FRONTIERS IN MICROBIOLOGY

NEW PERSPECTIVES IN CLINICAL MICROBIOLOGY

Brumfitt W, ed Hamilton-Miller JMT, assist. ed: New perspectives in clinical microbiology. 1978. ISBN 90-247-2074-5

Tyrrell DAJ, ed: Aspects of slow and persistent virus infections. 1979. ISBN 90-247-2281-0

Brumfitt W, Curcio L, Silvestri L, eds: Combined antimicrobial therapy. 1979. ISBN 90-247-2280-2

van Furth R, ed: Developments in antibiotic treatment of respiratory infections. 1981. ISBN 90-247-2493-7

van Furth R, ed: Evaluation and management of hospital infections. 1982. ISBN 90-247-2754-5

Kuwahara S, Pierce NF, eds: Advances in research on cholera and related diarrheas 1. 1983. ISBN 0-89838-592-X

Ristic M, Ambroise-Thomas P, Kreier J, eds: Malaria and babesiosis: Research findings and control measures. 1984. ISBN 0-89838-675-6

Kuwahara S, Pierce NF, eds: Advances in research on cholera and related diarrheas 2. 1984. ISBN 0-89838-680-2

Takeda Y: Bacterial diarrheal diseases. 1984. ISBN 0-89838-681-0

Hill MJ, Borriello SP, Hardie JM, Hudson MJ, Lysons RJ, Tabaqchali S, eds: Models of anaerobic infection. 1984. ISBN 0-89838-688-8

Taylor-Robinson D, ed: Clinical problems in sexually transmitted diseases. 1985. ISBN 0-89838-720-5

Borriello SP, Hardie JM, Drasar BS, Duerden BI, Hudson MJ, Lysons RJ, eds: Recent advances in anaerobic bacteriology. 1987. ISBN 0-89838-847-3

de Clercq E, ed: Frontiers in microbiology. 1987. ISBN 0-89838-959-3

FRONTIERS IN MICROBIOLOGY

From Antibiotics to AIDS

editor:
ERIK DE CLERCQ
Rega Institute, Department of Microbiology and Biochemistry, School of Medicine, Katholieke Universiteit Leuven, Belgium

1987 **MARTINUS NIJHOFF PUBLISHERS**
a member of the KLUWER ACADEMIC PUBLISHERS GROUP
DORDRECHT / BOSTON / LANCASTER

Distributors

for the United States and Canada: Kluwer Academic Publishers, P.O. Box 358, Accord Station, Hingham, MA 02018-0358, USA
for the UK and Ireland: Kluwer Academic Publishers, MTP Press Limited, Falcon House, Queen Square, Lancaster LA1 1RN, UK
for all other countries: Kluwer Academic Publishers Group, Distribution Center, P.O. Box 322, 3300 AH Dordrecht, The Netherlands

Library of Congress Cataloging in Publication Data

Frontiers in microbiology.

(New perspectives in clinical microbiology)
Proceedings of the International Symposium on Frontiers in Microbiology, held Nov. 1986 in Leuven, and sponsored by the National Fonds voor Wetenschappelijk Onderzoek and other organizations.
1. Medical microbiology--Congresses. I. De Clercq, Erik. II. International Symposium on Frontiers in Microbiology (1986 : Louvain, Belgium) III. Fonds national de la recherche scientificqu (Belgium) IV. Series.
[DNLM: 1. Allergy and Immunology--congresses. 2. Microbiology--congresses. 3. Virology--congresses.
QW 4 F935 1986]
QR46.F798 1987 616'.01 87-18548

ISBN 0-89838-959-3

DEDICATION

This book is dedicated to the memory of Professor Piet DE SOMER who was one of the great pioneers in the most diverse fields of microbiology (antibiotics, vaccines, interferon, antiviral and antitumor research).

He was born December 22, 1917 in Niel (near Antwerp), obtained his MD degree at the Katholieke Universiteit Leuven in 1942 and began his scientific career under Prof. R. Bruynoghe in the Laboratory of Bacteriology of the then bilingual (French-Flemish) University of Leuven. After his initial studies on the Rhesus factor, he shifted to research on antibiotics and engineered a fermentation plant in Genval for the production of penicillin, which was obtained in crystalline form at the end of 1947.

Based on this success, the pharmaceutical company, RIT, was built in 1950. It was further enlarged in 1958 and later incorporated in Smith, Kline & French Laboratories to become Smith Kline-RIT. In 1954, Dr. P. De Somer founded in Leuven the Rega Institute as a center for innovative research in microbiology. Ever since, he would remain the Director of the Institute and the Chairman of the Department of Microbiology. P. De Somer was appointed Professor in 1955. In the 1950s he guided the work that led to the discovery of various new antibiotics, such as virginiamycin (in 1954) and clometacillin (1962). In the late 1950's Prof. P. De Somer and his team achieved the large-scale production of the oral polio vaccine, which allowed Belgium to be the first country in the world to be freed from poliomyelitis. This was followed in the 1960's by the development of a vaccine against rubella.

From the end of the 1950's, Prof. De Somer took on the study of interferon. With his colleagues he pioneered on the elucidation of the mode of action of this molecule, its induction by polyanionic substances, and the large-scale production of human fibroblast interferon, hence making the latter available for clinical use. Thanks to De Somer, the Rega Institute became one of the world's leading centers for interferon research. The scope of this research has since 1980 been extended to cytokines (lymphokines, cell-growth factors) and the recently created unit of "genetic engineering" has played a major role in the identification of these cytokines.

In the 1970's an intensive antiviral research program, mostly in collaboration with synthetic chemistry laboratories all over the world, was launched, and this yielded a number of potent and selective antiviral agents, such as bromovinyldeoxyuridine (in 1979) and hydroxyphosphonylmethoxypropyladenine (1986), which hold great promise for the treatment of various virus infections.

From the earlier work of Prof. P. De Somer on antibiotics originated the more recent studies on the role of intestinal bacteria in the biotransformation of cholesterol, bile acids, steroid hormones and fatty acids. These studies clearly cross the barriers from microbiology to several other disciplines such as physiology and pharmacology. The truly interdisciplinary approach which has marked Prof. P. De Somer's scientific career is also reflected in the fields of immunology and immunopathology, as many of the fundamental principles of tumor and transplant immunity have been established in the Rega Institute during its now 33-year history.

P. De Somer

While being a great and innovative scientific leader, Prof. P. De Somer gained even more fame and prestige as the first rector of the Flemish Katholieke Universiteit Leuven. After being appointed in 1968 as prorector of this University, he was re-elected as rector in 1971, 1976 and 1981. At the verge of the last year of his last 5-year term as President of our University, he died on June 17, 1985 at the age of 68.

To commemorate the man from whom may be epitaphed that he subsequently founded a Pharmaceutical Company (in 1950), a Research Institute (in 1954), and a University (in 1968), we, his successors at the Rega Institute and Department of Microbiology, have invited to Leuven on 20-22 November 1986 a number of the world's leading authorities in the field of Microbiology.

The purpose of this reunion was to review the progress made in such fast developing areas as : chemotherapy (J.-M. Ghuysen : β-lactam antibiotics; P.A.J. Janssen : antimycotic drugs; R.T. Walker: antiviral drugs); gastrointestinal microbiota (D.C. Savage) and their role in the metabolism of both endogenous (T. Midtvedt) and exogenous (P. Goldman) substances; natural resistance mechanisms (S. Baron), interferons (J.T. Vilcek) and interleukins (C.A. Dinarello); tumor and virus immunity (A.C. Allison) and transplant immunity (based on total lymphoid irradiation: S. Strober); and virus research [M.R. Hilleman: vaccines, F. Rapp: virus latency, and R.C. Gallo: acquired immune deficiency syndrome (AIDS)]. Along the topics presented by these distinguished scientists, posters were presented by the staff members of the Rega Institute and in the Symposium Proceedings the contributions of the invited lecturers are followed by short communications pertaining to the recent scientific achievements of the Rega Institute. In this way, we wanted to pay tribute to Professor Piet De Somer, this Great Man of Science who left his hallmarks in the frontier fields of Microbiology and without whom the Rega Institute would never have existed or ever be the same again.

Prof. Erik De Clercq

R. Dillemans

PREFACE

The International Symposium on Frontiers in Microbiology has been dedicated to Prof. P. De Somer, whom I succeeded shortly after his death on 17 June 1985 as Rector of this now more than 560-year old University. When Prof. De Somer became the head of the University he started to remodel it, giving our old Alma Mater a more transparent administrative structure, strengthening its scientific and cultural autonomy, and establishing close links with the most prestigious national and foreign institutions. This made De Somer to one of the greatest, if not the greatest, of rectors in the history of Belgian Universities.

He was a great leader, a perfect organizer, a clever negotiator, and a brilliant orator. In his speeches one immediately sensed his intuitive cognition and witty evaluation of the values of life. He knew perfectly well how to persuade the unwilling and to disenchant the illusionist. Sometimes a visionary himself, he would not pursue his ideas unless there was a chance of success.

As innovative Prof. P. De Somer was in providing to this University a new face, or should I say facelifting, as international is his reputation as the founder, and, since its inception, only director, of the Rega Institute. Built now more than 30 years ago, the Rega Institute has remained one of the world's leading centers in microbiological research.

At this Institute, several new antibiotics were discovered, such as clometacillin, a semi-synthetic penicillin, and virginiamycin, which is still widely used as an animal feed additive. From virginiamycin new studies originated that led to the elucidation of the role of intestinal bacteria in the biotransformation of cholesterol and bile acids. In the Rega Institute polio vaccine was for the first time produced at a large scale, allowing Belgium to be the first country in the world where poliomyelitis was eradicated. It was also in the Rega Institute that the investigations started which ultimately led to the development of the first rubella vaccine.

The Rega Institute has been, and still is, the sanctuary of fundamental research in immunology and virology; where the basic principles of tumor and transplant immunity have been established, where interferon has flourished to the stage that from an almost elusive molecule it became available for extensive clinical trials, and where, more recently, a number of new antiviral compounds were born, which hold great promise for the treatment of various virus infections, including herpes simplex, varicella, zoster, cytomegalo, adeno and AIDS.

Prof. P. De Somer not only founded and directed the Rega Institute, he also was the driving force behind his research team which he led with a never waning enthusiasm and perspicuity. At both the University and Institute his advice was eagerly looked for. At International Congresses, his authoritative views were higly appreciated and often taken for granted.

De Somer had the innate gift to discriminate the important from the trivial and to intuitively discover new alleys of scientific research. He shunned bigotry and pretense, was adverse to mediocrity, and from his coworkers and their work he expected and requested excellence. He was loyal to his commitments and enchanted by innovative ideas. Many scientists from all over the world have been attracted to come to work in the Rega Institute as either pre- or postdocs. Back home, they have all become leading experts in their field.

Those that have received Prof. De Somer's scientific legacy at the Institute are left with the task to continue the tradition of excellence; excellence in research, excellence in teaching, excellence in serving mankind. Therefore all forces should be joined and united so that the magnificent patrimony inherited from our beloved Rector should not get lost.

With this Symposium we want to commemorate the man from whom may be said that he subsequently founded a Company, an Institute and a University. In the very first place, Prof. De Somer was a scientist, and as a scientist he climbed up from the laboratory bench to a top authority in science management. The Rega Institute, his creation, has played a pioneering role in the most varying areas of Microbiology.

This Symposium has brought together in the heart of Leuven the world's leading authorities in Microbiology. Most of our guests are among De Somer's best friends, and we are honored that they have made the long journey to Leuven to join us at this unique occasion and to pay tribute to the man to whom many researchers owe their dedication to science, many professors their attachment to the University, and the University itself its new youth, strength and wisdom.

Prof. Roger Dillemans
Rector K.U.Leuven

ACKNOWLEDGEMENTS

This Symposium was sponsored by

Nationaal Fonds voor Wetenschappelijk Onderzoek
Onderzoeksfonds K.U.Leuven
Rectoraat K.U.Leuven
Stichting Rega V.Z.W.
Ministerie van Onderwijs (Dienst Wetenschappelijk Onderzoek)
Alumni Lovanienses

Abbott Diagnostics
Janssen Research Foundation
Smith Kline - RIT
Upjohn
Merck Sharp & Dohme
Institut Mérieux Benelux
Kredietbank
Generale Bank
Lilly Benelux
IRE - Medgenix
Bristol-Myers
Agfa-Gevaert

Contributions were received from IBM, Flow Laboratories, Van der Heyden, Essex (Belgium), HVL, Pasture, International Medical Products, Beun-de Ronde, Boehringer Pharma, Amersham, Fonteyn, Gibco, LKB Instruments, Beecham, Millipore and Gera-Filtest

---

The Editor gratefully acknowledges the excellent editorial assistance of Christiane Callebaut who prepared the whole typescript of these Proceedings.

Photographs and illustrations : L. Palmaerts.

ORGANIZATION

This Symposium was organized by the Rega Institute and the Department of Microbiology of the Katholieke Universiteit Leuven, in collaboration with the Publication Relations Department of the K.U.Leuven

Honorary Committee

Prof. R. Dillemans (President, K.U.Leuven)
Prof. K. Tavernier (Executive Director, K.U.Leuven)
Prof. R. Borghgraef (Chairman, Biomedical Sciences, K.U.Leuven)

Organizing Committee

Prof. E. De Clercq (Chairman), K.U.Leuven
Prof. A. Billiau, K.U.Leuven
Prof. H. Eyssen, K.U.Leuven
Prof. M. Vandeputte, K.U.Leuven
Prof. H. Vanderhaeghe, K.U.Leuven
Prof. J. Desmyter, K.U.Leuven
Prof. J. Vandepitte, K.U.Leuven

Advisory Committee

Prof. W. Fiers (R.U.Gent)
Prof. L. Thiry (Pasteur Institute, U.L.Bruxelles)
Dr. J. Content (Pasteur Institute, U.L. Bruxelles)
Prof. A. Lacquet (Koninklijke Academie voor Geneeskunde van België)
Dr. E. De Maeyer (Orsay, France)
Dr. H. Schellekens (Rijswijk, The Netherlands)
Dr. D.A.J. Tyrrell (Salisbury, England)
Prof. H.J. Eggers (Köln, F.R. Germany)

TECHNICAL ASSISTANCE

The contributions from the Rega Institute were achieved with the excellent technical assistance of

Ann Absillis
Stanny Asselberghs
Luc Bassi
Erik Beuken
Etienne Clerinx
Rene Conings
Francine Cornette
Frieda De Meyer
Guido De Pauw
Chris Dillen
Pierre Fiten
Erik Fonteyn
Jozef Goebels
Dirk Henot
Gustave Hermans
Marcel Huybrechts
Luk Kerremans
Jean-Pierre Lenaerts
Rita Merckx
Jan Mertens
Willy Put
Alfred Ronsmans
Isabelle Ronsse
Omer Rutgeerts
Constant Segers
Miette Stuyck
Robert Van Damme
Lizette van Berckelaer
Ria Van Berwaer
Lieve Vandenbulcke
Marcel Vanderbruggen
Ria Vanlaer
Francis Vanlangendonck
Anita Van Lierde
Luc Verlooy
Luc Willebrords
Greet Wuyts
Willy Zeegers

CONTENTS

SECTION IV : TUMOR IMMUNITY
SECTION CHAIRMAN : Prof. M. Vandeputte

SECTION V : NEW HORIZONS IN VIRUS RESEARCH
SECTION CHAIRMAN : Prof. E. De Clercq

B. CONTRIBUTIONS FROM THE REGA INSTITUTE

I. RESEARCH UNIT OF PROF H. VANDERHAEGHE

V. RESEARCH UNIT OF PROF. E. DE CLERCQ

REVIEW LECTURES BY INVITED SPEAKERS

SECTION I : CHEMOTHERAPY

SECTION CHAIRMAN : PROF. H. VANDERHAEGHE

H. Vanderhaeghe

INTRODUCTION TO SECTION I : ANTIBACTERIAL, ANTIFUNGAL AND ANTIVIRAL CHEMOTHERAPY

H. VANDERHAEGHE
Rega Institute, Katholieke Universiteit Leuven, B-3000 Leuven, Belgium

I have the privilege to be the chairman of the first section of this Symposium because I am Professor De Somer's oldest collaborator who is still working in the Rega Institute. I will highlight some of the remarkable achievements of Prof De Somer early in his scientific carreer.

After obtaining his M.D. degree, P. De Somer did not go in medical practice but started research in the Laboratory of Bacteriology under Professor R. Bruynoghe. In 1944 he was a laureate of the Academy of Medicine with a study on infectious mononucleosis, and in 1945, again, he was a laureate of the Academy of Medicine with a study on the Rhesus factor.

In 1946, together with Professor C. de Duve, who would later obtain success and fame in the field of biochemistry, Professor De Somer started to investigate, in a room of the Papéteries de Genval (it could not be called a laboratory), the surface production of penicillin. After a few months, they arrived at the conclusion that this method was not appropriate and that deep fermentation was necessary. This was a rather bold conclusion, since at that time most plants were still using surface fermentation. Thus, a small fermentation plant was designed and set up in Genval. It would later become the pharmaceutical firm R.I.T. This was at a time when there was shortage of materials like stainless steel, difficulties in obtaining chemicals and solvents, problems of communication and travelling, etc. Despite these adverse conditions, penicillin production was started in 1947 and by the end of the year crystalline penicillin was obtained. For the next few years P. De Somer together with F. Naudts pioneered on the improvement of penicillin production, and, in 1950, the plant was considerably enlarged. Simultaneously, P. De Somer continued his work, together with P. Van Dyck, in the old, but still existing, Laboratory of Bacteriology. This led to the discovery in 1953 of griseomycin and in 1954 of virginiamycin. Virginiamycin had an excellent activity against penicillin-resistant staphylococci which began to emerge at that time. Success in the clinic, however, was limited because of poor absorption of the antibiotic. Nevertheless, virginiamycin is an important antibiotic, since it has excellent growth-promoting activity for poultry and swine and it is still used as a food additive.

The lecturers that take part in this first section are J.-M. Ghuysen, P.A.J. Janssen and R.T. Walker.

JEAN-MARIE GHUYSEN is Professor of Microbiology at the University of Liège. He obtained the degree of Pharmacist at the U. Liège in 1947 and a doctorate's degree in Chemistry in 1951. Shortly thereafter, he started to work in the Laboratory of Bacteriology under Professor M. Welsch. It should be mentioned that in the period between the two World Wars the director of this laboratory, Professor Gratia, had already worked on antibiotics. His product, actinomycetine, was too toxic for human use, but was interesting from a scientific viewpoint and further studied by Welsch and later by Ghuysen. J.-M. Ghuysen was appointed "chargé de cours" in 1961 and professor in 1966. He discovered that actinomycetin contained enzymes which lysed the bacterial cell wall. This provided important information as to the chemical composition of the bacterial cell wall. As penicillin is an inhibitor of the biosynthesis of some components of the bacterial cell wall, Prof. Ghuysen then turned to the mechanism of action of this antibiotic.

PAUL A.J. JANSSEN obtained his M.D. in 1951 at the University of Ghent where he also did postgraduate research in Pharmacology. After having gained additional experience in chemical pharmacology at the University of Köln (W.-Germany), Dr. Janssen started his own research program in the then small pharmaceutical company, Janssen Pharmaceutica, which was run by his father. In about 30 years more than 70,000 compounds were synthesized and examined, and 60 new therapeutic substances were introduced in human and veterinary medicine, making Janssen's research group the world's most successful pharmaceutical research organization. His achievements are described in a special issue of the Journal "Drug Development Research" (May-August 1986), Alan R. Liss (New York): "33 Years of Drug Development Research - A tribute to Paul Janssen at the occasion of his 60th birthday". Dr. Janssen holds several honorary doctorate degrees including a doctorate degree from the K.U.Leuven.

RICHARD T. WALKER studied chemistry at the University of Birmingham, where he obtained his Ph.D. in 1962, and became a lecturer in the Chemistry Department in 1968. Dr. Walker has worked with Professor H.G. Khorana at the University of Wisconsin (Madison) and has been a visiting faculty member at the Massachusetts Institute of Technology and visiting professor at the University of Osaka and at Hokkaido University in Sapporo (Japan). His field of research is the chemistry of nucleic acid derivatives. A great number of nucleoside analogues have been prepared in his laboratory and these compounds have been further examined for their antiviral and antitumor properties in the Rega Institute by Professor E. De Clercq. From this collaborative work has emerged BVDU (bromovinyldeoxyuridine), a potent and selective anti-herpes compound with promising clinical utility. Together with Prof. E. De Clercq, Dr. R.T. Walker has already directed two NATO/FEBS Courses on Antiviral Agents, and a third one is forthcoming in 1987. Dr. R.T. Walker is, since its inception, Editor of the Journal "Nucleic Acids Research".

β-LACTAM ANTIBIOTICS AS CARBONYL DONORS OF THE ACTIVE-SITE SERINE β-LACTAMASES, DD-PEPTIDASES AND LL-PEPTIDASES

J.-M. GHUYSEN
Service de Microbiology, Faculté de Médecine, Université de Liège, Institut de Chimie, B6, B-4000 Sart Tilman (Liège 1), Belgium

1. INTRODUCTION

The concept of antibacterial chemotherapy goes back to the discovery of the sulfa drugs in the 1930's. This concept, however, was definitely established only when the first cures were achieved by penicillin in 1941. Penicillin opened the antibiotic era in which we are living now.

Penicillin has given rise to a flourishing family of drugs (penicillins, cephalosporins, monobactams) designated collectively as β-lactam antibiotics (Fig. 1). All of them originally came from microorganisms of the soil but many of them were rendered more effective by chemical modifications. The multiplicity of the β-lactam molecules - more than 100,000 - that the pharmaceutical industry has produced essentially by trial and error, has been dictated primarily by the various countermoves that the bacteria have made to resist our most effective armory of antibacterial chemotherapy. One of countermoves - and so far the most impressive one - has been the rapid spread of genes coding for various types of β-lactamases.

The β-lactamases degrade benzylpenicillin, and other β-lactam antibiotics discovered subsequently, into biologically inactive metabolites by opening the β-lactam ring through hydrolysis of the endocyclic amide bond. β-Lactamase genes are readily transferable elements and bacteria can create new β-lactamases by recombination of existing genes. As a result of the selective pressure exerted by the massive use of the β-lactam antibiotics, almost every bacterial species has now members that have acquired a β-lactamase gene and about 100 different types of β-lactamases have been detected.

While there is still plenty of evidence for the continuing development of resistance by β-lactamase production, other lines of defense have appeared. Tolerance to β-lactam antibiotics, though poorly understood, can be defined as diminished and/or delayed killing by bacterial growth inhibitory concentrations of the drug. Intrinsic resistance may be due to a decreased accessibility of the drug to its targets by modifications of the bacterial cell envelope and/or modifications of the targets themselves resulting in a decreased susceptibility to the drug.

In this war between man and bacteria, Professor Piet De Somer has been one of the great pioneers. As a result of his and C. De Duve's studies on penicillin production, a plant (RIT, now Smith Kline-RIT) was built in Genval (Belgium) where, in spite of the many technical problems still unsolved at that time, penicillin was manufactured commercially as early as 1947. Subsequently, P. De Somer's contributions and insights have been decisive in opening up several other areas of major importance. After he founded the Rega Institute in 1954, scientific achievements increased ever since, not only in the field of antibiotics but also in several other fields of microbiology. As to the antibiotic field, suffice it to mention

J.-M. Ghuysen

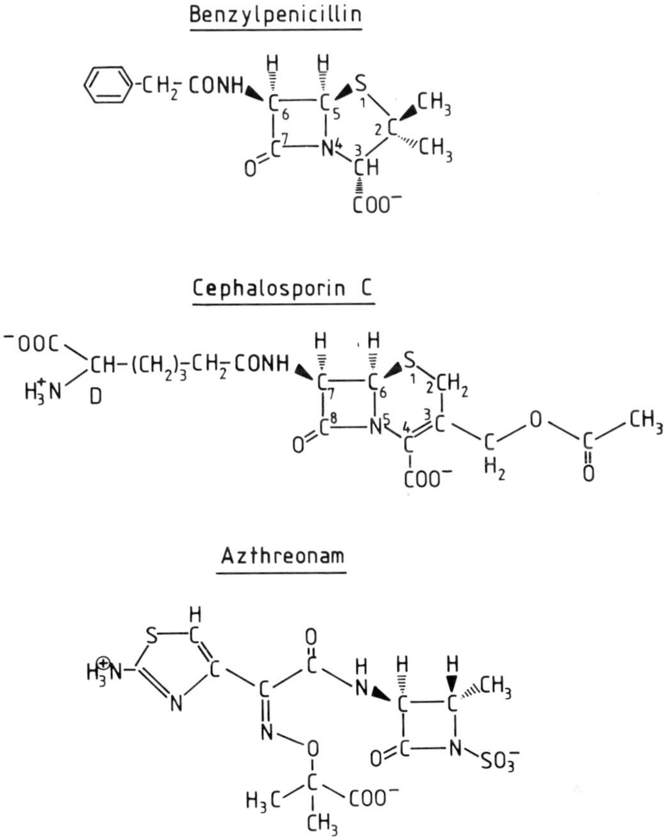

FIGURE 1. β-Lactam inactivators of the bacterial active-site serin DD-peptidases.

the discovery of griseomycin and virginiamycin and the design of one of the first semi-synthetic penicillins, clometocillin.

The aim of this article is to discuss advances made in the understanding of the catalytic mechanisms of two groups of enzymes whose common property is to recognize the β-lactam antibiotics : the DD-peptidases which are the target of penicillin action and the β-lactamases.

2. THE DD-PEPTIDASES/PENICILLIN BINDING PROTEINS

The biochemistry of the penicillin targets started some twenty years ago. In the late 60's, the primary structure of the bacterial wall peptidoglycan - which for many years had been known to be the site of the lethal lesion caused by penicillin - had been elucidated (1) (Fig. 2) and the pathway of wall biosynthesis had been unravelled (2) (Fig. 3). Killing of bacteria by penicillin was attributed to the selective inactivation of a membrane-bound transpeptidase, whose normal function is to incorporate the lipid-transported disaccharide peptide peptidoglycan precursors, as

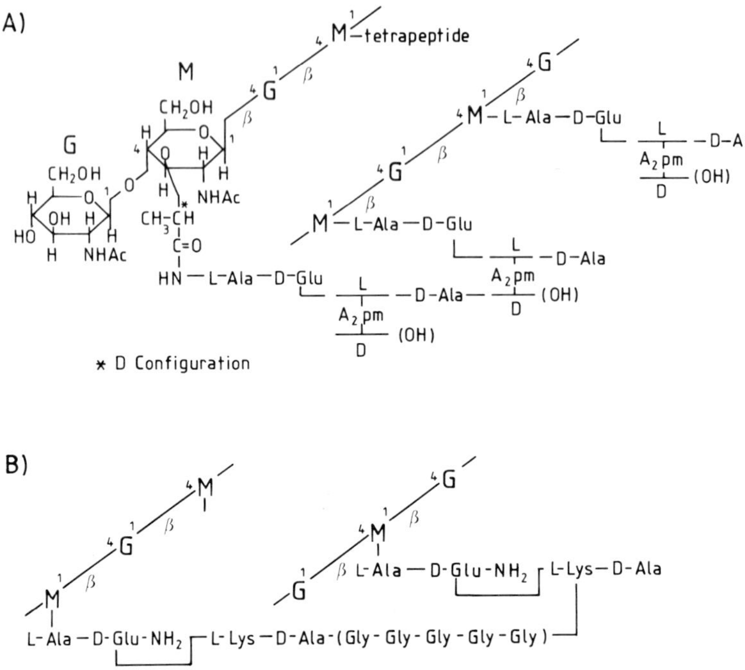

FIGURE 2. Primary structure of the wall peptidoglycans in Escherichia coli (A) and Staphylococcus aureus (B).

they emerge from the plasma membrane, to the preexisting wall peptidoglycan. Inactivation of the transpeptidase by penicillin was interpreted as a suicide process. The enzyme, mistaking the penicillin molecule for a D-alanyl-D-alanine-terminated peptidoglycan precursor (Fig. 3) undergoes derivatization in the form of a highly stable adduct. As a corollary, the transpeptidase behaves as a "penicillin binding protein".

As the research was progressing, the complexity of the reactions involved in wall peptidoglycan crosslinking became more and more evident and the unitary transpeptidase model had to be revised (3). All the bacteria possess, bound to the plasma membrane, multiple penicillin binding proteins (PBSs) of a molecular mass ranging from 25,000 to 100,000 or more (they are currently numbered in the order of decreasing size). Each bacterial species has its own assortment of PBPs. The several PBPs present in a single cell show varying susceptibility to β-lactam antibiotics so that, depending on the antibiotic concentration and the time of exposure used, the extent of inactivation of each of the PBPs may vary widely. The PBPs fulfil distinct cellular functions (in Escherichia coli, PBPs 1A/1B, 2 and 3 are related to cell elongation, shape maintenance and cell septation, respectively), do not show the same degree of essentiality for bacterial growth and may play compensatory roles.

The high-Mr PBPs (1A/1B, 2 and 3) have been assigned a role of prime importance in the killing of E.coli by the β-lactam antibiotics (3). But this is a rather exceptional situation. Often, comparison between the an-

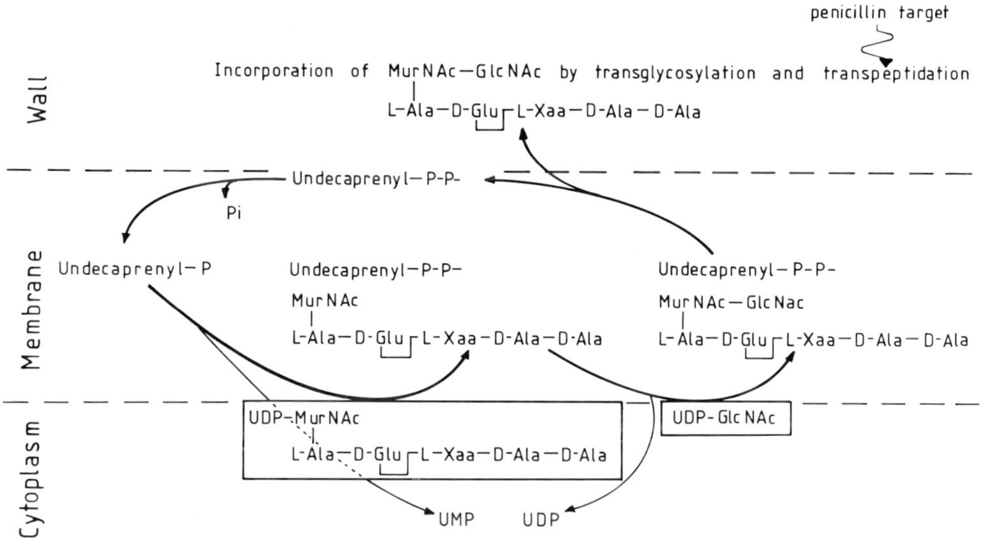

FIGURE 3. Biosynthesis of the bacterial wall peptidoglycans. Xaa = meso-diaminopimelic acid (as it occurs in E.coli) or L-lysine (as it occurs in S.aureus). Addition of the five glycine residues to the ε-amino group of the L-lysine residue (see Fig. 2) takes place at the level of the lipid intermediate (not shown).

tibacterial effectiveness of a β-lactam antibiotic and the concentration needed to derivatize some percentage (i.e. 50 %, 90 %, etc.) of one or more of the PBPs does not produce reasonable correlations and defining the killing target of a β-lactam antibiotic is difficult. There are cases where the first symptom of growth inhibition by penicillin has been related to a characteristic degree of saturation of each of the PBPs present, suggesting that the antibacterial effect arises from a summation of the extent of inactivation of several PBPs or from the abnormal relative concentrations of the PBPs left in a free form in the antibiotic-treated cells (4).

The PBPs to which catalytic activity could be assigned, catalyze various acyl transfer reactions. By transpeptidation (reaction A in Fig. 4), they perform closure of the bridges through which peptide units substituting adjacent glycan chains become linked together. This reaction is essential for the polymerization and expansion of the wall peptidoglycan in the form of an insoluble network. By carboxypeptidation, PBPs limit the amounts of donors available for transpeptidation (reaction B in Fig. 4) and/or produce new amino acceptor sites in the preformed peptidoglycan network (reaction C in Fig. 4). The significance of these hydrolytic activities is not clear. They probably control the level of cross-linking in the completed peptidoglycan and serve a role in wall remodelling throughout the bacterial life cycle.

Depending on whether the acceptor is the ω-amino group of a peptide or water, the transfer reaction leads to an increased or a decreased ex-

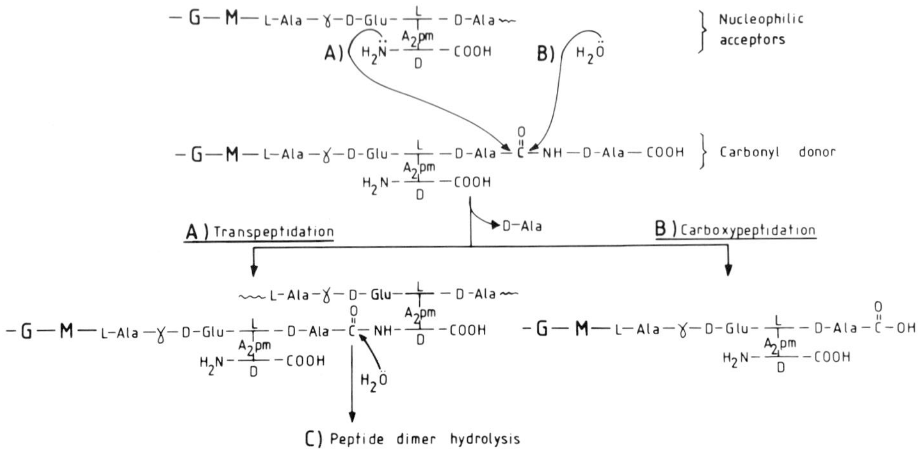

FIGURE 4. DD-peptidases/penicillin-binding proteins-catalyzed acyl transfer reactions with natural peptide donor and acceptor substrates.

tent of cross-linking. Yet, in all cases, the scissile peptide bond in the carbonyl donor is located in the α-position with respect to a free carboxylate and extends between two carbon atoms having the D configuration. This unique optical specificity justifies the name "DD-peptidase" given to the PBPs. Their selective inactivation by β-lactam derivatization explains the remarkable selective toxicity of these antibiotics.

Isolated DD-peptidases/PBPs may have high activity on well-defined peptides. These peptides may be of natural origin (as those shown in Fig. 4) or especially designed to function as donor and acceptor analogues (as those shown in Fig. 5). Other PBPs (of high molecular mass) catalyze, in concert, peptide cross-linking by transpeptidation and glycan synthesis by transglycosylation. Their bifunctionality is revealed by the products that they generate upon incubation with the lipid-linked precursor N-acetylglucosaminyl-N-acetylmuramyl(pentapeptide)-diphosphoryl-undecaprenol (Fig. 6). These reactions proceed with a low turnover number of 0.01 s^{-1} or less, suggesting that the _in vitro_ conditions poorly mimick the _in vivo_ situation. As shown with PBP3 of _E.coli_, removal by gene fusion of the amino terminal 240 amino acid region which, presumably, contains the penicillin-insensitive transglycosylate site, results in a truncated, 348 amino acid polypeptide that binds penicilin (5).

The DD-peptidase/PBPs are not easily accessible to biochemical studies. They occur in a small number of copies per cell, from perhaps 100 to 2000, and their membraneous nature poses obvious problems (though some of them can be detached from the plasma membrane by careful proteolytic action without apparent alteration of the catalytic properties). DNA recombinant techniques as means to prepare large quantities of water-soluble DD-peptidase/PBPs are actively investigated. In addition, such enzymes are excreted spontaneously during the growth of bacteria (i.e. actinomycetes strains).

The gene encoding the extracellular DD-peptidase/PBP of _Streptomyces_

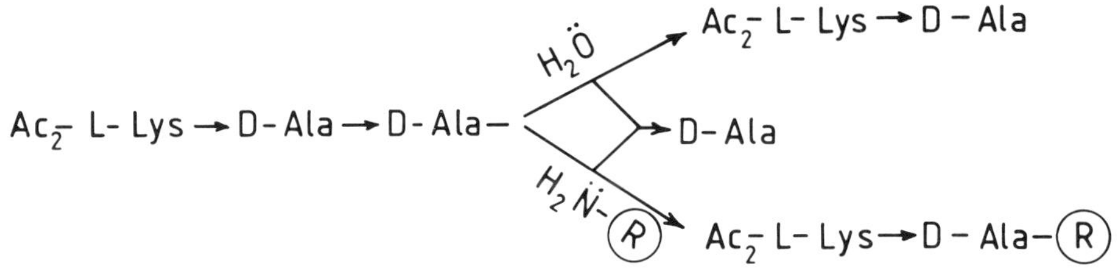

FIGURE 5. DD-peptidases/penicillin-binding proteins-catalyzed acyl transfer reactions with peptide donor and acceptor analogues.

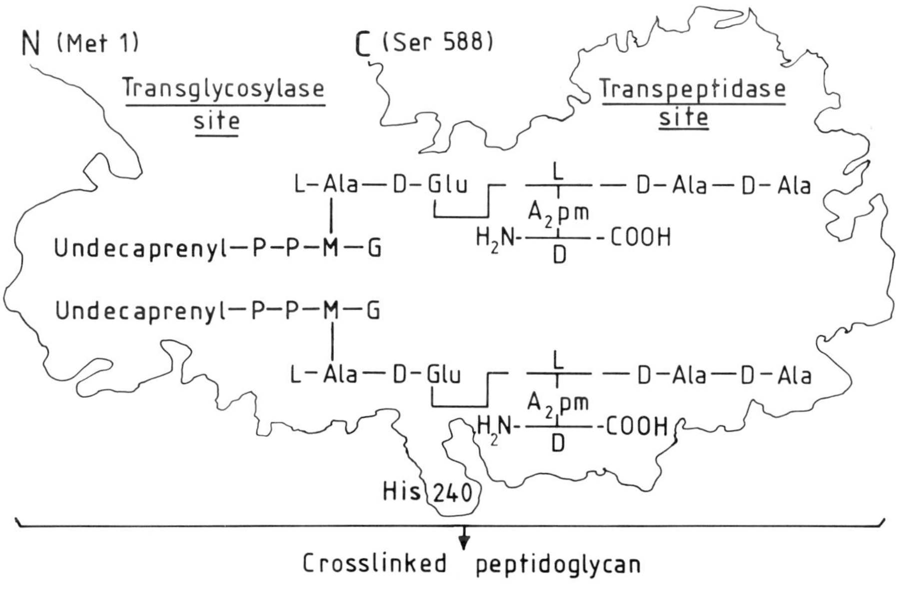

FIGURE 6. Lipid-linked disaccharide (pentapeptide)-diphosphoryl-undecaprenol as substrate of the bifunctional penicillin-binding proteins. The transglycosylase site catalyzes glycan elongation (with release of the undecaprenyl-P-P). The transpeptidase site catalyzes cross-linking between peptide units.

R61 has been cloned with amplified expression of the excreted protein (6). The gene codes for a 406 amino acid precursor which possesses both a cleavable N-terminal signal peptide and a cleavable C-terminal extension. This C-terminal segment has no runs of hydrophobicity typical of a membrane anchor. But, by analogy with E.coli PBP no 5, it may be hypothesized that, should it not be cleaved during maturation, this segment would function as a halting signal through which the enzymes would become membrane-bound by a mechanism not previously described. Whatever the mechanism and physiological significance of excretion, this water-soluble 349 amino acid enzyme has served its role in the unravelling of the enzymatic, biochemical and structural properties of the DD-peptidases/PBPs.

3. DD-PEPTIDASES AND β-LACTAMASES-CATALYSED ACYL TRANSFER REACTIONS

In recent years we have witnessed one of those remarkable convergences which not unfrequently lead to episodes of high drama in science. The convergence is that, following initial studies made in 1976 (7) and 1979 (8), all the DD-peptidases/PBPs and the majority of the β-lactamases which have so far been characterized are active-site serine enzymes which can be regarded as counterparts of the LL-peptidases of the trypsin family. They all operate by an acyl enzyme mechanism.

3.1. General reaction and basic equations (9)

The DD-peptidases and β-lactamases-catalysed reactions proceed *via* the formation of a covalent intermediate in which the electrophilic carbon of the carbonyl donor substrate becomes ester-linked to the hydroxyl group of the enzyme's serine residue (Fig. 7). Enzyme acylation produces a leaving group or product P_1. In turn, transfer of the acyl moiety to an exogenous nucleophilic acceptor HY causes release of $R-\overset{O}{\overset{\|}{C}}-Y$, or product P_2, and enzyme regeneration.

This three-step reaction can be presented by

$$E + D \underset{}{\overset{K}{\rightleftharpoons}} E.D \xrightarrow[P_1]{k_{+2}} E\text{-}D^{\star} \xrightarrow[HY]{k_{+3}} E + P_2 \qquad \underline{1}$$

where E = enzyme; D = carbonyl donor; HY = exogenous nucleophile; E.D = Michaelis complex; $E\text{-}D^{\star}$ = serine ester-liked acyl enzyme; K = dissociation constant; k_{+2} and k_{+3} = first-order rate constants; P_1 = leaving group and P_2 = second reaction product. It is assumed that [D] >> [E], E.D is in rapid equilibrium with free enzyme and free donor, and HY (i.e. water) is at a saturating concentration.

The values of k_{cat} and Km are related to K, k_{+2} and k_{+3} by

$$k_{cat} = \frac{k_{+2}k_{+3}}{k_{+2} + k_{+3}} \qquad Km = \frac{Kk_{+3}}{k_{+2} + k_{+3}}$$

so that k_{+2}/K (i.e. the second-order rate constant of enzyme acylation by the carbonyl donor) is equal to k_{cat}/Km.

The proportion of total enzyme $[E]_o$ which occurs as acyl enzyme $[E\text{-}D^{\star}]_{ss}$ at the steady state of the reaction is

$$\frac{[E]_o}{[E\text{-}D^{\star}]_{ss}} = 1 + \frac{k_{+3}}{k_{+2}} + \frac{Kk_{+3}}{k_{+2}[D]} \qquad \underline{2}$$

The time necessary for the acyl enzyme to reach a certain percentage of its steady state level is

$$t = \frac{-\ln\left(1 - \dfrac{[E\text{-}D^{\star}]}{[E\text{-}D^{\star}]_{ss}}\right)}{k_{+3} + k_a} \qquad \underline{3}$$

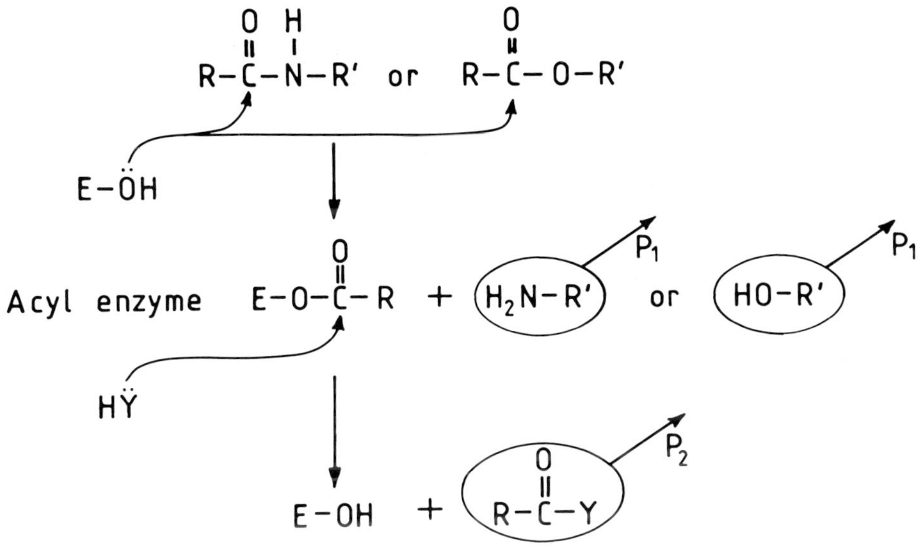

FIGURE 7. Active-site serine enzyme-catalyzed acyl transfer reaction between an amide (or ester) carbonyl donor and an exogenous nucleophilic acceptor. R-CONH-R' or R-COO-R' = amide or ester carbonyl donor; E-OH = enzyme with the hydroxyl group of the serin residue; H_2N-R' or HO-R' = leaving group; HY = exogenous acceptor (H_2O, an amino compound, etc.).

where $k_a = \frac{k_{+2}}{1 + (K/[D])}$ is the pseudo-first-order rate constant of acyl enzyme formation at a given concentration of carbonyl donor.

For $k_{+3} >> k_{+2}$, the steady-state proportion of enzyme present as acyl enzyme is low ($k_{cat} = k_{+2}$ and Km = K). For $k_{+2} >> k_{+3}$, this proportion is high ($k_{cat} = k_{+3}$ and Km = Kk_{+3}/k_{+2}) and the accumulated acyl enzyme can be trapped, isolated and characterized. If k_{+3} has a very small absolute value, the acyl enzyme is highly stable and the donor is a mechanism-based inactivator.

The presence of competing nucleophiles (alcohol, hydroxylamine, other amino compounds) in the aqueous reaction mixture modifies the fate of the donor since the enzyme activity undergoes partitioning at the level of the acyl enzyme between hydrolysis, alcoholysis, aminolysis or transpeptidation. An alternate nucleophile may also increase the rate of consumption of the donor (i.e. k_{cat}) if, in water, the acyl enzyme accumulates ($k_{+2} > k_{+3}$).

3.2. DD-peptidases and β-lactamases-catalysed reactions with non-cyclic peptide and ester carbonyl donors

The substrate profiles of some DD-peptidases and β-lactamases have been explored using several peptides and depsipeptides as donors (Fig. 8) and by carrying the reactions both in water and in aqueous solutions con-

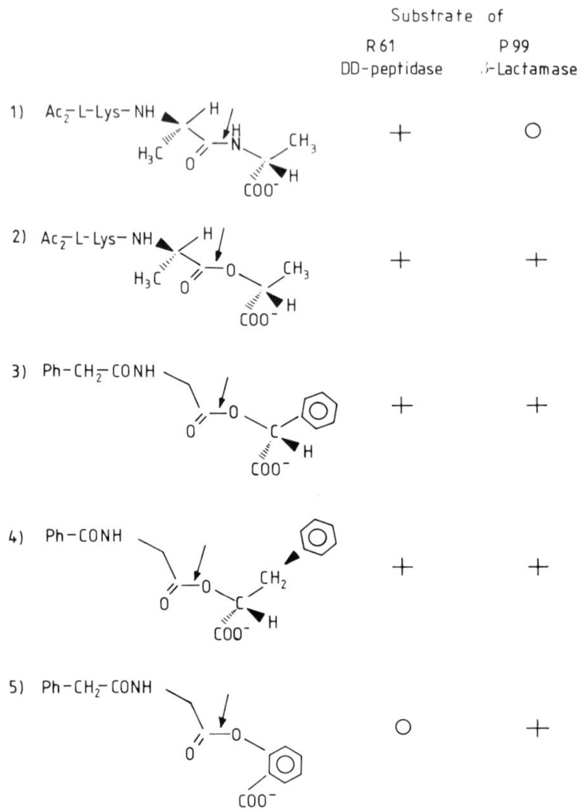

(1) Ac_2-L-Lys^2—D-Ala^1—D-Ala^1; (2) Ac_2-L-Lys—D-Ala—D-Lac;

(3) Phenylacetylglycyl—D—mandelate; (4) Hippuryl (D) phenyllactate;

(5) Phenylacetylglycylsalicylate.

FIGURE 8. Amide and ester compounds as carbonyl donors of some DD-peptidases and β-lactamases. The arrows indicate the scissile bond. (1) : Ac_2-L-Lys-D-Ala-D-Ala; (2) Ac_2-L-Lys-D-Ala-D-lactate; (3) phenylacetylglycyl-D-mandelate; (4) hippuryl-(D)-phenyllactate; (5) phenylacetylglycylsalicylate.

taining alternate acceptors.

Not surprisingly, the conclusion of these studies is that a peptide or a depsipeptide of high donor activity for a DD-peptidase (such as Ac_2-L-Lys-D-Ala-D-Ala or Ac_2-L-Lys-D-Ala-D-lactate) closely resembles the C-terminal portion of a peptidoglycan precursor. With the DD-peptidases studied, the k_{+3}/k_{+2} ratio is such that enzyme acylation is usually slower or much slower than enzyme deacylation, the efficacy of binding of D-alanyl-D-alanine-terminated peptides is relatively weak ($Km \cong K$ = 0.1-10 mM; a situation which favors maximization of the overall reaction rates) and the

k_{+2} values of enzyme acylation may range from $\cong 1\ s^{-1}$ to $\cong 50\ s^{-1}$. Replacement of the scissile peptide bond by an ester bond (as it occurs in the depsipeptides) usually results in an increased rate of enzyme acylation (k_{+2}).

Relatively little is known on the β-lactamases-catalysed acyl transfer reactions involving non-cyclic donors. But three observations have been made. 1) All the β-lactamases appear to be devoid of peptidase activity. 2) Some β-lactamases (such as that produced by *Enterobacter cloacae* P99) have esterase activity (10); hence β-lactamases and DD-peptidases exist which catalyse acyl transfer reactions from the same depsipeptide donors. 3) Other β-lactamases (such as that produced by *Bacillus licheniformis*) have no or little esterase activity; β-lactam compounds are, apparently, the only donors on which they operate.

As expected from an acyl enzyme mechanism, the presence of alternate nucleophiles causes partitioning of the acyl enzymes formed during interaction between non-cyclic (peptide and/or ester) donors and the DD-peptidases or β-lactamases. The following examples illustrate the underlying mechanisms.

The use of D-alanyl-D-alanine-terminated peptides as donors by the DD-peptidases reveals a subtle "push and pull" interplay at the level of the acyl enzyme, between the leaving group D-Ala, water and an exogenous amino compound NH_2-R structurally related to the wall peptidoglycan interpeptide bridge. Depending on the DD-peptidases, the relative acceptor activity may be D-Ala << 55.5 M H_2O << mM NH_2-R or 55.5 M H_2O << D-Ala << mM NH_2-R (Fig. 9).

The DD-peptidases of the first type hydrolyze the donor to completion without any significant accumulation of acyl enzyme and without any sign of interference by the leaving group D-Ala (which accumulates in the reaction mixture). Millimolar concentrations of an exogenous, suitable peptide acceptor channel part of the donor to the transpeptidated product at the expense of its hydrolysis (Fig. 9A). These DD-peptidases (for example the PBPs 4, 5 and 6 of *E.coli* and the *Streptomyces* R61 DD-peptidase) catalyze concomitant carboxypeptidation and transpeptidation reactions on a competitive basis.

The DD-peptidases of the second type (for example the 26,000-Mr, membrane-bound DD-peptidase of *Streptomyces* K15) (11) behave differently. The leaving group D-Ala effectively suppresses the acceptor activity of water, performs attack of the acyl enzyme and regenerates the initial carbonyl donor. Though the DD-peptidase turns over and a non-negligible proportion of it occurs as acyl enzyme, little of the carbonyl donor is consumed. In turn, this "pace-making" mechanism can be overcome by supplementing the reaction mixture with a suitable peptide acceptor. This peptide, at millimolar concentrations, largely saturates the enzyme (and is the only acceptor HY of reaction *1*). Consequently, the carbonyl donor is transpeptidated until completion, no acyl enzyme accumulates and the DD-peptidase functions as a strict transpeptidase (Fig. 9B). Experimental evidence also shows that the peptide acceptor does not behave as a simple alternate nucleophile. It influences the initial binding between the enzyme and the donor and enhances the efficacy of the ensuing enzyme acylation step.

These properties offer a possible explanation for the observations that certain PBPs have no detectable or weak enzymic activity *in* vitro. The high-Mr PBPs 1A/1B, 2 and 3 of *E.coli*, already mentioned, may lack hydrolytic potency because they possess a "pace-making" mechanism like that described above. They may have low intrinsic transpeptidase activity because in the assays so far devised, they do not find the right amino acceptor needed for an efficient acylation of the protein by the donor.

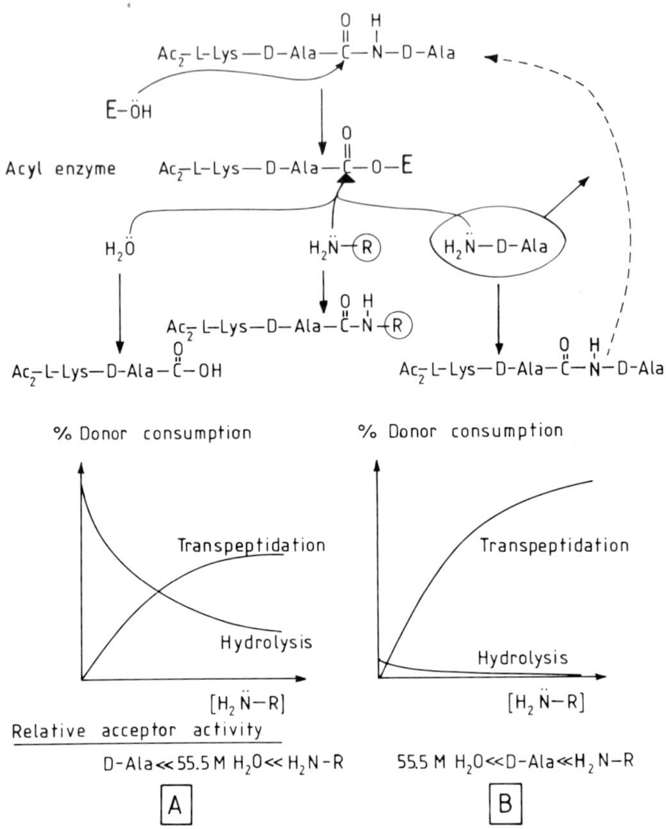

FIGURE 9. Competition between the leaving group D-Ala, water and an exogenous amino acceptor at the level of the acyl enzyme formed during DD-peptidase-catalyzed acyl transfer reaction with Ac_2-L-Lys-D-Ala-D-Ala as carbonyl donor. A) The DD-peptidase catalyzes concomitant carboxypeptidation and transpeptidation reactions. B) The DD-peptidase catalyzes, almost exclusively, a transpeptidation reaction. E-OH = enzyme with the hydroxyl group of the serine residue.

3.3. DD-peptidases and ß-lactamases-catalyzed reactions with cyclic (ß-lactam) carbonyl donors

ß-Lactam compounds may have high donor activity for the ß-lactamases and the DD-peptidases if the two following minimal requirements are fulfilled. A negative charge must occur at the 3' in the penicillins and 4' in the cephalosporins (this group is usually a carboxylate but a sulfate is found in the monobactams) and the acyl side chain borne by the ß-lactam ring must be on the ß face in a cis-position to the thiazolidine ring in the penicillins or dihydrothiazine ring in the cephalosporins (Fig. 1).

Enzyme acylation of ß-lactam compounds produces a leaving group which remains covalently attached to the enzyme ester-linked electrophilic portion of the donor. Consequently, the enzyme's active site remains occupied by a leaving group which cannot leave. In this respect, the reactions catalyzed by the β-lactamases contrast with those catalyzed by the DD-peptidases (Fig. 10).

The β-lactamases have adapted themselves to the situation created by a non-diffusable leaving group. Water is an excellent attacking nucleophile of the acyl (penicilloyl, cephalosporoyl, etc.) enzyme which, usually, is very short-lived. The β-lactamases can hydrolyze β-lactam antibiotics with very high k_{cat} values (up to 1000 s^{-1}).

In contrast, β-lactam compounds are mechanism-based inactivators of the DD-peptidases. Constant k_{+3} of reaction 1 may have a very small absolute value. However, the efficacy of the inactivation depends on the va-

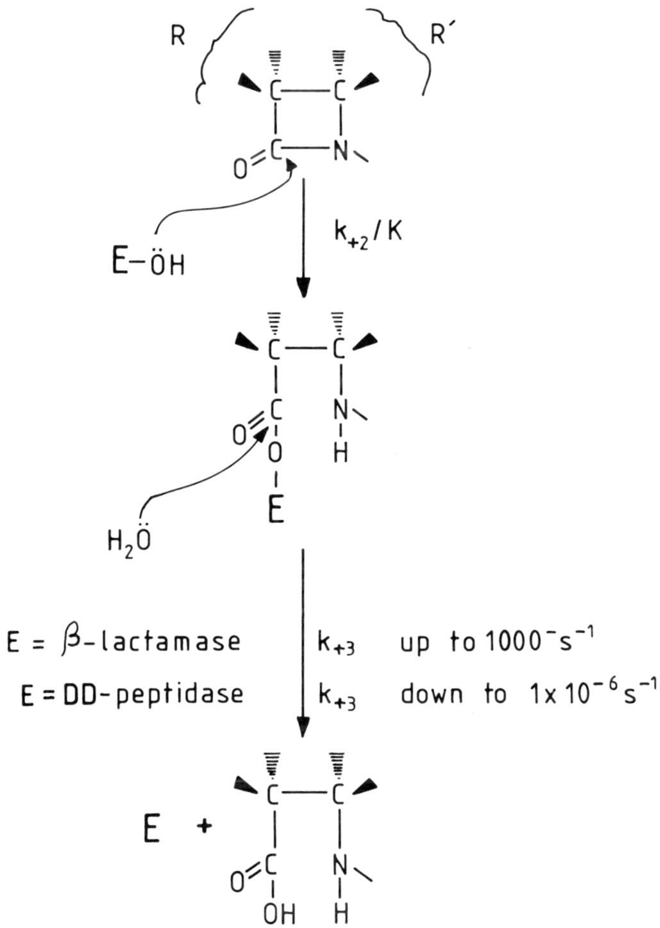

FIGURE 10. β-Lactam compounds as carbonyl donors of the β-lactamases and DD-peptidases. E-OH = enzyme with the hydroxyl group of the serine residue.

lues of all the kinetic parameters (9). The higher the second-order rate constant of enzyme acylation (k_{+1}/K; $M^{-1}s^{-1}$) and the smaller the first-order rate constant of acyl enzyme breakdown (k_{+3}; s^{-1}), the lower the β-lactam compound concentration ([D]) for which the term $Kk_{+3}/k_{+2}[D]$ of equation _2_ (which for $k_{+3} << k_{+2}$ simplifies to $[E]_o/[E\text{-}d^*]_{ss} = 1 + (Kk_{+3}/k_{+2}[D])$ becomes negligible. When this term is smaller than 0.01, virtually all the DD-peptidase is immobilized as acyl enzyme at the steady state of the reaction.

The time required for the reaction to reach the steady state is another important feature of the inactivation process. Assume four β-lactam compounds acting on a given DD-peptidase. They are used at the same concentration ([D] = 1 µM), the k_{+3} value for acyl enzyme breakdown is the same ($k_{+3} = 1 \times 10^{-4}s^{-1}$; i.e. the acyl enzyme has a half-life of 125 min) but the relevant k_{2+}/K values are 10, 100, 1000 and 10,000 $M^{-1}s^{-1}$. It follows from equations _2_ and _3_ that 9.1 %, 58 %, 90.9 % and 99 %, respectively, of the DD-peptidase is acylated at the steady state of the reaction and that it takes 698 min, 348 min, 70 min and 7.6 min, respectively, for the reaction to reach 99 % of the steady-state.

β-Lactam compounds have been extensively used to probe the active sites of the DD-peptidases/PBPs (12). The values of k_{+3} and k_{+2}/K (and sometimes the individual values of K and k_{+2}) have been determined accurately in some cases. An interesting conclusion of these studies is that a cyclic β-lactam compound is not necessarily a better acylating agent of the active-site serine residue than a non-cyclic peptide (or ester) donor. Moreover, as observed with non-cyclic donors, the efficacy of binding of the β-lactam compounds may be relatively weak (K = 0.1-10 mM). However, because of high k_{+2} values, the DD-peptidases/PBPs can be (almost) completely immobilized as acyl enzymes in a few seconds or minutes by β-lactam compound concentrations much smaller than the dissociation constant K.

Plots of extents of protein acylation _versus_ β-lactam compound concentrations give access to the β-lactam compound concentration ($[D]_{0.5}$) which is necessary to achieve 50 % saturation of a given DD-peptidase/PBP after a given time of incubation ($t_{0.5}$) and at a given temperature. Under certain conditions (9), these $[D]_{0.5}$ and $t_{0.5}$ values are related to the second-order rate constant of protein acylation by

$$([D]t)_{0.5} = 0.69 \frac{K}{k_{+2}} \qquad \underline{4}$$

Equation _4_ is valid if i) the incubation time is much smaller than k_{+3} (or $k_a >> k_{+3}$); ii) [D] << K; and iii) the β-lactam compound concentrations used are such that a 100 % level of protein acylation would be achieved at the steady-state of the reaction. That these conditions are fulfilled can be checked graphically; a 23-fold (or a 7.2-fold etc.) increase in the β-lactam concentration should be sufficient to cause a change in the extent of enzyme acylation from 10 % to 90 % (or from 20 % to 80 %, etc.). Hundreds of $[D]_{0.5}$ values concerning the interactions between β-lactam compounds and PBPs have been published. These data may give at least rough estimates of the values of the second-order rate constant of protein acylation (although in many instances, the experimental conditions are not described in sufficient detail to be sure that equation _4_ applies).

3.4. <u>Properties of the acyl enzymes formed by reaction between β-lactam carbonyl donors, the DD-peptidases and the β-lactamases</u>

The above discussion is an oversimplification. The acyl enzymes formed by reaction between some DD-peptidases and β-lactam compounds may be relatively short-lived. Thus the benzylpenicilloyl PBP4 of *Staphylococcus aureus* has a half-life of about 1.5 min. Conversely, the acyl enzymes formed by reaction between β-lactamases and β-lactam compounds may be long-lived. This is the case for the pairs cefoxitin-RTEM β-lactamase (k_{+3} = $5 \times 10^{-3} s^{-1}$) and cloxacillin-*E.cloacae* β-lactamase (k_{+3} = $1.5 \times 10^{-4} s^{-1}$). Finally, long-lived acyl enzymes (whether they are formed with a DD-peptidase or a β-lactamase) may undergo intramolecular rearrangements. Enzyme regeneration is then a complex pathway.

The first acyl enzyme rearrangement reported in the literature describes the rupture of the C_5-C_6 bond in an acyl (benzylpenicilloyl) DD-peptidase (13) (Fig. 11). This reaction, in which water donates the proton on C_6, has two consequences. The leaving group of the enzyme acylation step can now leave the enzyme's active site (to give rise to N-formyl-penicillamine) and the new phenylacetylglycyl enzyme (which is identical to that formed by reaction with phenylacetylglycyl-D-mandelate) is susceptible to concomitant and immediate attack by water and a structured amino compound. Hydrolysis of the benzylpenicilloyl enzyme with release of benzylpenicilloate and rupture of C_5-C_6 proceed at approximately the same low rates (= $0.5 \times 10^{-4} s^{-1}$) so that C_5-C_6 rupture is not an effective means for the enzyme to get relieved of the bound inactivator. Release of the phenylacetylglycyl moiety is pH-independent. Release of the benzylpenicilloyl moiety increases at high pHs. The second-order rate constant computed for the attack of the benzylpenicilloyl enzyme by OH^- ions is about 4 $M^{-1}s^{-1}$, a value which is not much higher than that observed for the hydrolysis of α-methylpenicilloate.

A second type of acyl enzyme rearrangement is observed with those cephalosporins which have a good 3' leaving group, for example an acetoxy substituent (14). Such groups are eliminated from the acyl enzyme before deacylation occurs and an exocyclic methylene is formed. This elimination may have important consequences with respect ot cephalosporin turnover by the β-lactamases and inactivation of the DD-peptidases since the acyl enzymes which have undergone elimination at position 3 are more hydrolytically inert than their parents. As shown with a DD-peptidase, the 3' substituent influences enzyme acylation in a way not predictable solely on the basis of the electronic effects and, conversely, the DD-peptidase may strongly influence the rates of the elimination reaction.

A third type of acyl enzyme rearrangement, which is an integral part of the β-lactamase inactivation mechanisms, involves ring opening between S_1 and C_5 in the penam sulfones and 6-β-bromo(iodo)penicillanate or between O_1 and C_5 in clavulanate (15). Ring opening launches a series of rearrangements which proceed ultimately to enzyme inactivation. The established common feature is that of a branched pathway where β-elimination of the acyl enzyme leads to a hydrolytically inert acyl enzyme $E\text{-}d^i$ and/or further modification of some residues of the enzyme's active site.

$$
\begin{array}{ccccccc}
E + D & \overset{K}{\rightleftharpoons} & E.D & \xrightarrow{k_{+2}} & E\text{-}D^{*} & \xrightarrow{k_{+3}} & E + P \\
 & & & & \downarrow k_{+4} & & \\
 & & & & E\text{-}D^{i} & &
\end{array}
$$

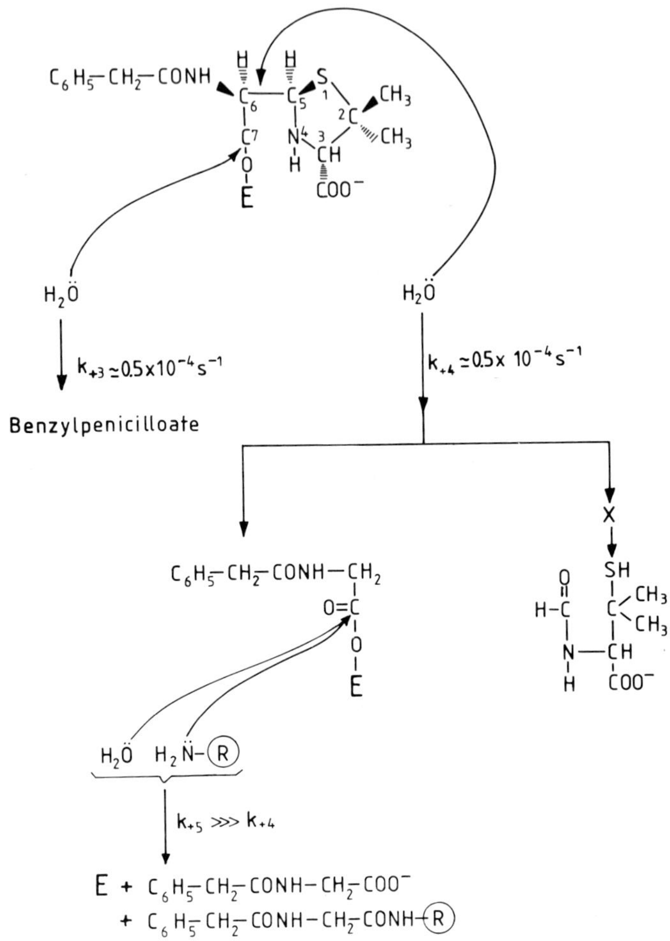

FIGURE 11. Rearrangement of the acyl enzyme formed by reaction between benzylpenicillin and a DD-peptidase. Rupture of the C_5-C_6 bond (k_{+4}) causes release of the leaving group of the enzyme acylation step.

For $k_{+3} \ll k_{+4}$, the fate of the ß-lactamase is inactivation. Under other conditions, the ß-lactamase can be totally inactivated only if $[D]/[E]_o \geq (k_{+3} + k_{+4})/k_{+4}$. The final composition of the reaction mixture depends on both the concentration of the penam and the $[D]/[E]_o$ ratio value. For $[D]/[E]_o < (k_{+3} + k_{+4})/k_{+4}$, the final mixture contains free enzyme, $E\text{-}D^i$ and hydrolyzed penam. For $[D]/[E]_o > (k_{+3} + k_{+4})/k_{+4}$, the final mixture contains $E\text{-}d^i$ and both hydrolyzed and intact penam.

6-ß-Iodopenicillinate has been used as a means to probe the active sites of β-lactamases. In all cases, rearrangement of the acyl enzyme gives rise to a dihydrothiazine chromophore (16). But depending on the β-lactamases, the value of second-order rate constant of enzyme acylation varies from 1 x 10^6 $M^{-1}s^{-1}$ to 100 $M^{-1}s^{-1}$. Moreover, k_{+3} may not be negligible in which case complete enzyme inactivation requires high molar ratios of 6-β-iodopenicillinate to enzyme.

6-β-Iodopenicillanate has also been used as a DD-peptidase labelling reagent. Rearrangement of the acyl enzyme also gives rise to the dihydrothiazine chromophore. Though a weak acylating donor (k_{+2}/K = 0.75 $M^{-1}s^{-1}$), 6-β-iodopenicillinate has the advantage to form an adduct which is much more stable than that formed by reaction with benzylpenicillin.

4. STRUCTURE OF, AND EVOLUTIONARY RELATIONSHIP BETWEEN DD-PEPTIDASES AND β-LACTAMASES

Advances have been made in three areas.

Gene cloning and sequencing (and sometimes direct analysis of the expressed proteins) has given access to the primary structure of ten β-lactamases (from plasmid pBR322, *E.coli*, *Citrobacter freundii*, *Enterobacter* P99, *Salmonella typhimurium*, *Bacillus licheniformis*, *Bacillus cereus*, *Staphylococcus aureus*, *Streptomyces albus* G and *Streptomyces cacaoi*) (17), three DD-peptidase/PBPs of low molecular mass (from *E.coli*, *B.subtilis* and *Streptomyces* R61) (6,18) and four bifunctional PBPs of high molecular mass, (PBPs 1A, 1B, 2 and 3 of *E.coli*) (19). The proteins or the penicillin-binding (transpeptidase) domains of the bifunctional PBPs are 250- to 350-amino acid polypeptide chains. Using a program which compares protein sequences on the basis of the relative amino acid substitution frequencies found among families of homologous proteins, some of these proteins (or domains) form pairs or groups which match very well with respect to the amino acid sequences. However, when the pairs or groups thus defined are compared to each other, the overall similarity becomes marginally significant or vanishes.

Enzymic degradation of long-lived acyl enzymes formed with β-lactam compounds and analysis of the β-lactam-labeled peptide fragments has led to the identification of the active-site serine residue. This serine residue is always close to the amino terminus of the protein (at position 62 in the *Streptomyces* R61 DD-peptidase) or the penicillin-binding (transpeptidase) domain of the bifunctional PBPs. Moreover, in all cases, a lysine residue occurs at the third position on the carboxyl side of the active-site serine (Ser*-Xaa-Xaa-Lys).

The exocellular DD-peptidase of *Streptomyces* R61 and several β-lactamases have been crystallized. The tertiary structure and active site of the DD-peptidase are known at 2 Å resolution (20). X-ray analysis at 3.5 Å of the β-lactamases of *Bacillus licheniformis* (20) and *B.cereus* (21) (these two β-lactamases have highly homologous amino acid sequences) has provided clear images of the polypeptide folding. Though the *Streptomyces* DD-peptidase and the baccilus β-lactamases lack structural relatedness in the primary structure, they have very similar tertiary structures. They contain eight helices and a β-sheet of five strands protected on the front face by two helices and on the back face by another helix (Fig. 12). Matching of the helices and sheet strands between the three structures is so great that it is impossible to argue against an evolutionary relationship.

Alignment of the primary structures (with appropriate deletions or insertions) of the seventeen penicillin-recognizing enzymes for which the data are available, generates boxes which contain identical and/or functionally homologous residues (Fig. 13). These boxes occur at similar posi-

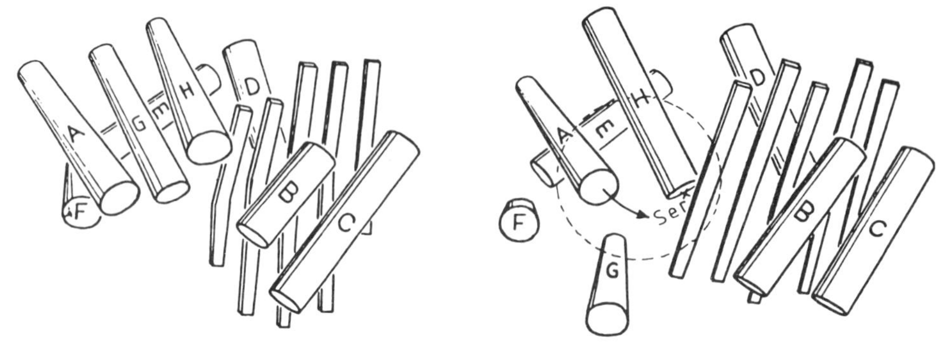

Bacillus licheniformis
β-lactamase

Streptomyces R 61
DD-peptidase

FIGURE 12. Spatial disposition of secondary structures in the Bacillus licheniformis β-lactamase and the Streptomyces R61 DD-peptidase. The eight α helices (cylinders) are marked A through H. The β sheet (five strands) is protected by helices B and C on one face and by helix D on the other. The active-site serine (star) is at the N-terminus of helix H. The active site area is circled.

Enzymes	Box I	Box II	Box III	Box IV	Box V	Box VI	Box VII	
a	G_{42} D_{45}	R_{57} F R V G S* V T K	Y_{90}	T_{101} V R	D_{225}	W_{271}	G_{297} H T G	Streptomyces R61 DD-peptidase
b	G_{44} D_{47}	L_{59} F E L G S* V S K I	Y_{92}	T_{97} A K G	D_{217}	W_{260}	H_{314} K T G	β-lactamases
c	G_{45} D_{50}	R_{65} F A F A S* T S K P Y M I M Y	Y_{105}	T_{109} E K L	E_{166}	W_{210} L	D_{233} K S G T	
d	G_{24} D_{29}	L_{43} F P M C S* V F K	Y_{74}(?)	T_{90} G K	E_{152}	W_{194}	D_{217} K T G	
e	G_{23} D_{28}	R_{42} F A Y G S* T F K	Y_{83}(?)	T_{95} G K	E_{260}	W_{200}	D_{223} K S G	
f	G_{42} D_{48}	R_{67} Y S P A S* T F K	H_{107}	D_{112} L R	D_{157}	L_{196}	A_{209} K T G	
g	G_{20} D_{29}	R_{31} L P I A S* M T K	$Y_{61 or 68}$	L_{81} R K	D_{490}	W_{207}	L_{226} K T G	Low MW PBPs
h	G_{28} D_{37}	R_{39} R D P A S* L T K	N_{80}	F_{90} L K	D_{175}	W_{203}	I_{212} K T G H	
i	G_{444} D_{447}	L_{460} R Q V G S* N I K	Y_{511}	Q_{523} S K	D_{588}	W_{697} 702	G_{715} K T G	High MW PBPs
j	G_{489} E_{492}	R_{505} R S I G S* L A K	Y_{557}	L_{569} T R	D_{647}	W_{672}	G_{697} K T G	
k	G_{300} D_{305}	V_{325} Y P P A S* T V K	Y_{367}	H_{375} G R	D_{490}	W_{512}	A_{543} K S G	
l	G_{288} E_{292}	V_{302} F E P G S* T V K	Y_{334}	V_{344} A R	E_{456}	L_{476}	I_{493} K T G T	

FIGURE 13. Conserved boxes along the primary structures of penicillin-recognizing enzymes. a) Streptomyces R61 DD-peptidase; b) enterobacterial β-lactamases; c) B.licheniformis, B.cereus, pBR322 and S.aureus β-lactamases; d) Streptomyces albus G β-lactamase; e) Streptomyces cacaoi β-lactamase; f) Salmonella typhimurium Oxa-2 β-lactamase; g) B.subtilis PBP5; h) E.coli PBP5; i) E.coli PBP1A; j) E.coli PBP1B; k) E.coli PBP2; l) E.coli PBP3. The boxes in i to l refer only to the penicillin-binding (transpeptidase) domains of the PBPs.

tions along the amino acid sequences (22) and some of them are known to occupy critical positions in the tertiary structure of the *Streptomyces* DD-peptidase (23). It has been proposed (22) that this conserved pattern probably expresses a common polypeptide scaffolding.

Assuming that hundreds, perhaps thousands, of different DD-peptidases and β-lactamases occur in the bacterial world, the accumulated information is still very limited. Yet, it has led to significant conclusions and allows predictions to be made.

The β-lactamases, DD-peptidases and penicillin-binding/transpeptidase domains of bifunctional PBPs have derived by divergent evolution from a common ancestor gene that must have coded for an essential DD-transpeptidase. These enzymes offer an example of persistence of similarities between three-dimensional structures while i) the similarity between primary structures has progressively disappeared as a function of the evolutionary distance; and ii) the active sites have been modified to give rise to multiple penicillin-binding DD-peptidase of varying functionality and essentiality and multiple penicillin-hydrolysing β-lactamases of varying anti-β-lactam antibiotics spectra.

The tertiary arrangement of packed α-helices and β-strands shown in Fig. 12 is unprecedented. Though structural changes have occurred as the sequences have diverged, the active-site serine DD-peptidase and β-lactamases form a very large family of enzymes which are characterized by a similar fold of the polypeptide chains.

Differences in the catalyzed reactions can, at least in part, be viewed as the expression of the variations seen in the amino acid composition of some of the conserved boxes. As a corollary, these amino acids are targets of choice to be studied by site-directed mutagenesis.

5. CATALYTIC MECHANISM OF THE β-LACTAMASES, DD-PEPTIDASES AND LL-PEPTIDASES OF THE TRYPSIN FAMILY, AN UNIFIED VIEW

Understanding the established catalytic properties of an enzyme at atomic resolution can be approached only by cross-fertilization between various disciplines and technologies : tertiary structure determination, gene analysis and site-directed mutagenesis, molecular modelling and theoretical chemistry. As described above, recent advances in the field of the DD-peptidases and β-lactamases have brought the goal closer.

The DD-peptidases and β-lactamases as well as the LL-peptidases of the trypsin family catalyze rupture of a peptide (amide, ester) bond in a carbonyl donor. Such a bond has a permanent dipole which confers a partial positive charge on the carbonyl carbon atom. Bond rupture requires polarization of the C=O by an electrophile, attack of the carbonyl carbon atom by a nucleophile and proton donation on the nitrogen (oxygen) atom. The reaction as it is catalyzed by the active-site serine LL-peptidases (and using the numbering of chymotrypsin) is currently attributed to the concerted action between the conserved triad Ser 195 ... Hist 57 ... Asp 102 acting as proton abstractor and donator and a pair of hydrogen bonds from the backbone NH of Gly 193 and Ser 195 acting as electrophile.

The tertiairy structure of the penicillin-recognizing enzymes is not known to a resolution that permits identification of the amino acid side chains possibly involved in catalysis in concert with the active-site serine. But histidine is not a conserved residue and one histidine-free β-lactamase (from *Streptomyces albus* G) is known.

Since histidine is not always involved in enzyme-catalyzed rupture of a peptide (amide, ester) bond, the role of the diad His 57 ... Asp 102 in chymotrypsin has been addressed by theoretical studies (24). *Ab initio* calculations have led to an unified concept according to which the essential components of the catalytic machinery of the β-lactamases, DD-peptidases and LL-peptidases are a serine residue, a polarizing agent and a water molecule. This water molecule acts as the proton carrier while the diad His 57 ... Asp 102, characteristic of the LL-peptidases, simply orients the interacting system and is dispensable. The functioning of the active-site serine β-lactamases, DD-peptidases and LL-peptidases thus depends only on the initial geometry of the interacting system that is generated by the carbonyl donor bound to the enzyme active site. Depending on the R and R' groups which flank the scissile bond (Fig. 14), the dihedral

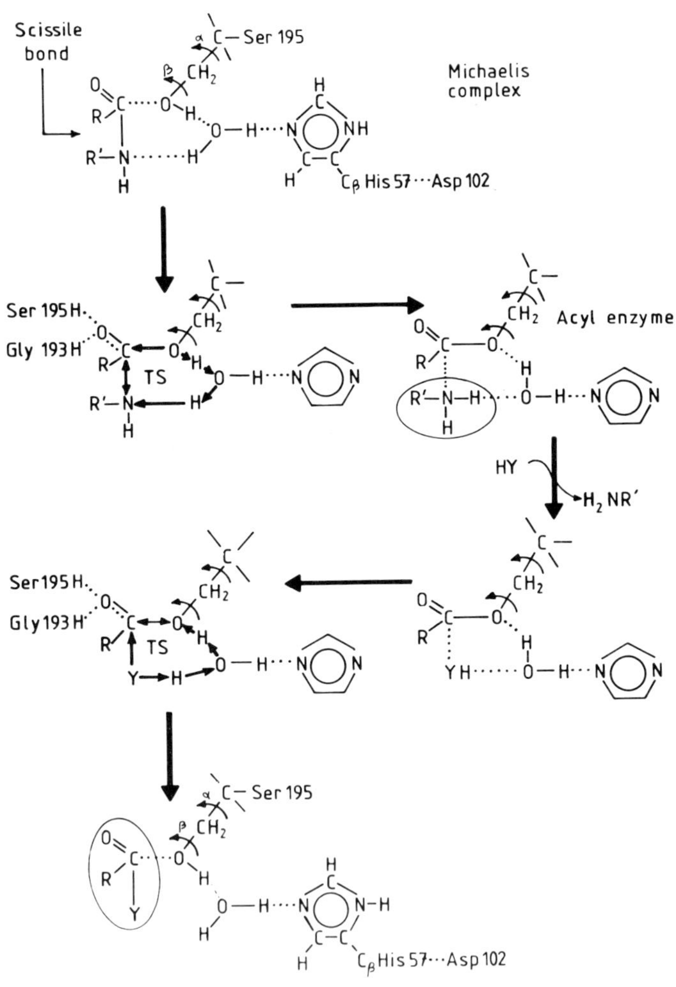

FIGURE 14. Proton shuttle during chymotrypsin-catalysed rupture of the scissile bond of a susceptible amide carbonyl donor. TS = transition state. A water molecule serves a protein carrier. The diad His 57 ... Asp 102 simply serves in orienting the interacting system.

angles α and β of the side chain of the active-site serine adopt varying orientations which greatly influence the polarization of the C=0 bond and, consequently, the sense and ease of the proton shuttle. Under optimal geometrical conditions, the whole interacting system rearranges itself in such a way that enzyme acylation (and enzyme deacylation) occur at high rates without any significant energy barrier.

The above concept has a corollary. β-Lactam compounds may have donor activity for the mammalian LL-peptidases (trypsin, chymotrypsin, elastase, plasmin, thrombin) if their structures match the active sites. Assuming that the LL-peptidases differ from the DD-peptidases in having endopeptidase *versus* carboxypeptidase activity and in preferring L *versus* D configuration, cephalosporins have been modified accordingly and converted into effective mechanism-based inactivators of the LL-peptidases (25). These cephalosporins are 4' esters or amides and bear the acyl substituent in the 7α position so that they better mimick the mammalian substrates. Inactivation occurs by generation of a stable acyl enzyme. Reaction 1 and equations 2, 3 and 4 apply.

In the case of porcine pancreatic elastase and as shown by refined X-ray crystallography studies (25), the acyl enzyme formed with the long-lasting inactivator 3-acetoxymethyl-7-α-chloro-3-cephem-4-carboxylate-1,1-dioxide tert-butyl ester undergoes elimination of both the 7' chlorine and the 3' substituent. In this case, His 57 rotates away from its normal position in the enzyme's active site and becomes exposed to the exocyclic methylene. Slow alkylation of His 57 by Michael addition ($t1/2 = 2$ hours) provides the acyl enzyme with a second covalent bond which is not subject ot hydrolysis (or reversal by hydroxylamine). This particular cephalosporin is a potent inactivator of the human leukocyte elastase. The value of the second-order rate constant of enzyme acylation is 161,000 $M^{-1}s^{-1}$ (with $K = 1.8 \times 10^{-7}$ M and $k+2 = 0.029\ s^{-1}$).

Hence, 45 years after the first use of penicillin as a drug, β-lactam compounds - which so far have served their therapeutical role as antibacterial agents by inactivating the DD-peptidases - can now be remodelled to attack other enzyme targets of great medical importance.

ACKNOWLEDGEMENTS

It gives me great pleasure to thank my colleagues who contributed to the work described here : P. Charlier, J. Coyette, O. Dideberg, G. Dive, C. Duez, J. Dusart, J.M. Frère, B. Joris, J. Lamotte-Brasseur, M. Leyh-Bouille, M. Nguyen-Distèche from the University of Liège, Belgium and J. Kelly and J. Knox from the University of Connecticut, Storrs, CN, USA.

The work in Liège has been supported by an *Action concertée* from the Belgian Government (convention 79/84-I1), the *Fonds de la Recherche Scientifique Médicale*, Brussels (contract no 3.4507.83) and a Convention tripartite between the *Région wallonne*, Continental Pharma and the University of Liège.

REFERENCES

1. Ghuysen JM. Bact. Rev. 32:425-464, 1968.
2. Blumberg PM, Strominger JL. Bact. Rev. 38:291-335, 1974.
3. Spratt BG. J. Gen. Microbiol. 129:1247-1260, 1983.
4. Williamson R, Tomasz A. Eur. J. Biochem. 151:475-483, 1985.
5. Hedge PJ, Spratt BG. FEBS Lett. 176:179-184, 1984.
6. Duez C, Piron-Fraipont C, Joris B, Dusart J, Urdea MS, Martial JA, Frère JM, Ghuysen JM. Eur. J. Biochem., in press, 1987.

7. Frère JM, Duez C, Ghuysen JM, Vandekerckhove J. FEBS Lett. 70:257-260, 1976.
8. Knott-Hunziker V, Waley SG, Orlek BS, Sammes PG. FEBS Lett. 99:59-60, 1979.
9. Ghuysen JM, Frère JM, Leyh-Bouille M, Nguyen-Distèche M, Coyette J. Biochem. J. 235:159-165, 1986.
10. Rasmussen JR, Strominger JL. Proc. Natl. Acad. Sci. USA 75:84-88, 1978; Pratt RF, Govardhan CP. Proc. Natl. Acad. Sci. USA 81:1302-1306, 1984.
11. Nguyen-Distèche M, Leyh-Bouille M, Pirlot S, Frère JM, Ghuysen JM. Biochem. J. 235:167-182, 1986.
12. Frère JM, Joris B. CRC Crit. Rev. Microbiol. 11:299-396, 1985.
13. Frère JM, Ghuysen JM, Degelaen J, Loffet A, Perkins HR. Nature 258: 168-170, 1975.
14. Boyd DB. J. Med. Chem. 27:63-66, 1984.
15. Cartwright SJ, Waley SG. Med. Res. Rev. 3:341-382, 1983.
16. De Meester F, Frère JM, Waley SG, Cartwright SJ, Virden R, Lindberg F. Biochem. J. 239:575-580, 1986.
17. Ambler R. Phil. Trans. R. Soc. Lond. B289:285-301, 1980; Dale JW, Goldwin D, Mossakowski D, Stephenson P, Wall S. FEBS Lett. 191:39-44, 1985; Dehottay P et al. Eur. J. Biochem., submitted; Lenzini M. et al., unpublished data.
18. Broome-Smith J, Edelman A, Spratt BG. in The Target of Penicillin (Höltje JV, Hackenbeck R, Labischinski H, Eds). Walter De Gruyter, Berlin, pp 403-408, 1983; Todd JA, Roberts AN, Johnstone K, Piggot PJ, Winter G, Ellar DJ. J. Bacteriol. 167:257-264, 1986.
19. Broome-Smith JK, Edelman A, Yousif S, Spratt BG. Eur. J. Biochem. 147:437-446, 1985; Asoh S, Matsuzawa H, Ishino F, Strominger JL, Matsuhashi M, Ohta T. Eur. J. Biochem. 160:231-238, 1986; Nakamura M, Maruyama IN, Soma M, Kato J, Suzuki H, Hirota Y. Mol. Gen. Genet. 191:1-19, 1983.
20. Kelley JA, Dideberg O, Charlier P, Wéry JP, Libert M, Moews PC, Knox JR, Duez C, Fraipont C, Joris B, Dusart J, Frère JM, Ghuysen JM. Science 231:1429-1437, 1986.
21. Samraoui B, Sutton BJ, Todd RJ, Artymiuk PJ, Waley SG, Philips DC. Nature 320:378-380, 1986.
22. Joris B et al., unpublished data.
23. Kelley J et al., unpublished data.
24. Dive G et al., unpublished data.
25. Doherty JB, Ashe BM, Argenbright LW, Barker PL, Bonney RJ, Chandler GO, Dahlgren ME, Dorn CP Jr, Finke PE, Firestone RA, Fletcher D, Hagmann WK, Mumford R, O'Grady L, Maycock AL, Pisano JM, Shah SK, Thomson KR, Zimmerman M. Nature 322:192-194, 1986; Navia MA, Springer JP, Lin T-Y, Williams HR, et al. Science, in press, 1987.

NEW ANTIFUNGAL AGENTS : A RESULT OF TARGET-ORIENTED DRUG DESIGN

P.A.J. JANSSEN
Janssen Pharmaceutica, B-2340 Beerse, Belgium

1. INTRODUCTION

Until 1966, the choice of antifungal agents for the management of fungal disease has been relatively limited. Indeed, apart from Whitfield's ointment and potassium iodide, only polyene antibiotics (amphotericin B and nystatine), griseofulvin, tolnaftate and flucytosin were available for widespread use (1). Both the spectrum of activity and the side effect profile of these agents were far from optimal.

In 1966, exactly 20 years ago, the first broad spectrum imidazole derivative was synthesized. Its name was etonam (R 10 100) (2). This topically active imidazole was rapidly followed by the well known agents miconazole and clotrimazole. These topical imidazoles were a significant improvement when compared with the already existing antifungals. Indeed, their spectrum included the vast majority of pathogenic fungi of the skin and the mucous membranes : dermatophytes and yeasts (3). In certain patients with systemic mycoses, the intravenous form of miconazole has proven to be a valuable alternative for the rather toxic amphotericin B (4).

In contrast with the achievements in antibacterial chemotherapy, several decennia of antifungal drug research have not produced a single broad-spectrum orally-active antifungal agent. As a consequence, the chemical research has focussed on the synthesis of a compound with improved oral absorption without, however, loosing the already acquired advantage of broad-spectrum activity. The result of this drug design became evident in 1976, with the advent of ketoconazole (R 41 400) (5).

2. KETOCONAZOLE : ANTIFUNGAL AND PHARMACOLOGIC TURNING TABLE

Shortly after entering clinical trials, it became evident that ketoconazole was a most promising agent. Excellent results were obtained in the treatment of paracoccidioidomycosis and histoplasmosis (6). Later, blastomycosis and systemic candidosis were added in this list (7). The fame of the drug came especially from its results in chronic mucocutaneous candidosis (CMC). Studies in vaginal candidosis had shown cure rates of 90 % when the drug was given at a dose of 400 mg for 5 consecutive days (8). Equally promising results were obtained in pityriasis versicolor and dermatomycoses (9).

As a consequence of these findings, ketoconazole became rapidly available for widespread use. The broad usage of the drug initially confirmed the activity and side effect profile obtained during the first clinical trials. Apart from nausea, headache and pruritus, ketoconazole appeared to be well tolerated. Only after some time rare but more serious side effects were reported. The most important one was hepatitis (10). This phenomenon appeard to be a metabolic drug idiosyncrasy especially observed in elderly female patients on long-term treatment. The overall incidence of the hepa-

P.A.J. Janssen

titis was 1 in 10,000 to 1 in 15,000 patients. When patients with onychomycosis were considered, this incidence rose to 1 in 2,000 patients. For patients with vaginal candidosis and pityriasis versicolor (5 to 10 days therapy), the incidence was estimated to be between 1 in 500,000 to 1 in a million (9).

A second type of unexpected "side" effects consisted of decreased libido and gynecomastia (9). These side effects were virtually never seen at the usual therapeutic dose of 200 mg, but were occasionally reported in (coccidioidomycosis) patients receiving high multiple daily doses of ketoconazole (11). To understand this phenomenon it is important to go back to the biochemical mechanism of action of imidazoles in general and ketoconazole in particular.

The azole antifungals investigated so far are potent and selective cytochrome P-450 inhibitors. Cytochrome P-450 isozymes (P-450) are iron containing hemoproteins catalyze the monooxygenation of a broad spectrum of lipophylic substances. Examples are shown in Table 1.

TABLE 1.

Cytochromes P-450 are involved in the synthesis and metabolism of endobiotics such as

Fatty acids	Mineralocorticoids
Sterols	Vitamin D
Sex steroids	Leukotrienes
Glucocorticoids	Prostaglandins

They play a key role in the metabolism of xenobiotics.

The name P-450 is derived from the fact that in its reduced form and in the presence of carbon monoxide this pigment gives maximum absorption at about 450 nm. The active site of P-450 isozymes contains an iron protoporphyrin IX moiety (Fig. 1). The iron is penta- or hexacoordinate, four of the ligands being contributed by the tetradentate prophyrin ring.

The fifth ligand is a thiolate iron contributed by a cysteine residue of the apoprotein. In the substrate-free ferric state the 6th coordination position of the iron is occupied by an easily exchangeable ligand, perhaps water.

The catalytic sequence for P-450 enzymes involves different steps (Fig. 2; for a review see ref. 12): (1) binding of the substrate (RH); (2) transfer of an electron with reduction of the iron atom; (3) binding of molecular oxygen to the 6th coordination position of the heme iron; (4) transfer of a second electron; (5) cleavage of the oxygen-oxygen bond with formation of a molecule water; (6) transfer of the second oxygen to the substrate and (7) dissociation of the hydroxylated product (ROH).

The following equation shows a typical reaction in which P-450 serves as the oxygenating catalyst :

$$RH + O_2 + 2e^- + 2H^+ \rightarrow ROH + H_2O.$$

A number of azole antifungals such as miconazole and ketoconazole (Fig. 3) show high affinity for microsomal P-450 in Saccharomyces cerevisiae and Candida albicans (13). Evidence is available indicating that the N_3 of the imidazole ring binds to the 6th coordination position of the iron atom in-

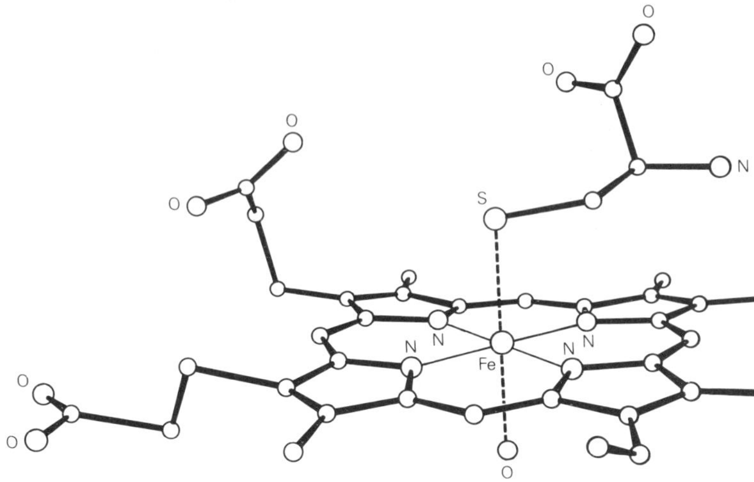

FIGURE 1. Proposed perspective drawing of the iron protoporphyrin in the resting form. The fifth ligand is a thiolate anion contributed by a cysteine residue of the apoprotein. The sixth ligand might be H_2O.

hibiting in this way oxygen binding and activation and thus hydroxylation of the substrate.

In Fig. 4 a proposed perspective drawing of the ketoconazole P-450 complex is given; the unhindered nitrogen is bound to the 6th position of the heme iron. The hydrophobic nonligating portion of the azole antifungal is bound to the apoprotein. A number of studies suggest that the hydrophobic interactions between azole antifungals and the apoprotein might be at the origin of their affinity and selectivity (14,15).

In yeast and fungi a P-450 isozyme is involved in the biosynthesis of ergosterol, the main sterol present in their membranes. Indeed, the 14α-demethylation of the precursor, lanosterol, is, as shown in Fig. 5, in the presence of NADPH and molecular oxygen, catalyzed by a P-450 isozyme, the $P\text{-}450_{14DM}$ (13,14).

A number of studies (13,14) have shown that miconazole, ketoconazole and other azole antifungals inhibit at nanomolar concentrations the ergosterol biosynthesis in C.albicans. Similar effects were observed for example with ketoconazole in Aspergillus fumigatus (16) and with miconazole in Trichophyton mentagrophytes (17).

When C.albicans were incubated for 24 h in the presence of 7.5×10^{-9} M ketoconazole a complete inhibition of ergosterol synthesis was observed. Fifty % inhibition was already achieved after 1 h contact with 5×10^{-9} M ketoconazole (18,19). In all studies done (13), the inhibition of ergosterol synthesis coincided with an accumulation of 14α-methylsterols (Fig. 6).

The role of sterols is to modulate membrane fluidity and to stabilize membranes (13). The fluidity of the lipid bilayer regulates the membrane's properties as well as membrane-bound enzymes. Ergosterol is needed for growth and plays an important role in yeast mitochondria. Indeed, high levels of ergosterol are found associated with mitochondrial enzymes such as

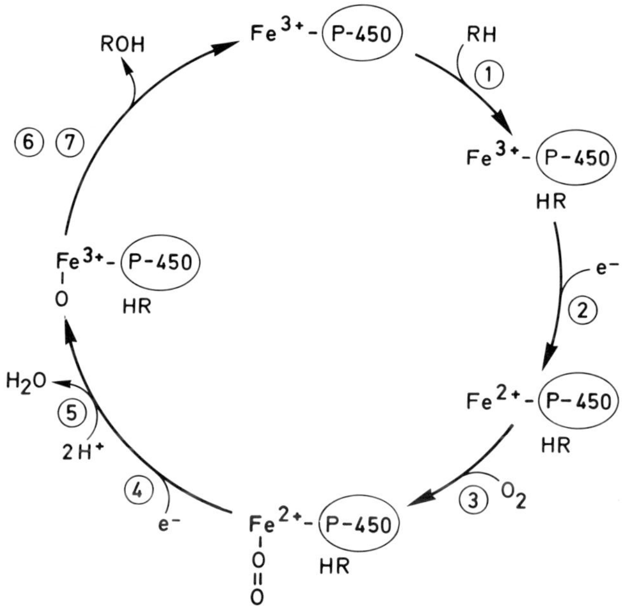

FIGURE 2. The catalytic sequence of a cytochrome P-450 catalyzed hydroxylation of the substrate RH.

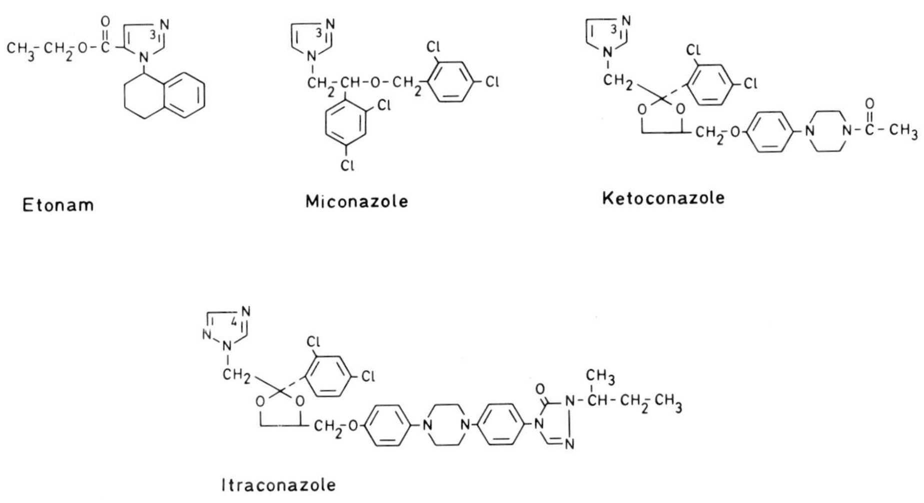

FIGURE 3. Chemical structure of N-substituted azole antifungals.

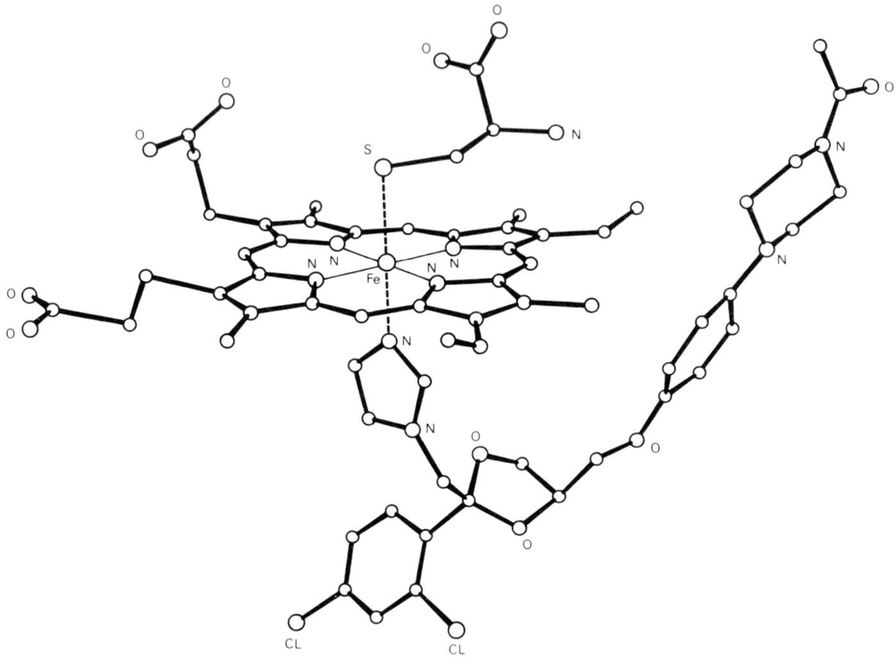

FIGURE 4. Proposed structure of the cytochrome P-450-ketoconazole complex. The unhindered nitrogen (N_3) is bound to the 6th coordination position of the heme iron.

cytochrome oxidase. Therefore, from a depletion in ergosterol will result changes in membrane permeability, membrane-found enzymatic activities, development of mitochondria and cell proliferation. Evidence has been collected showing the necessity to remove the 14α-methyl group from i.e. 24-methylenedihydrolanosterol (Fig. 6) to obtain a functional sterol (13). Thus the induced accumulation of 14α-methylsterols will contribute to the deteriorating effects of azole antifungals on fungal cells.

A consequence of ergosterol depletion and/or 14α-methylsterol accumulation is the activation of the chitin synthase. This enzyme is responsible for the synthesis of chitin, a major component of the septa and hyphal tip (Fig. 7a). It has been proven quantitative, as well as by using fluorescence microscopy, in C.albicans (13) and A.fumigatus (16) that the depletion of ergosterol and accumulation of 14α-methylsterols create unbalanced conditions in the cell from which a generalized activation of the chitin synthase may originate (Fig. 7b). Hence, abnormal cells ensue with too high a content of chitin.

In summary (Fig. 8), the interaction of azole antifungals with $P\text{-}450_{14DM}$, gives rise to an inhibition of ergosterol synthesis. This, in turn, leads to a number of changes in membrane and membrane-bound enzyme properties, thus accounting for their antifungal activity.

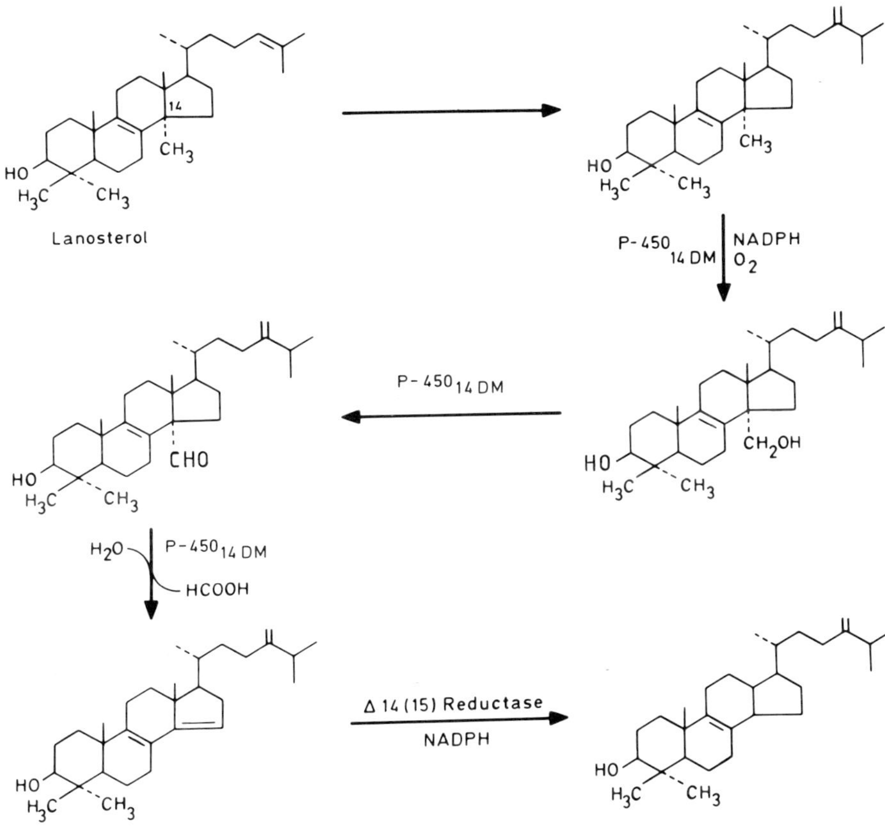

FIGURE 5. The 14α-demethylation of lanosterol. $P\text{-}450_{14DM}$ = the cytochrome P450 isozyme involved in the catalysis of the 14α-demethylation.

The role of P-450 isozymes in cellular metabolism in mammalian cells calls for an extensive evaluation of the interaction of azole antifungals with mammalian P-450 isozymes. The P-450-dependent 14α-demethylase, similar to the enzyme present in yeast and fungi, has a key function in the synthesis of cholesterol. However, the 14α-demethylation of lanosterol by subcellular fractions of liver is much less sensitive to azole antifungals than the demethylation by similar fractions of yeast. Some results are summarized in Fig. 9. This difference in sensitivity between yeast and mammalian P-450-dependent 14α-demethylases was also observed by comparing the effects of azole antifungals on ergosterol and cholesterol synthesis in intact C.albicans and mammalian cells, respectively (13,15).

Of interest is ketoconazole's interaction with P-450 isozymes of testis microsomes. Indeed at high concentrations this imidazole decreases the absorbance at 450 nm of the reduced P-450-CO complex. An IC_{50} value of about 4×10^{-7} M and 1×10^{-6} M was found with piglet or dog testis microsomes, respectively (19,20). The microsomal synthesis of androgens is ca-

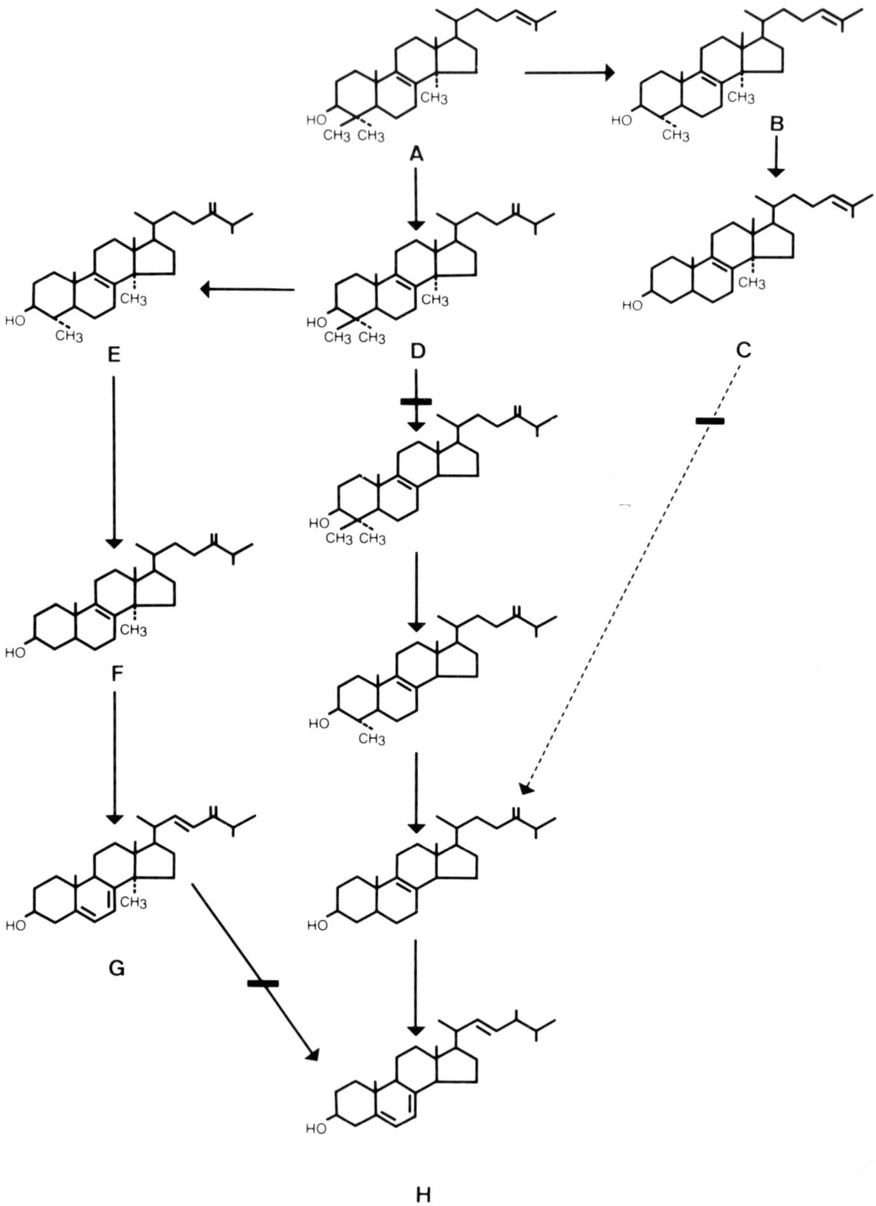

FIGURE 6. Simplified biosynthetic pathway of ergosterol in yeast and fungi. A : lanosterol; B : 4,14-dimethylzymosterol; C : 14α-methylzymosterol; D : 24-methylene-dihydrolanosterol; E : obtusifoliol; F : 14-methylfecosterol; G : 14α-methyl-ergosta-8,24(28)-dien-3β-ol; H : ergosterol.

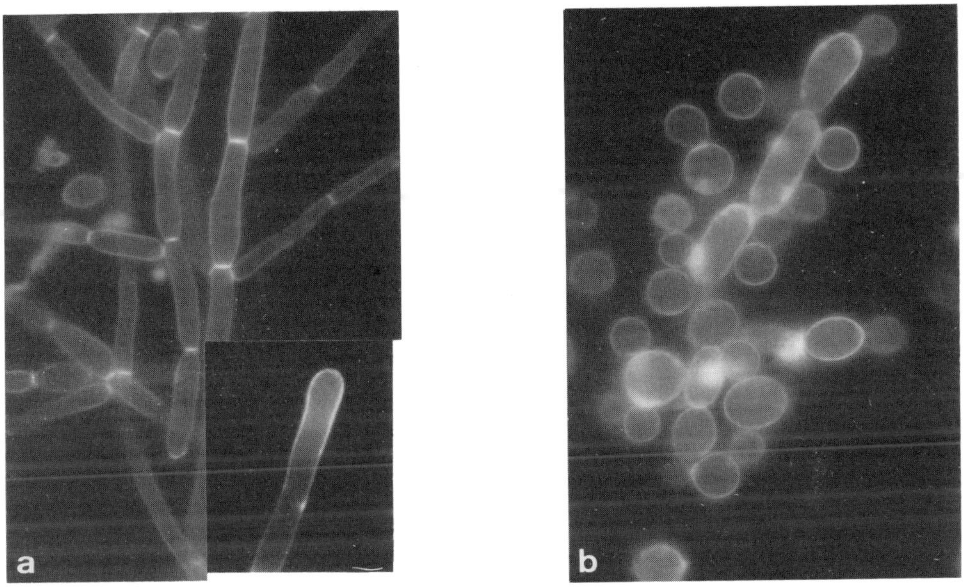

FIGURE 7. Fluorescence microscopy of C.albicans grown for 24 h and then stained with Direct Yellow 62 (Janssen Chimica) : a, control cells; b, cells treated with 5 x 10^{-9} M ketoconazole. The insert shows the hyphal tip taken from a control culture.

talyzed by two P-450-dependent enzyme systems, the 17α-hydroxylase and the 17,20-lyase (Fig. 10). Our studies have shown that ketoconazole interferes with the 17,20-lyase resulting, at concentrations higher than those needed to inhibit ergosterol synthesis, in an inhibition of testosterone biosynthesis (21,22). This effect makes ketoconazole, at high doses, a candidate for the treatment of some androgen-depending diseases, mainly prostate carcinoma.

The investigations on the effects of azole derivatives on the different types of mammalian P-450 has not only resulted in new antifungals with improved selectivity towards fungal P-450 but also in new possibilities for therapeutic applications of azole derivatives in other disease, i.e. prostatic cancer (23), hirsutism (24), acne (25) and psoriasis (26). These findings have made ketoconazole a real pharmacologic turning table for the testing of the role of P-450 in these conditions.

3. ITRACONAZOLE : AN EXAMPLE OF TARGET-ORIENTED DRUG DESIGN

When looking at ketoconazole's therapeutic and safety profile it is evident that space is left for improvements. Certain fungi such as Aspergillus spp. are not very sensitive to ketoconazole. The increasing incidence of aspergillosis in immunocompromised patients, however, necessitates the availability of a more potent drug against Aspergillus (27). Especially in dermatology, certain infections such as tinea capitis (due to M.canis) and toenail infections still constitute a major therapeutic problem which is not solved by ketoconazole (27). Pharmacokinetics and distribution of ketoconazole show that plasma levels are usually higher than

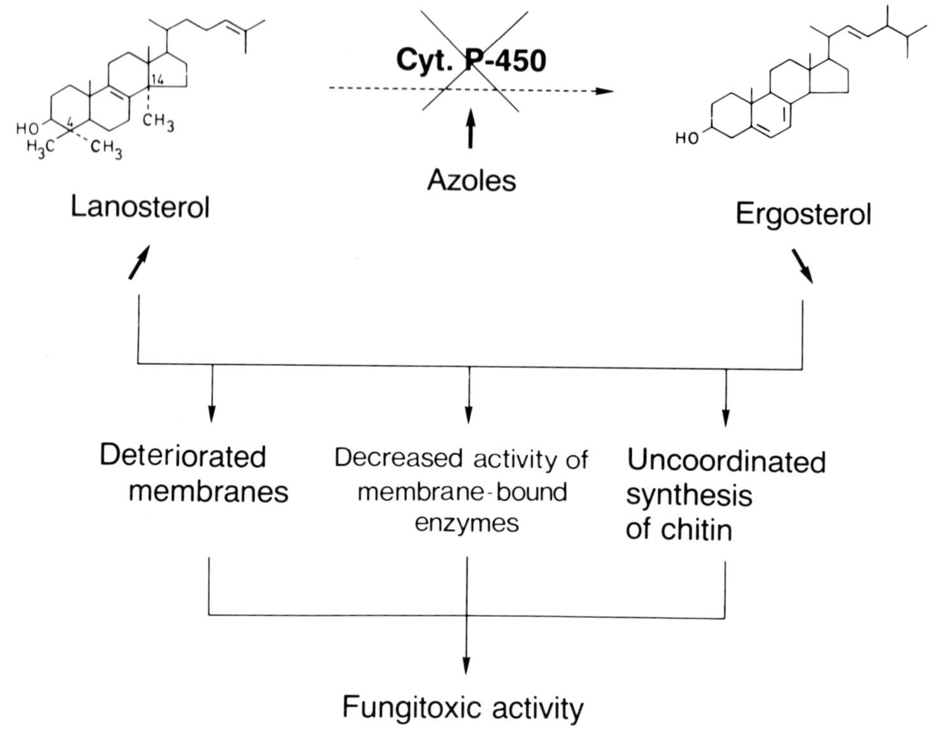

FIGURE 8. Summary of the mode of action of azole antifungals.

the levels observed in the tissues where fungal infections usually prevail (skin, mucosa, lungs) (28). The comments above on cytochrome P-450 suggest that an improved selectivity in the biochemical mechanism of action would also be beneficial to the safety profile of a systemic antimycotic.

As a consequence, drug design has focussed on the development of a molecule which included Aspergillus in its spectrum, which had an even more pronounced effect on dermatophytes (especially M.canis) and which showed a reversed plasma level/tissue level ratio. These molecules were evaluated in a variety of in vitro and in vivo systems to evaluate the effects on several mammalian P-450 systems. This complex process has resulted in the synthesis and development of itraconazole (R 51 211) (29). Itraconazole is a fairly large molecule with a highly lipophilic character. Its spectrum encompasses all common human pathogenic fungi including Aspergillus. This spectrum has been confirmed in animal models, and, in addition, these models have demonstrated that itraconazole, in contrast to ketoconazole, is also capable of curing meningeal infections due to Cryptococcus neoformans (30).

The biochemical mechanism of action of itraconazole is similar to that of the other azole antifungals. As shown in Fig. 11, the N_4 of the triazole ring of itraconazole binds to the 6th coordination position of the heme iron in a similar fashion as does the N_3 of the imidazole ring of

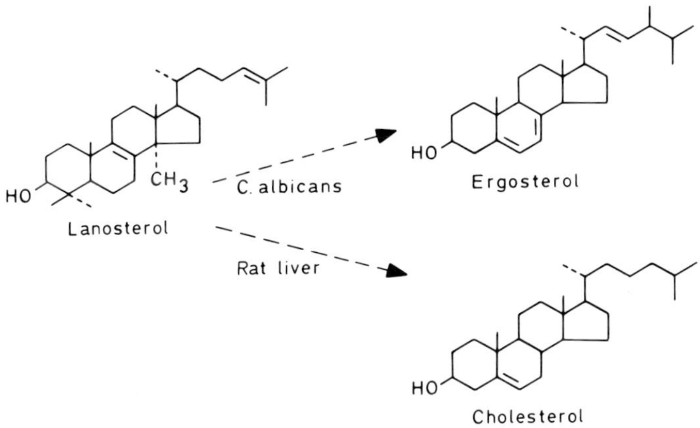

Azole derivative	I_{50}-value (M)	
	Rat liver	Candida albicans
ITRACONAZOLE	7×10^{-6}	7×10^{-8}
KETOCONAZOLE	2×10^{-6}	7×10^{-8}
MICONAZOLE	6×10^{-6}	8×10^{-8}

FIGURE 9. Effects of azole derivatives on ergosterol or cholesterol synthesis in subcellular fractions (microsomes + cytosol) of C.albicans or rat liver. [^{14}C]-mevalonate was used as substrate (incubation time : 3 h).

ketoconazole. Furthermore, the hydrophobic nonligating portion of itraconazole binds more tightly to the apoprotein of the C.albicans microsomal P-450 isozyme than does the corresponding portion of ketoconazole. This higher affinity is reflected by the inhibitory effect of itraconazole on the P-450-dependent 14α-demethylation of lanosterol or 24-methylenedihydrolanosterol, the precursors of ergosterol. Indeed 50 % inhibition was already reached after 1 h contact of intact C.albicans cells with 5.5×10^{-10} M of itraconazole (19), whereas for ketoconazole 5×10^{-9} M were needed (18). Itraconazole is also 10 times more active in Pityrosporum ovale (31). The difference in activity was even more pronounced when the effects of these azoles on sterol synthesis in A.fumigatus were studied. At least 100-fold higher concentrations of ketoconazole than of itraconazole were needed to achieve 50 % inhibition of ergosterol synthesis (16).

Of great interest is the fact that in contrast with ketoconazole, itraconazole has almost no effect on the P-450 isozymes of testis microsomes. Also, this triazole derivative is almost devoid of effects on the P-450 isozymes of liver microsomes, adrenal microsomes and mitochondria (15). Neither does itraconazole inhibit the P-450-dependent aromatase in

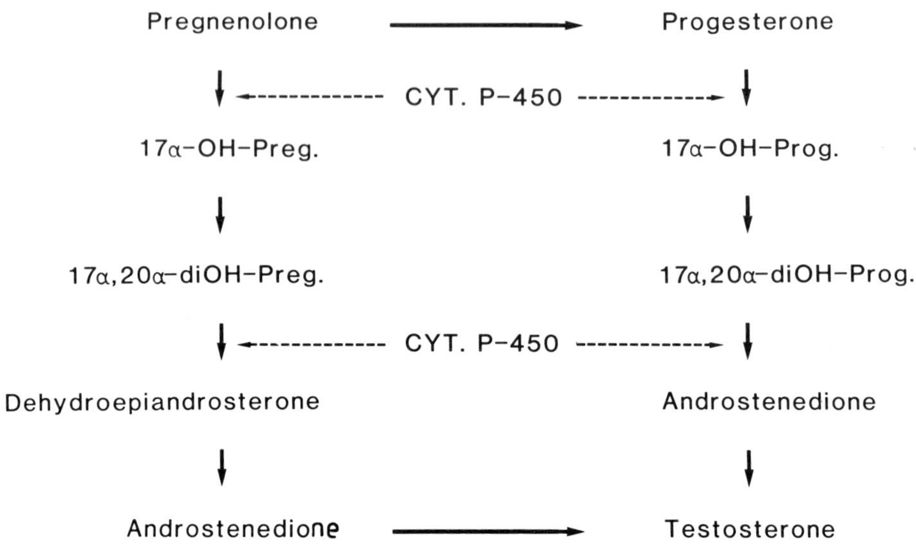

FIGURE 10. Androgen synthesis in testis microsomes.

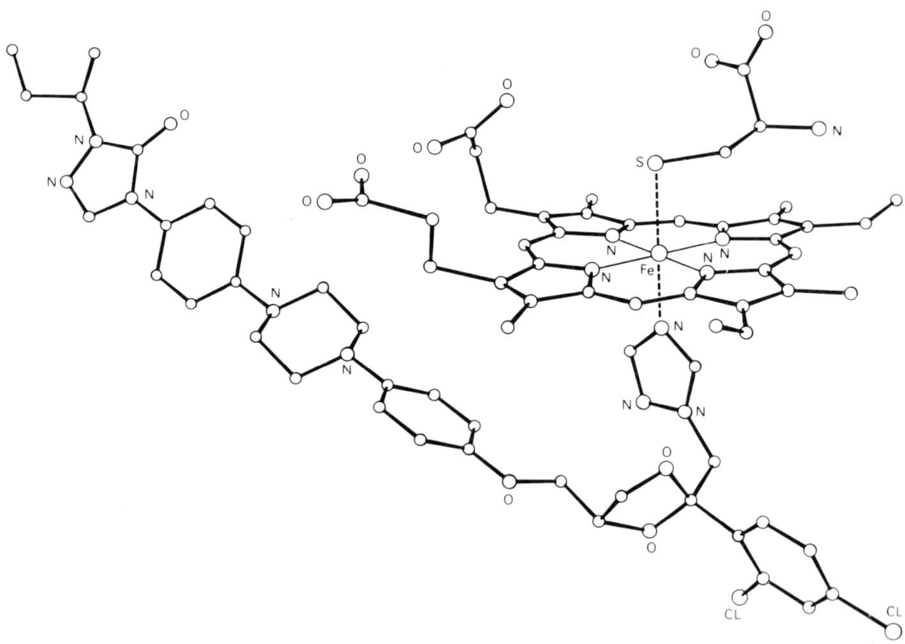

FIGURE 11. Proposed structure of the cytochrome P-450-itraconazole complex. The N_4 of the triazole ring is bound to the 6th coordination position of the heme iron. Figs. 1, 4 and 11 were calculated by H. Moereels from the Department of Theoretical Medicinal Chemistry.

human placenta microsomes nor the activation of vitamin D_3 in kidney mitochondria (unpublished results).

Another major difference between itraconazole and ketoconazole is certainly the pharmacokinetic profile. As mentioned before, ketoconazole usually achieves plasma drug levels higher than tissue drug levels (skin, mucosa, lungs). This situation is completely reversed in the case of itraconazole (32). Indeed, tissue levels with itraconazole are more elevated than the corresponding plasma levels. In addition, recent kinetic studies in skin and vaginal mucosa have indicated that, when treatment with itraconazole is discontinued, the half-life in tissue is much longer than that in plasma. This study has led to two conclusions : (1) skin and vaginal mucosa must be regarded as separate compartments from which itraconazole, once entered, is not redistributed to the systemic circulation; (2) treatment of a few days in reality give antifungal tissue levels for a much longer time than the actual period of therapy. During the last 4 years, several clinical studies with itraconazole have been completed in dermatology, gynaecology and internal medicine. These studies have been conducted in 22 countries on the 5 continents (27). The results obtained in vaginal candidosis point to the predictive value of animal tests. Indeed, the rat model for vaginal candidosis had suggested that itraconazole is about 5 times more active in vaginal candidosis than ketoconazole. Looking at the clinical results in vaginal candidosis with different therapy schedules it is obvious that at total dose of 400 mg itraconazole generates almost the same results as a total dose of 2000 mg of ketoconazole (400 mg/day for 5 days) (33).

Apart from this finding, the clinical studies have indicated that, especially in chronic vaginal candidosis, treatment of the partner may result in a lower recurrence rate (34). In pityriasis versicolor, ketoconazole had already shown to be highly active when one tablet was given daily for 10 consecutive days. Itraconazole proved at least as active as ketoconazole in this disease, since 200 mg given daily for 5 to 7 days led to a cure in more than 90 % of the patients (35).

Itraconazole has been tested extensively in dermatophytoses of the skin. In these infections, the high affinity of the drug has played a significant role. Indeed, as mentioned above, therapy for a few days result in high antifungal drug levels in the skin which remain substantial even after plasma drug levels have returned to zero. For this reason itraconazole has been compared with griseofulvin (100 mg *versus* 500 mg daily) in the treatment of tinea corporis and tinea cruris (2 weeks therapy) and palmo-plantar infections (4 weeks therapy) (36). The outcome of these studies has revealed that 2 or 4 week treatment periods are clearly not sufficient for griseofulvin, but result in 80 to 90 % response with itraconazole. This brings systemic antifungal therapy in line with systemic antibiotic treatment for which adequate predictable responses are obtained with relatively short treatment courses.

Even more dramatic results with itraconazole have been obtained in the field of systemic fungal disease. For paracoccidioidomycosis and histoplasmosis it was not surprising that itraconazole showed excellent effects (37). Indeed, miconazole and especially ketoconazole had already proven to be highly active in these infections (1). Itraconazole results in the same high cure rates but with lower daily doses and shorter treatment courses : 200 to 400 mg ketoconazole daily for 6 months *versus* 100 to 200 mg itraconazole daily for 3 months.

Recently, it has been demonstrated that the same holds for North American blastomycosis (38). Thus, 400 mg ketoconazole daily for 6 months resulted in 80 % response. The same response was obtained with 200 mg itraconazole given daily for 3 months.

As for ketoconazole, the results obtained with itraconazole in systemic candidosis are highly dependent upon the patient's immune status (39). Neutropenic patients tend to show less response than patients with a normal neutrophil count. In addition, preliminary data suggest that *Torulopsis glabrata* may be more difficult to treat with itraconazole than *C.albicans* infections. Of course, this is not different from the results obtained with other azole derivatives.

Itraconazole represents a major therapeutic break-through in some systemic mycoses such as sporotrichosis, chromomycosis, aspergillosis and possibly even meningeal infections caused by *Cryptococcus neoformans*. Since 1903, potassium iodide has been considered as the treatment of choice in lymphangetic sporotrichosis. Eighty years later, in 1983, Dr. Angela Restrepo had the first therapeutic success with itraconazole in a patient with cutaneous and lymphangetic sporotrichosis (40). Since then approximately 50 patients with sporotrichosis have received itraconazole (100 to 200 mg daily for 3 months) and about 90 % of them responded well to treatment. Another relatively rare but very disabling infection is chromomycosis. Thus far the treatment of choice has been a combination of 5-fluorocytosine and thiabendazole given for about 1 year. With the disappearance of thiabendazole as drug for human use in many countries, treatment of chromomycosis had become a serious problem. Itraconazole, 200 mg daily for 3 to 6 months, has yielded excellent results in infections caused by *Cladosporium carionii* and adequate results in infections caused by *Fonsecaea pedrosoi* (41). Also, for the extremely rare chromomycosis-like infections such as the infections caused by *Exophiala spinifera*, itraconazole has created a new therapeutic hope (42).

For coccidioidomycosis only a limited number of patients have thus far received itraconazole therapy, but some rather dramatic results have already been obtained, i.e. for infections with bone involvement (43).

In the field of opportunistic fungal infections some new trends are being observed. Until 4 to 5 years ago, the main fungal pathogen responsible for death in hematological units was *Candida albicans* and, to some extent, also other *Candida spp*. During the last years, however, there has been a shift from these yeast-induced fatalities towards *Aspergillus*-induced death in neutropenic patients. A possible reason for this may be the more prolonged periods of neutropenia with the more aggressive immunosuppressive therapy combined with the antimicrobial prophylaxis against bacteria and *Candida*. Aspergillosis is also more frequently reported in patients after organ transplants and in patients with chronic obstructive pulmonary disease. Another new trend is certainly the development of meningeal mycosis in AIDS patients. Especially *C.neoformans* plays a major role in this respect. The available therapeutic means against these new challenges are very limited and, consequently, itraconazole has been carefully evaluated in both aspergillosis and meningeal cryptococcosis. The effects on aspergillosis should be evaluated according to the type of infection (44). Acute invasive aspergillosis in neutropenic patients has responded to itraconazole treatment in about 50 % of the cases.

The semi-invasive type of aspergillosis which is often observed in patients with chronic obstructive pulmonary disease responds much better to itraconazole, with about 70 % of the patients showing a favorable response. The therapeutic effect of itraconazole in aspergilloma is difficult to evaluate. In a certain number of patients favorable clinical effects have been observed and in a small number a mycological negativation has been seen after 16 to 22 months of therapy. Certainly more work has to be done to define the true value of itraconazole in treatment of aspergilloma.

Keratomycosis due to Aspergillus spp. is a fairly uncommon disease in Europe but is one of the main causes of blindness in certain parts of India, especially Southern India. In contrast to the previous types of aspergillosis, the causative organism in the majority of these cases is not A.fumigatus but A.flavus. Itraconazole at 200 mg daily for only 2 to 3 weeks has been given to a number of patients with corneal ulcerations caused by Aspergillus. In more than 70 % of the cases this has resulted in mycological negativation combined with improved visual capacity.

Until now, 18 patients with cryptococcal meningitis have been treated with itraconazole (39). Fifteen were positive for HIV (human immunodeficiency virus). It is still early to give an accurate assessment of the effect of itraconazole in this disease, but it seems that the response is related to the life expectancy at the start of treatment. Indeed, when patients survived for more than one month after initiation of itraconazole treatment, the therapeutic effect against the Cryptococcus infection became much more pronounced. Therefore it seems that itraconazole may have some therapeutic value in cryptococcal meningitis when treatment can be started early enough in the course of the disease.

The last mycotic area for which itraconazole has been considered is antifungal prophylaxis in immunocompromised patients. The first studies indicated that the drug may have life-saving potential in these patients, especially in units having a high Aspergillus-induced mortality. These studies also indicated that poor absorption in this type of patients may pose a significant problem. Several of the mycosis-related fatalities were associated with low itraconazole levels in the plasma (45).

At present, more than 4,000 well documented itraconazole patients records have been collected. A safety and side effect analysis of this material suggests an overall side effect incidence of less than 5 %, with nausea (1 %), headache (0.4 %), pyrosis (0.4 %) and dizziness (0.3 %) being the most frequently reported side effects. Thus far serious drug-associated toxicity has not yet been observed.

4. CONCLUSION

The period from 1966 till 1986 has been of major importance in antifungal drug development; a development which had previously remained in the shadow of the progress made with antibiotics. Miconazole was the starting point, ketoconazole the pharmacologic turning point and itraconazole as the result of target-oriented drug design. This demonstrates that long-term dedication is the only way leading to successful drug research.

5. SUMMARY

Until 20 years ago, the antifungal armamentarium was limited to polyene antibiotics, griseofulvin, tolnaftate and flucytosin. In 1966 however, the first broad spectrum imidazole derivative etonam (R 10 100) was synthesized, immediately followed by the well known imidazoles miconazole and clotrimazole. These agents have initiated a drastic change in the topical antimycotic therapy. Ten years ago, in 1976, the first orally active broad spectrum antifungal, ketoconazole (R 41 400), was synthesized. Again, this agent has created a revolution in the therapy of infections unresponsive to topical agents, as well as the treatment of systemic mycoses.

The advantages and disadvantages of ketoconazole have resulted in an intensified drug design program from which itraconazole (R 51 211) was selected for further development. This choice was based on the spectrum of activity (including _Aspergillus_) as well as some major pharmacokinetic properties, e.g. a favorable tissue level/plasma level ratio and a long half-life. In addition, itraconazole has a much greater affinity for the fungal cytochrome P-450 than for the mammalian cytochrome P-450, resulting in an improved safety profile. Clinical results now collected for almost 4,000 patients point to the high potential of itraconazole for the treatment of fungal diseases.

REFERENCES

1. Levine HB. _in_ Ketoconazole in the Management of Fungal Disease (Levine HB, Ed). Acid Press, Hong Kong, pp 47-53, 1982.
2. Vanbreuseghem R, Van Cutsem J, Thienpont D. Chemotherapia 12:107-122, 1967.
3. Van Cutsem JM, Thienpont DC. Chemotherapy 17:392-404, 1972.
4. Borgers M, Vanden Bossche H. _in_ Ketoconazole in the Management of Fungal Disease. (Levine HB, Ed). Adis Press, Hong Kong, pp 25-47, 1982.
5. Heeres J, Backx LJJ, Mostmans JH, Van Cutsem J. J. Med. Chem. 22: 1003-1005, 1979.
6. Symoens J, Moens M, Dom J, Scheijgrond H, Dony J, Schuermans V, Legendre R, Finestine N. Rev. Infect. Dis. 2:674-691, 1980.
7. National Institute of Allergy and Infectious Diseases Mycoses Study Group. Ann. Intern. Med. 103:861-872, 1985.
8. Proost J, Cauwenbergh G. Curr. Ther. Res. 35:747-752, 1984.
9. Cauwenbergh G. _in_ Oral Therapy in Dermatomycoses : A Step Forward. Proceedings of a Symposium, Frankfurt, February 1985. The Medicine Publishing Foundation, Symposium Series No. 16:27-43, 1985.
10. Janssen PAJ, Symoens J. Am. J. Med. 74(1B):885, 1983.
11. Graybill JR, Galgiani J, Stevens DA, Dismukes W, Cloud G, Sugar A, Alsip S, Craven P, Huppert M, Bowles C. _in_ Coccidioidomycosis (Einstein HE, Catanzaro A, Eds). The National Foundation for Infectious Diseases, Washington D.C., pp 474-482, 1985.
12. Janssen PAJ, Vanden Bossche H. Arch. Pharmacol. Chem., in press, 1987.
13. Vanden Bossche H. _in_ Current Topics in Midical Mycology (McGinnis MR, Ed). Springer-Verlag, New York, vol. 1, pp 313-351, 1985.
14. Yoshida Y, Aoyama Y. _in_ In Vitro and In Vivo Evaluation of Antifungal Agents (Iwata K, Vanden Bossche H, Eds). Elsevier Science Publishers, Amsterdam, pp 123-134, 1986.
15. Marichal P, Gorrens J, Vanden Bossche H. J. Med. Vet. Mycol. 23:13-21, 1985.
16. Morita T, Nozawa Y. J. Invest. Dermatol. 85:434-437, 1985.
17. Vanden Bossche H, Lauwers W, Willemsens G, Marichal P, Cornelissen F, Cools W. Pestic. Sci. 15:188-198, 1984.
18. Vanden Bossche H, Willemsens G, Marichal P, Cools W, Lauwers W. _in_ Mode of Action of Antifungal Agents (Trinci APJ, Ryley JF, Eds). Cambridge University Press, Cambridge, pp 321-341, 1984.
19. Vanden Bossche H, Bellens D, Cools W, Gorrens J, Marichal P, Verhoeven H, Willemsens G, De Coster R, Beerens D, Haelterman C, Coene MC, Lauwers W, Le Jeune L. Drug Develop. Res. 8:287-298, 1986.

20. Vanden Bossche H, Lauwers W, Willemsens G. in EORTC Genitourinary Group, Monograph 2, Part A. Therapeutic Principles in Metastatic Prostate Cancer (Schroeder FH, Richards B, Eds). Alan R. Liss, New York, pp 187-196, 1985.
21. Vanden Bossche H, De Coster R, Amery WK. in Pharmacology and Clinical Uses of Inhibitors of Hormone Secretion and Action (Furr B, Wakeling AE, Eds). Ballière-Tindall, London, pp 288-307, 1986.
22. Vanden Bossche H, Lauwers W, Willemsens G, Cools W. in Microsomes and Drug Oxidations (Boobis AR, Caldwell J, de Matteis F, Elcombe CR, Eds). Taylor and Francis, London, pp 63-73, 1985.
23. Amery WK, De Coster R, Caers I. Drug Developm. Res. 8:299-307, 1986.
24. Carvalho D, Pignatelli D, Resende C. Lancet ii:560, 1985.
25. Ghetti P, Patrone P, Tosti A. Arch. Dermatol. 122:629, 1986.
26. Rosenberg EW, Belew PW. Arch. Dermatol. 118:370, 1982.
27. Cauwenbergh G, De Doncker P. Drug Developm. Res. 8:317-323, 1986.
28. Merkus JMWM, Bisschop MPJM. in Oral Therapy in Vaginal Candidosis. (Eliot BW, Ed). The Medicine Publishing Foundation, Symposium Series No. 13, pp 15-21, 1984.
29. Van Cutsem J, Van Gerven F, Zaman R, Heeres J, Janssen PAJ. in Proc. 13th International Congress of Chemotherapy, Vienna, Austria, August 28-September 2, 1983.
30. Van Cutsem J, Janssen PAJ. Bull. Soc. Française Mycol. Méd. 15:31-44, 1986.
31. Marichal P, Gorrens J, Van Cutsem J, Van Gerven F, Vanden Bossche H. J. Med. Vet. Mycol. 24:487, 1986.
32. Heykants J, Michiels M, Meuldermans W, Monbaliu J, Lavrijsen K, Van Peer A, Levron JC, Woestenborghs R, Cauwenbergh G. in Recent Trends in the Discovery, Development and Evaluation of Antifungal Agents (Fromtling et al., Eds). J.R. Prous Science Publishers, Barcelona, in press, 1987.
33. Peeters F, Van Der Pas H, Proost J, Janssens D, Snauwaert E, Van Cutsem J, Cauwenbergh G. Curr. Ther. Res. 39:496-504, 1986.
34. Calderon RA. Rev. Infect. Dis., suppl., in press, 1987.
35. Delescluse J, Cauwenbergh G, Degreef H. Brit. J. Dermatol. 114:701-703, 1986.
36. Cauwenbergh G, De Doncker P, Stoops K, De Dier A, Goyvaerts H, Schuermans V. Rev. Infect. Dis., suppl., in press, 1987.
37. Galimberti R. Rev. Infect. Dis., suppl., in press, 1987.
38. Dismukes W. in Proc. 26th ICAAC, New Orleans, September 29-October 1, 1986.
39. Cauwenbergh G. in Proc. Annual Meeting of the Danish Society for Mycopathology, Copenhagen, October 3-5, 1986.
40. Restrepo A, Robledo J, Gomez I, Tabares AM, Gutierrez R. Arch. Dermatol. 122:413-417, 1986.
41. Borelli D. Rev. Infect. Dis., suppl., in press, 1987.
42. Spraker MK, Van Dyke TE, Jones HE, Artis WM. in Proc. IXth International Congress of the International Society for Human and Animal Mycology, Atlanta, Georgia, May 19-45, 1985.
43. Graybill JR, Stevens DA, Galgiani JN, Dismukes WE, Cloud G, Ganer A, Arathoon E, Fetchick R, Diaz M, the NIAID Mycoses Study Group. in Proc. 26th Interscience Conference on Antimicrobial Agents and Chemotherapy, New Orleans, September 29-October 1, 1986.
44. Cauwenbergh G. in Proc. XXth Scientific Meeting of the German Mycological Society, Freiburg, May 22-24, 1986.
45. Tricot G, Joosten E, Boogaerts MA, Vandepitte J, Cauwenbergh G. Rev. Infect. Dis., suppl., in press, 1987.

R.T. Walker

ANTIVIRAL CHEMOTHERAPY : DESIGN OR SERENDIPITY ?

R.T. WALKER
Department of Chemistry, University of Birmingham, P.O.Box 363, Birmingham B15 2TT, United Kingdom

There are currently in the world, and particularly in Belgium, many people who owe their good eyesight and in some cases, even their lives, to the decision made by Professor Piet De Somer that the work being done by Prof. Erik De Clercq at the Rega Institute on antiviral chemotherapy was worth supporting and that the discoveries made should be exploited. From this joint research work involving chemists in the United Kingdom, and, more recently, those in the Rega Institute, together with the expertise in other areas from several other laboratories around the world, has arisen the compound BVDU [(E)-5-(2-bromovinyl)-2'-deoxyuridine] (1), the most potent member of an ever-growing family of 5-vinylpyrimidine-2'-deoxynucleosides which exhibit anti-herpesvirus activity (2).

The discovery of the antiviral properties of BVDU was entirely serendipitous as is, to the best of my knowledge, the discovery of all other compounds with antiviral properties, at least if one regards biological screening as giving serendipitous results.

I wish to use this opportunity to examine the current position of antiviral chemotherapy and to pose and try to answer the following questions:

1. Why are there currently so few antivirals licensed for use when so many people die from, or are severely inconvenienced by, viral disease ?
2. Is there likely to be such a thing as a safe antiviral ?
3. In view of the cost of discovery, development, testing and determining the safety of a drug, are viral diseases worthwhile targets for pharmaceutical companies ? On the other hand are we, as scientists, morally justified in allowing discoveries which are made, not to be exploited just because there is no direct financial benefit ?
4. What is the future for antiviral chemotherapy and what can we do to hasten the change from serendipity to rational design ?

In trying to answer some of these questions in detail, I will use three examples from among the very few antiviral compounds which are known and I will introduce these at the beginning, so that their current status can be assessed. The compounds are amantadine (**1**), acyclovir (**2**) and BVDU (**3**).

Amantadine (3) is the only compound to have been generally licensed for treatment of a viral disease caused by a virus other than a member of the herpesvirus group. It is also the only non-nucleoside compound to have been licensed. Influenza A and B viruses were shown to be inhibited by amines and ammonium compounds as long ago as 1962 and amantadine was selected as a compound with a suitable *in vivo* activity against influenza A virus. Despite the fact that the only serious competitor to emerge in the

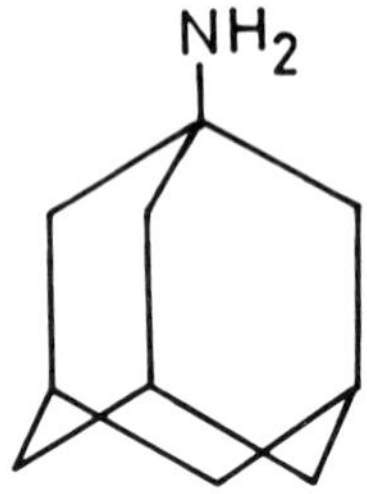

1

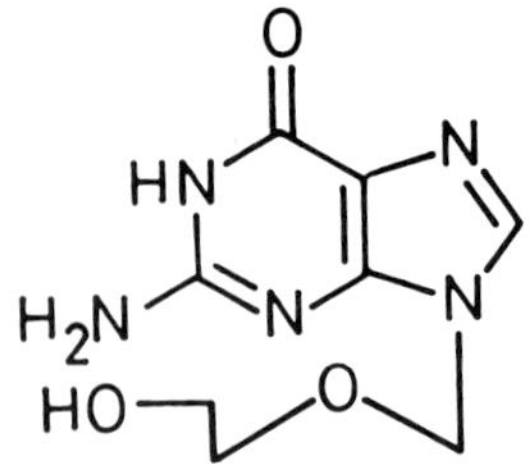

2

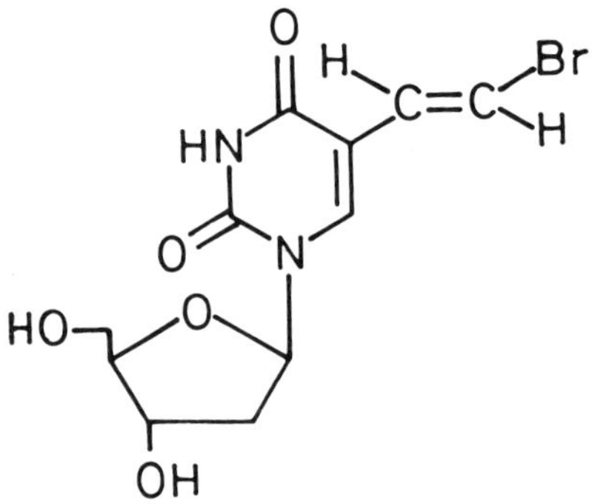

3

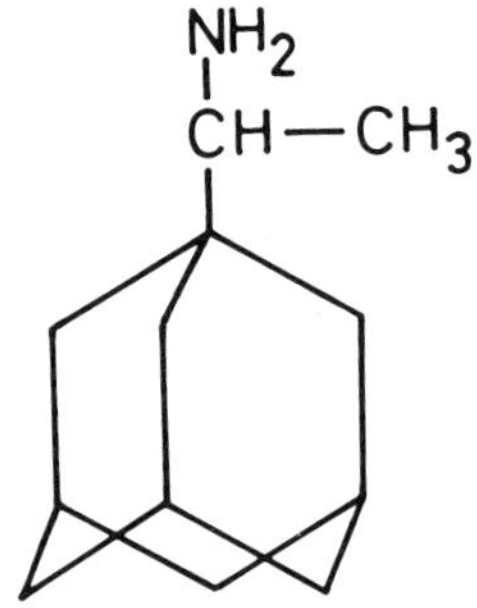

4

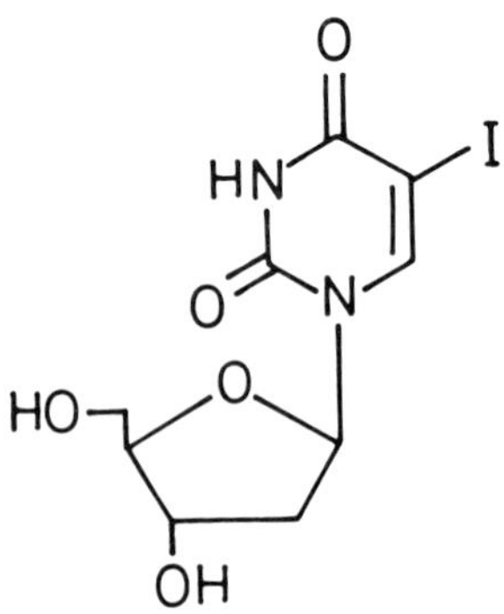

5

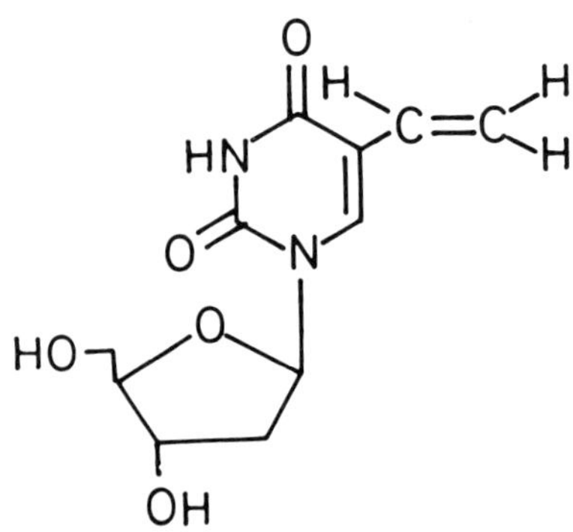

6

last 25 years is the close related rimantadine (4), and the commercial potential for a compound capable even of reducing slightly the economic losses caused by influenza A would be considerable, the compound has never been widely used although its prophylactic efficiency has been well demonstrated, because of fears voiced about its toxicity. The precise mode of action is not clear and indeed is currently the topic of some research. Originally it was thought to act by affecting the pH of cells such that the acidic pH required for fusion of the viral and vacuolar membranes after penetration of the virus into cell by viropexis is not maintained. However, recently (4) it has been argued that the specific and high antiviral activity of amantadine when compared to other amines cannot possibly be due entirely to this effect, and in cell culture, two concentration-dependent inhibitory actions against virus replication seem to take place. At a concentration > 0.1 mM, a non-specific inhibition is seen which is not peculiar to amantadine but is a property of many amines and is probably due to elevation of the pH of the cell. However, at concentrations more realistically related to the _in vivo_ concentrations required to inhibit viral infection (0.1-0.5 μM), a strain-specific inhibiting effect of amantadine is observed. Recent work with viral mutants has shown that the target is the viral membrane M_2 protein, the product of a spliced transcript of genome RNA segment 7.

Acyclovir [9-(2-hydroxyethoxymethyl)guanine] (2) (5) is the first (and so far the only representative) of the new generation of antiviral compounds which are much more effective and much less toxic than the previously available compounds, which has been licensed for clinical use. Indeed, more is probably known about the mode of action or at least the necessary metabolic pathways of these newer antiviral agents than has been learnt in the previous 20 years when compounds like 5-iodo-2'-deoxyuridine (5) have been available. The selectivity indexes observed _in vitro_ with these new compounds (6) are often in the range of 10^3-10^4, and this appears to be due in whole or in part to the required activation of the compound to the monophosphate which is achieved only by the virus-encoded thymidine kinase and not by the normal cellular enzymes. Needless to say, acyclovir, like the other compounds of this generation was not designed for the purpose it is now used for, as its phosphorylation pattern is so bizarre that even with hindsight it is not plausible to concoct a convincing explanation for its design.

Thus, the efficacy and specificity of acyclovir is due to its phosphorylation by the herpesvirus-encoded thymidine kinase (TK), whereas at any reasonable concentration it is not a substrate for the cellular enzyme. That the viral TK can exhibit such specificity when its normal substrate is thymidine, and here we have a purine deriative with C-2' and C-3' of the sugar moiety missing, defies any rational explanation until something is known about the mechanism of action of this enzyme. Subsequent metabolism of acyclovir monophosphate to the diphosphate by a cellular GMP kinase (7) is at least within the realms of reason (although it probably defies Murphy's Law). The triphosphate which is produced by some nebulous cellular enzyme is a good substrate for the virus-specified DNA polymerase, and a poor substrate for the cellular enzyme, and, when incorporated into DNA, acyclovir, of course, has to act as a chain terminator, and presumably this is how it prevents viral replication. Thus the compound is activated only in the virus-infected cell and only interferes with viral replication, which accounts for its selectivity.

The synthesis of _(E)-5-(2-bromovinyl)-2'-deoxyuridine_ (3) was not designed and this review gives me the opportunity of committing the original series of chances and mishaps to paper for the first time.

In the early 70's, the group lead by Prof. A. Stanley Jones and myself was interested in synthesizing analogues of thymidine which contained potential radiation (u.v. or μ)-sensitive groups at their 5-position. It was hoped to incorporate these analogues into DNA in place of thymidine and then by irradiation cause the production of derivatives incompatible with the stability of the DNA double-helix. The application to local irradiation of tumors is obvious. At this stage the design stops. The compounds chosen for synthesis were 5-vinyl-2'-deoxyuridine (6) and 5-ethynyl-2'-deoxyuridine (7). The synthesis of these compounds has been reviewed in detail elsewhere (2), and I do not wish to give the impression that we were the only people working in this field; indeed, we were not even the first to synthesize these compounds.

5-Vinyl-2'-deoxyuridine was eventually synthesized and proved to be incorporated into DNA (8). However, it was so toxic *in vitro*, causing chromosomal damage (9) that it clearly could not be used for our original purpose. Subsequently, it was found that *in vivo* this compound appeared to have no biological activity at all, and we will return to this aspect later (10).

We thought that one possible route to 5-ethynyl-2'-deoxyuridine (7) would be to add bromine across the double bond of the 5-vinyl compound (6) and then to remove 2 moles of HBr. At the base level, indeed bromine could be added and one mole of HBr spontaneously eliminated to give (*E*)-5-(2-bromovinyl)uracil (BVU) (11). However, the second mole of HBr could not be removed under any conditions. In dispair, we turned to a synthesis of 5-ethynyluracil which we had attempted years previously and which had failed. This time it worked (12), and the deoxynucleoside when prepared was shown by Prof. E. De Clercq to have significant antiviral activity although it was too toxic (it was subsequently shown to be a thymidylate synthetase inhibitor) (13). The next compound examined was BVDU, and the rest of the story follows.

(*E*)-5-(2-bromovinyl)-2'-deoxyuridine (BVDU) has a metabolic pathway somewhat similar to acyclovir (7). Again selective phosphorylation of the nucleoside by the virus-encoded thymidine kinase is responsible for the specificity of the compound which is particularly effective against herpes simplex virus type 1 (HSV-1) and varicella-zoster virus (VZV). Before one gets the idea that this thymidine kinase is likely to phosphorylate anything and everything which can vaguely be called a nucleoside and which has a 5'-hydroxyl function, then it should be made quite clear that this is not so. The lack of antiviral properties of many nucleosides is probably due to the fact that they are not substrates for this enzyme and, indeed (*Z*)-5-(2-bromovinyl)-2'-deoxyuridine (8) is much less active as an antiviral agent, primarily because it is not a substrate for the thymidine kinase (14). It is interesting to see that the viral TK apparently has a much lower specificity than the cellular enzyme and it is clearly important to clone, sequence and determine the three-dimensional structure and mode of action of as many different thymidine kinases as possible so that some attempt at substrate design can be made. BVDU 5'-monophosphate (BVDUMP) is further phosphorylated to the diphosphate by the same viral enzyme (a fact which I find only slightly less surprising than the phosphorylation of acyclovir by a thymidine kinase). However, the facts appear to be clear and explain why BVDU is essentially ineffective against HSV-2, because BVDUMP is a poor substrate for the HSV-2 TK. Once again the nebulous cellular triphosphate-making enzyme is conjured up, and here the problems start. BVDU 5'-triphosphate (BVDUTP) is both a substrate and a com-

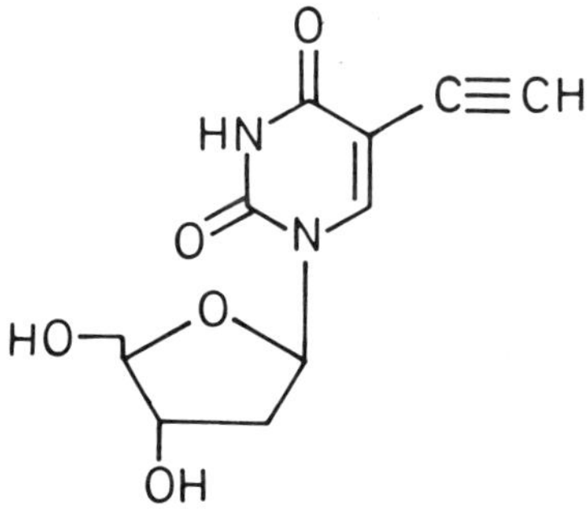

7

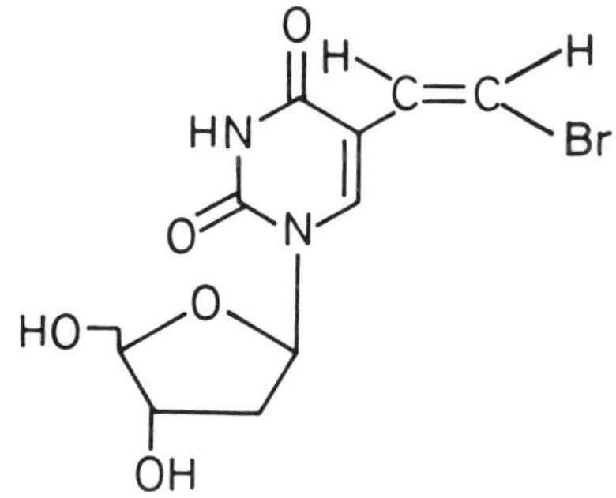

8

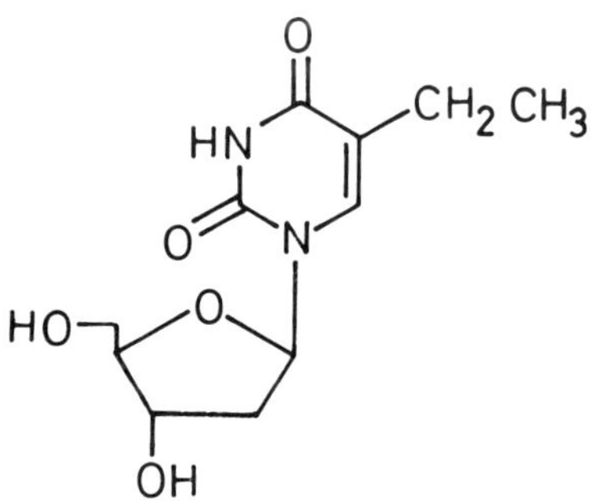

9

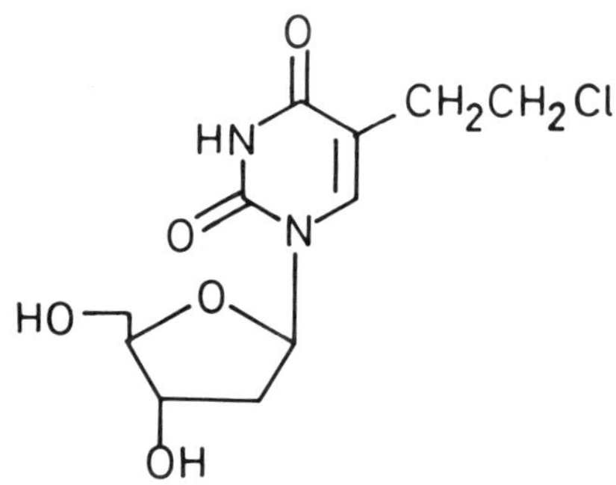

10

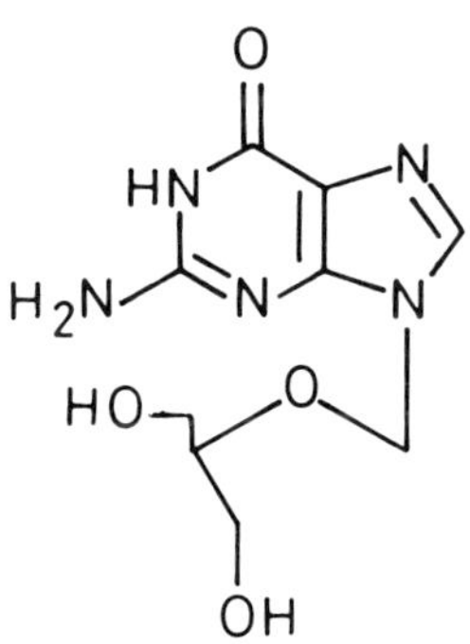

11

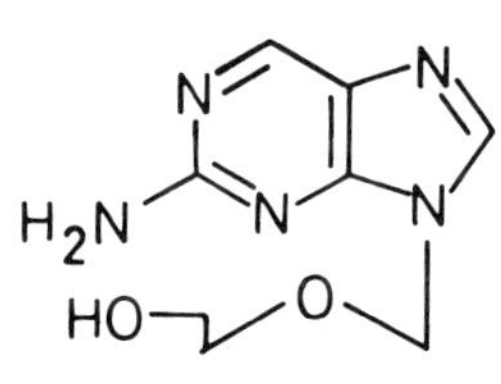

12

petitive inhibitor of dTTP, for the virus-encoded DNA polymerase as well as for the cellular enzyme. Until recently, it has always seemed to be a tenet of faith among virologists that incorporation of an analogue into viral DNA is enough to explain the antiviral properties of these compounds. This explanation has been thought as sufficient for 5-iodo-2'-deoxyuridine and 5-ethyl-2'-deoxyuridine (9) for years. This may be correct but I think more evidence is needed before scientists have to assume this postulate proven. There are certainly many in vitro studies and also some in vivo studies which suggest that incorporation of thymidine analogues into DNA, especially at the minute amounts required here for BVDU to demonstrate its antiviral efficacy, does not impair DNA replication or transcription at all. Of course such an event might potentially be mutagenic, but one needs more than the occasional mutagenic event to explain the antiviral properties of BVDU at the levels used in vitro.

Thus we now have some of the details of three representative antiviral compounds. One, amantadine, has been licensed but is not really used; one, acyclovir, is licensed and used and a rational explanation for its action has been devised; and one, BVDU, which shows great promise based on in vitro, as well as in vivo data, but for which further work is needed to know how it works.

Before reverting to try and answer the questions posed at the beginning of this talk, let me just complete the background by listing those agents which have been licensed for clinical use (Table 1).

TABLE 1. Drugs approved for human use against viral infection.

1. Amantadine (Influenza, virus A)
2. 5-Iodo-2'-deoxyuridine, (HSV-1)
3. 5-Trifluoromethyl-2'-deoxyuridine, (HSV-1)
4. Adenine arabinoside, (HSV-1)
5. Acyclovir, (HSV-1, HSV-2, VZV)
6. 5-Ethyl-2'-deoxyuridine, (HSV-1) [W.-Germany only]
7. 5-Iodo-2'-deoxycytidine, (HSV-1) [France only]
8. Ribavirin, (Respiratory syncytial virus)

It is clear that currently there are so few antivirals licensed because of a combination of many factors. These include the general belief held until recently that because of the close and specific association of a virus with its host, it would be impossible to prevent replication of the former without damaging the latter. The advent of acyclovir has shown this view to be clearly wrong but it takes a long time to change such a widely held belief. The regulations for developing a potentially new drug are, rightly, very stringent. Particularly in the antiviral field because of the close and specific association already mentioned, the relevance of much of the required toxicological evaluation can be questioned but until guidelines which will safeguard the general population from another disaster like the thalidomide case can be agreed upon, the present rules which have stood the test of time have to remain.

On the other hand, we, as scientists, need to question how safe is safe ? It is unlikely that any drug can possibly be effective and totally safe. In the current climate which is dominated by lawyers who wish to seize on the possibility of any side effect of medication being exploited for the profit of the supposed victim (and, of course, also themselves), we have almost reached the stage where a company would prefer to not enter a field and let people continue to suffer and on occasions die by "Act of God" rather than run the risk of expensive litigation and all the attendant bad publicity affecting sales of other products. Eventually this problem is going to have to be faced and as scientists we ought to try and influence the politicians in what we feel is the right direction.

Although few people would go through a year without receiving at least one viral infection, it is often unlikely to be more than an inconvenience, resulting perhaps in a child missing a few days of school, an adult missing a few days of work or continuing to work despite the infection and being less efficient and also spreading the disease. In industrial countries, few people nowadays die of viral infection and hence any drug really has to be seen to be safe although the advent of AIDS may chance our normal attitude to viral infections, which seems to be that if one waits long enough they will cure themselves. In any case, because there are so few remedies for viral infections, viral diseases are often not fully identified and when they are, the procedure is so lengthy that the infection is often under normal control by the time the identity is established. One reason cited for the lack of interest in amantadine is the fact that it is only effective against influenza virus A strains *in vivo*, and thus strain typing would be necessary before use. However, as amantadine is most effective when used prophylactically and the widespread influenza infections are clearly typed in advance, this seems to be more of an excuse than a valid reason. Yet, for any antiviral chemotherapy regime to assert a maximum effect, rapid and accurate diagnosis is required and this would entail a complete review of the attitude of the medical profession to viral infections.

Thus the population in general has to come to a consensus as to whether they want most virus diseases treated or not other than by vaccination (which often causes precisely the same problems, as recently seen in the U.K. with the pertussis vaccine). There is never going to be an absolutely safe drug and for most viral diseases one will have to trade the inconvenience caused by an infection with the often ill-defined chance of some side effects. If this is accompanied by automatic litigation wherever some possible side effect is spotted, then I suggest that we are unlikely to see many antiviral drugs developed in the future.

The really devasting viral diseases are normally associated with Third World Countries. Here, the problems are somewhat different in that antiviral chemotherapy is expensive and therefore these countries cannot afford such medication and hence there is little incentive for companies to do research in this area. Secondly, as used to be the case in the now developed countries, viral disease is often a result of poor nutrition and poor hygiene and it could be argued that it is better to spend the money on the cause not the symptoms of the problem.

Thus, for producing a reasonable profit from the very high investment required for research into synthesis, evaluation, licensing and marketing an antiviral, there are very few viral diseases which appear worthwhile targets. I suppose that ever since the discovery of acyclovir herpesvirus infections are top of nearly everyone's list. As most herpes infections are accompanied by latency, this is almost a viral disease designed by a pharmaceutical company ! As distinct from smallpox, I think it is safe to say that the herpesvirus will always be with us and always likely to be reactivated. There must be very few people even in the developed countries who have not been infected by HSV-1, HSV-2 or VZV and are therefore potential targets for recurrent attacks from within. This together with the associated problems of vaccination make the herpes and influenza viruses the most attractive targets for antiviral chemotherapy. Thereafter we have the rotaviruses, rhinoviruses, hepatitis virus and the retroviruses as possibilities, but it is my opinion that in terms of potential commercial success and chances of design based on present knowledge, herpesviruses offer possibilities orders of magnitude higher than the others.

Following this diversion into the socio-economic-political sphere, let us return to the world of facts and science and examine the future for antiviral chemotherapy and, in particular, the chances for rational design.

We know that viruses are not the invincible force they were once thought to be. The advent of acyclovir (2) and BVDU (3) has shown that antivirals can be made and specific targets do exist. In discussing targets, I make no apology for using the comprehensive list discussed in detail by Professor Prusoff (15) three years ago. I have no intention here of going into this list in detail. Suffice it to say that as our knowledge of virus replication increases, it becomes clearer that there are many events in the replication pathway which are unique to the virus and which therefore present possible targets for attack which should not necessarily be accompanied by host cell toxicity.

TABLE 2. Viral targets.

1. Extracellular attack by antibody, enzymes or chemicals
2. Adsorption of virus to cell surface
3. Transport across cell wall
4. Uncoating of virus
5. Transport of viral genome into cytoplasm or nucleus
6. Transcription or reverse transcription
7. Methylation of RNA
8. Protein synthesis or processing
9. Virus-encoded enzymes
10. Virion assembly
11. Virion release

We need to know much more detail about most of the processes listed here for most virus classes before we can attempt the design of antivirals, but the targets are undoubtedly there and this alone should give us hope. I would like to concentrate on one target which is listed as "enzymes". This refers to those enzymes required for virus replication which are coded for by the viral genome. As we learn more about the coding capacity of the viral genomes, undoubtedly more of these enzymes will be identified, but, as these enzymes inevitably use nucleoside derivatives as substrates, it is not surprising that it is against this target that the nucleoside chemist should direct his synthetic expertise. This explains, of course, why herpesvirus, which codes for so many enzymes is susceptible to so many nucleoside analogues because the substrate specificities of the host and viral enzymes are different and are there to be exploited.

The enzymes encoded for by some herpesviruses are listed in Table 3.

TABLE 3. Herpesvirus-encoded enzymes.

	Herpesvirus-encoded enzymes				
	HSV-1	HSV-2	VZV	CMV	EBV
Thymidine kinase	+	+	+	-	?
DNA polymerase	+	+	+	+	+
DNase	+	+	+	+	+
Uracil-DNA-glycosylase	+	+	?	?	?
dUTPase	+	+	?	?	?
Ribonucleotide reductase	+	+	?	?	+

The only enzymes which have been exploited so far are the thymidine kinase and DNA polymerase. As already mentioned, all nucleoside analogues which have so far been synthesized and show antiviral activity against herpesvirus, have to be a substrate for the thymidine kinase coded for by the enzyme. TK^- virus strains and TK-altered virus strains (altered so that the enzyme specificity is modified or rendered inactive but the gene is still present) are unaffected by the nucleoside analogues. It is not known how many nucleoside analogues which have no antiviral activity are inactive because they are not substrates for the TK. Usually no attempt is made to find why a compound is inactive, and at present it is not possible to bypass this step and get the 5'-phosphate into the cell directly (16). It is likely that for some while to come, the only way in which we will pick up a new lead is by *in vitro* screening. *In vivo* evaluation is too expensive, requires too much material, is too time consuming and often the results on animal models are open to too many different interpretations for it to be possible to do much more than rely on an *in vitro* evaluation. The relevance of the results so obtained to the curing of a clinical disease can be tenuous at best, and at the worst can mislead completely so that an active compound is missed.

A false positive result from an *in vitro* test may result in a waste of time and money but a false negative may mean the compound is lost for ever. Some examples can be given to illustrate this point. *In vitro*, 5-vinyl-2'-deoxyuridine (**6**) is a potent antiviral agent and also very toxic; *in vivo* it appears to be biologically inert (10). The reason seems to be that in most cell culture lines, the levels of nucleoside phosphorylase which results in the degradation of the nucleoside to the inactive pyrimidine base, are low, whereas *in vivo* they are quite high. 5-Vinyl-2'-deoxyuridine is a very good substrate for this enzyme (17), and hence *in vivo* it may be impossible to achieve the levels of deoxynucleoside required for phosphorylation to occur. BVDU offers a less extreme example. Although it is very active *in vivo*, the *in vitro* results would indicate that one should expect an *in vivo* activity some 10-100 times greater than that found. Once again, BVDU is a good substrate for nucleoside phosphorylase but it is still possible to achieve, for a sufficiently long time, the necessary levels of nucleoside *in vivo* so that phosphorylation can occur. The base, (*E*)-5-(2-bromovinyl)uracil (BVU), is metabolically inert, but *in vivo* it also shows antiviral activity (18), probably because nucleoside phosphorylase can transfer the carbohydrate moiety from another deoxynucleoside to BVU to form BVDU.

It is also possible to be embarassed by true positives. When BVDU was discovered, it was immediately realized that several other 5-(2-halogenovinyl)-2'-deoxyuridines and 2'-deoxycytidines may have similar *in vitro* activity (19). Yet, all subsequent *in vivo* work was done with BVDU for the very non-scientific reason that it was synthesized first and *in vivo* testing was too expensive because of the material it consumed. Even when it was realized that nucleoside phosphorylase was causing a substantial loss in *in vivo* activity, this was not enough of an incentive for the other (i.e. carbocyclic) analogues to be investigated *in vivo* in sufficient depth.

A false negative from an *in vitro* assay is altogether another problem. Once a compound has been tested and found to be inactive, it is unlikely to receive another chance. However, there are already several examples of nucleoside antiviral agents which *in vitro* show a very low activity but *in vivo* work very well. These false negatives have of course been detected because by analogy with compounds of known activity; one expected a positive result and, therefore, the *in vivo* test followed a negative *in vitro* result. An example of this is 5-(2-chloroethyl)-2'-deoxyuridine (**10**) which is not terribly active *in vitro* but effective *in vivo* against systemic HSV-1 infections at a 10-fold lower dose than either acyclovir or BVDU (20).

The discovery of acyclovir led to the synthesis of many analogues of which DHPG (**11**) was not more active *in vitro* but clearly more effective *in vivo* (6). A better example of discrepancy between *in vitro* and *in vivo* results is that of 6-deoxyacyclovir (**12**) (21). As acyclovir itself was found by chance, had 6-deoxyacyclovir happened to be the first compound made, the negative *in vitro* result would have given no incentive for *in vivo* exploration or indeed for the chemical synthesis of any other analogues. However, with the *in vivo* activity of acyclovir known, the *in vitro* data on 6-deoxyacyclovir could be disregarded; 6-deoxyacyclovir is a prodrug of acyclovir and much higher plasma levels of acyclovir can be achieved using this prodrug.

The only thing worse than getting a false negative *in vitro* result is not investigating a new compound at all. Nowadays it is incumbent upon all chemists who synthesize compounds which might have some activity, to get them adequately evaluated. Unless the synthesis is trivial, it is most unlikely that anyone will resynthesize a compound described in the literature unless the likelihood of it having biological activity is overwhelming. Thus, chemists should realize that they are doing a disservice to the community by synthesizing compounds and not having them evaluated for biological activity.

We now finally turn to the problems of designed synthesis in the antiviral field. Let us first consider the much simpler task and more predictable area of synthesizing analogues in an area where a lead compound has been established. Amantadine (**1**) was made this way, but as its mode of action was then unknown one can hardly claim a rational design, rather a slightly less-than-random screening process.

Acyclovir (**2**) and BVDU (**3**) have also been lead compounds and many hundreds of analogues of each have been synthesized. Although the mode of action of acyclovir-triphosphate as a chain terminator of viral DNA syn-

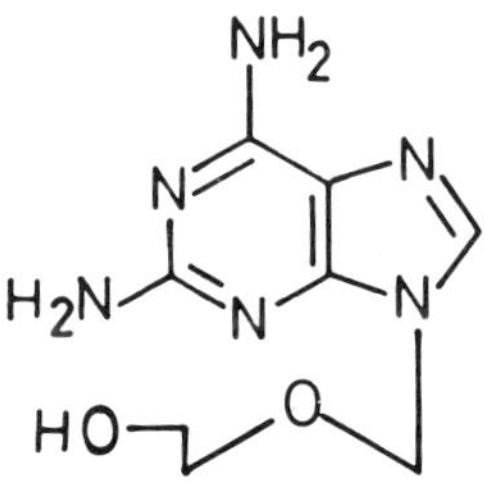

13

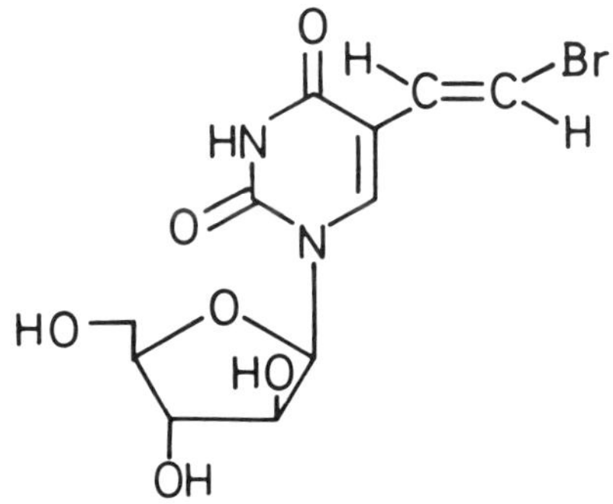

14

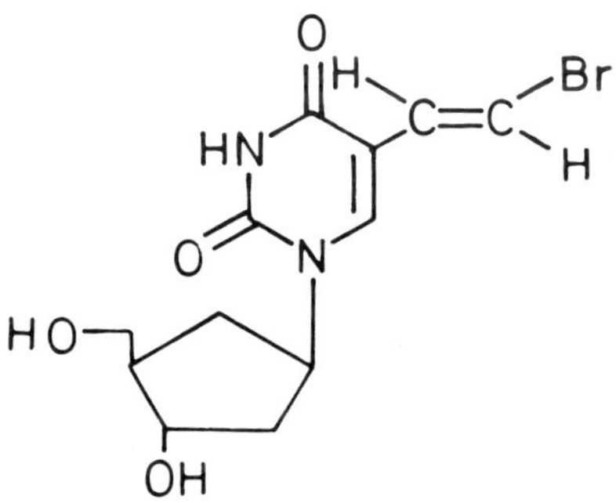

15

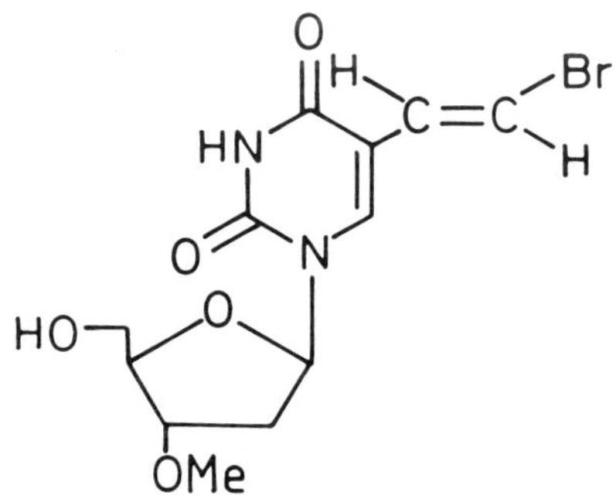

16

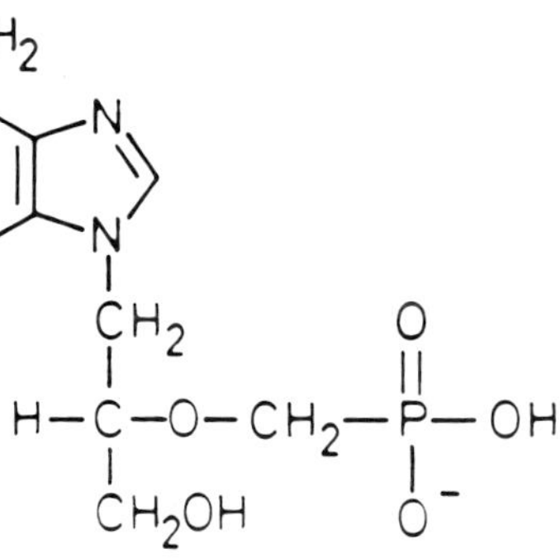

17

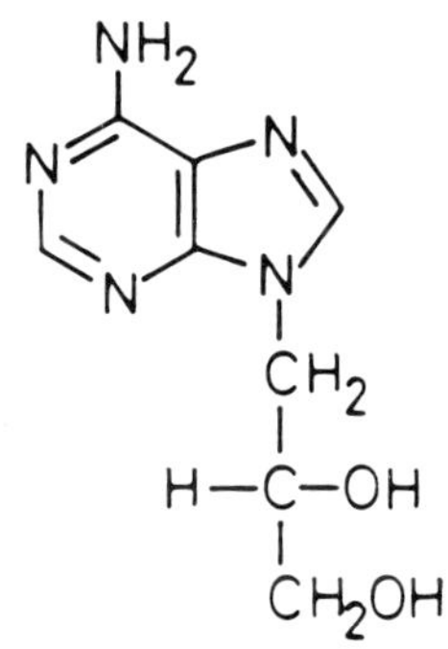

18

thesis is reasonably well established, this did not refrain several groups of chemists from adding the "3'-hydroxyl group" to give DHPG (11). This compound is also very active but appears to be too toxic. Most of the other analogues were prepared in an attempt to synthesize something which had not been made before and had no design in them. An exception is the prodrug approach to modify the heterocycle such that it afforded the 2,6-diamino derivative (13), which is converted by adenosine deaminase to yield acyclovir itself (22) or the 6-deoxy derivative (12) which is converted by xanthine oxidase to yield again acyclovir (21).

A similar story happened for the BVDU analogues. The ara-derivative (14) was an early active derivative to be found (23), but since then, few, if any, compounds with useful activity have been found. As the mode of action of BVDU itself is still being elucidated, this is perhaps not too surprising. Thus, two attempts at a rational design to improve the efficacy of BVDU have given less active compounds. First, the carbocyclic derivative (15) was made in an attempt to prevent the degradation of the compound by nucleoside phosphorylase. This, it clearly did, but, although the compound was surprisingly active *in vitro*, it had relatively low *in vivo* activity (24), which ought to tell us something about the metabolism of carbocyclic nucleosides. The second designed compound was the 3'-*O*-methyl derivative (16) which was synthesized in an attempt to prevent incorporation of BVDU into DNA and make it a chain terminator (25). This compound has no activity because it is not a substrate for the virus-encoded thymidine kinase. Thus even the limited tinkering with a lead compound has not produced very exciting results. Can we use our knowledge that BVDU has such pronounced antiherpes activity to design something which would work in the same way but have a different structure ? As we don't exactly know how BVDU works in suppressing viral replication, then clearly we cannot do this, but just assuming that we require a deoxynucleoside 5'-triphosphate to inhibit specifically the DNA polymerase coded for by the virus (and this assumption is probably incorrect), what could we do ?

Any nucleoside selected for this purpose has to be a substrate for four enzymes and should not be a substrate for four others as a minimum requirement (Scheme I). First, it has to avoid nucleoside phosphorylase (A), and must be a substrate for the viral TK [B] but not for the host cell TK (C), as affinity for the latter would probably confer toxicity. The 5'-monophosphate so formed should not to be a thymidylate synthetase inhibitor (D) (or again it will be toxic for the host cell) but has to be phosphorylated successively to the di- and triphosphate by [B'] and [E]. Presumably this triphosphate has to be a substrate or inhibitor of the virus-encoded DNA polymerase [F] and not be recognized by the cellular enzymes [G]. As yet, we really know nothing about the specificity, active site or mechanism of action of any of these enzymes; so we cannot possibly design a compound to be a substrate for four of them and avoid being recognized by the other four. From what we know on the mode of action of the antiherpes compounds, a specific inhibitor of the virus-encoded TK would almost certainly show antiviral properties. Here, we are only dealing with three enzymes : the viral and cellular TK and nucleoside phosphorylase, but no such inhibitor has yet been found. All nucleoside analogues appear either to be substrates of the virus-encoded TK or not to be recognized at all.

We have several lead antiherpes agents; we know probably as much about herpesviruses in molecular terms as any virus; herpesviruses are clearly worth exploiting commercially and many *in vitro* and *in vivo* assay systems are available. Yet, new discoveries are rare and designed agents

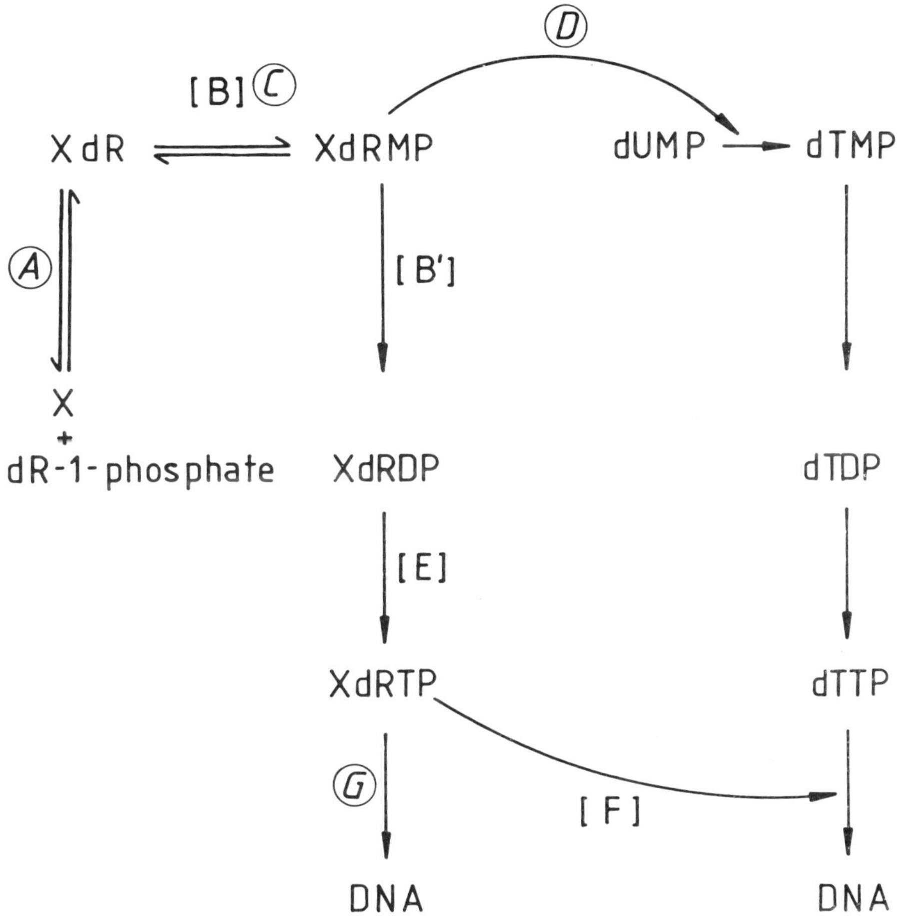

SCHEME I. Possible metabolic routes for an antiviral pyrimidine nucleoside analogue. [B], [B'], [E], [F] : required enzymes; (A), (C), (D), (G) : enzymes to be avoided.

very few. The recently described broad-spectrum anti-DNA virus agent (S)-9-(3-hydroxy-2-phosphonylmethoxypropyl)adenine [(S)-HPMPA] (17) (26) is such an example, however. The synthesis of this compound involved a rather small chemical modification of an already established wide-spectrum antiviral agent, (S)-DHPA (18). It certainly constitutes a new lead for anti-DNA virus agents.

What can one suggest for the future ? Advances have to be made on many fronts. For design to stand any chance, we must know more at the molecular level about how viruses replicate. This means we need to know their genomic structure, the structure of the protein products and the details of crucial steps unique to virus replication. As chemist, we rely on the virologists, molecular biologists and eventually X-ray crystallographers to get these data for all viruses of potential interest.

Then perhaps attempts at design can be made, at least then many of the excuses have been removed, and as the chemical synthesis should provide not too much of a problem it will then become apparent how good the chemist is at original design.

In the meantime, we have to rely on serendipity and working on leads thus obtained. This approach means that the testing must be reliable and relevant to the eventual treatment of a clinical disease. After all is said and done, there is little money to be made in curing fibroblast cell cultures of herpesvirus infection or ferrets of influenza virus !

This is where we return full circle to the contribution made to this field by Prof. Piet De Somer. It is not by chance, or should I say serendipity, that one of the most reliable and comprehensive antiviral evaluation laboratories is situated in the Rega Institute. Prof. De Somer had the foresight to realize that reliable, quantitative and interpretable antiviral assays were required and that in Prof. Erik De Clercq he had the man to exploit this need. Thus he backed this research with all the energy we would expect from such a man, and the results can, and will continue to be seen, as a monument to this great Man of Science.

REFERENCES

1. De Clercq E, Descamps J, De Somer P, Barr PJ, Jones AS, Walker RT. Proc. Natl. Acad. Sci. USA 76:2947-2951, 1979.
2. De Clercq E, Walker RT. Pharm. Ther. 26:1-44, 1984.
3. See reviews by Oxford JS. _in_ Problems of Antiviral Chemotherapy (Stuart-Harris Sir Charles H, Oxford JS, Eds). Academic Press, London, pp 231-264, 1983; _in_ Targets for the Design of Antiviral Agents (De Clercq E, Walker RT, Eds). Plenum Press, New York, pp 159-176, 1984.
4. Hay AJ, Wolstenholme AJ, Skehel JJ, Smith MH. EMBO J. 4:3021-3024, 1985.
5. Elion GB, Furman PA, Fyfe JA, de Miranda P, Beauchamp L, Schaeffer HJ. Proc. Natl. Acad. Sci. USA 74:5716-5720, 1977; Schaeffer HJ, Beauchamp L, de Miranda P, Elion GB, Bauer DJ, Collins P. Nature 272: 583-585, 1978.
6. De Clercq E, Walker RT. Prog. Med. Chem. 23:187-212, 1986.
7. De Clercq E. Biochem. Pharmacol. 33:2159-2169, 1984.
8. Walker RT, Jones AS, De Clercq E, Descamps J, Allaudeen HS, Kozarich JW. Nucleic Acids Res. Symp. Series no 8:s95-s102, 1980.
9. Cassiman JJ, De Clercq E, Jones AS, Walker RT, Van den Berghe H. Brit. Med. J. 283:817-818, 1981.
10. Walker RT, Balzarini J, Coe PL, De Clercq E, Harnden MR, Jones AS, Noble SA, Rahim SG. Nucleic Acids Res. Symp. Series no 11:215-218, 1982.
11. Bleackley RC, Jones AS, Walker RT. Tetrahedron 32:2795-2797, 1976.
12. Barr PJ, Jones AS, Walker RT. Nucleic Acids Res. 3:2845-2849, 1976.
13. De Clercq E, Descamps J, Verhelst G, Walker RT, Jones AS, Torrence PF, Shugar D. J. Infect. Dis. 141:563-574, 1980.
14. Jones AS, Rahim SG, Walker RT, De Clercq E. J. Med. Chem. 24:759-760, 1981; Cheng Y-C, Dutschman G, De Clercq E, Jones AS, Rahim SG, Verhelst G, Walker RT. Mol. Pharmacol. 20:230-233, 1981.

15. Prusoff WH, Lin T-S, Mancini WR, Otto MJ, Siegel SA, Lee JJ. in Targets for the Design of Antiviral Agents (De Clercq E, Walker RT, Eds). Plenum Press, New York, pp 1-27, 1984.
16. Bennett LL Jr, Brockman RW, Montgomery JA. Nucleosides & Nucleotides 5:117-124, 1986.
17. Desgranges C, Razaka G, Drouillet F, Bricaud H, Herdewijn P, De Clercq E. Nucleic Acids Res. 12:2081-2090, 1984; Desgranges C, Razaka G, Rabaud M, Bricaud H, Balzarini J, De Clercq E. Biochem. Pharmacol. 32:3583-3590, 1983.
18. De Clercq E, Desgranges C, Herdewijn P, Sim IS, Jones AS, McLean MJ, Walker RT. J. Med. Chem. 29:213-217, 1986.
19. Jones AS, Verhelst G, Walker RT. Tetrahedron Lett.: 4415-4418, 1979.
20. De Clercq E, Rosenwirth B. Antimicrob. Agents Chemother. 28:246-251, 1985.
21. Krenitsky TA, Hall WW, de Miranda P, Beauchamp LM, Schaeffer HJ, Whiteman PD. Proc. Natl. Acad. Sci. USA 81:3209-3213, 1984.
22. Good SS, Krasny HC, Elion GB, de Miranda P. J. Pharm. Exp. Ther. 227: 644-651, 1983.
23. Sakata S, Shibuya S, Machida H, Yoshino H, Hirota K, Senda S, Ikeda K, Mizuno Y. Nucleic Acids Res. Symp. Series no 8:s39-s42, 1980.
24. Herdewijn P, De Clercq E, Balzarini J, Vanderhaeghe H. J. Med. Chem. 28:550-555, 1985.
25. Jones AS, Sayers JR, Walker RT, De Clercq E. Unpublished results.
26. De Clercq E, Holy A, Rosenberg I, Sakuma T, Balzarini J, Maudgal PC. Nature 323:464-467, 1986.

SECTION II : INTESTINAL MICROFLORA

SECTION CHAIRMAN : PROF. H. EYSSEN

H. Eyssen

INTRODUCTION TO SECTION II : GASTROINTESTINAL MICROBIOTA AND THEIR ROLE IN THE METABOLISM OF ENDOGENOUS AND EXOGENOUS SUBSTANCES

H. EYSSEN
Rega Institute, Katholieke Universiteit Leuven, B-3000 Leuven, Belgium

I joined the laboratory of Professor P. De Somer in 1952 when I was still a medical student. In 1952 Professor De Somer was especially interested in the, then unexplained, extremely high toxicity of penicillin for guinea pigs and hamsters. A single dose of a few thousand units of penicillin killed 80 to 90 % of the animals within 2-4 days. Usually, 10-20 % of the guinea pigs survived and proved to be resistant to high and repeated doses of penicillin. The same toxic effects were seen after oral administration of bacitracin, a non-absorbed antibiotic. Tetracyclines also were highly toxic. Animals resistant to administration of penicillin, could be killed by a single dose of tetracycline, and *vice versa*. We postulated that antibiotics, by eliminating the so-called "normal" or indigeneous microflora (microbiota) caused a fatal enterocolitis by microorganisms not susceptible to the antibiotic. This, or at least very similar, types of enterocolitis have been noted after antibiotic therapy in man. In the 1950's and 1960's this enterocolitis was due to an overgrowth of staphylococci; in more recent years it can be ascribed to *Clostridium difficile* (after lincomycin therapy).

In other animals, i.e. piglets and chickens, small amounts of antibiotics were found to exert the opposite effect : they stimulated growth and promoted the efficiency of absorption of nutrients from the intestine. Apparently, some "normal" intestinal bacteria must have been harmful to the host. These observations led Piet De Somer to study the effects of the intestinal microflora in gnotobiotic animals. Gnotobiotic animals are originally completely germ-free animals that are deliberately contaminated with one or more strains of defined intestinal microorganisms to explore the influence of these bacteria on the physiology of the host. In our germfree animal unit we found that in newborn chickens *Streptococcus faecalis* and certain undefined filterable agents caused growth inhibition, because they developed in the small intestine and affected the efficiency of nutrient absorption. Small amounts of antibiotics in the food eliminated these harmful "normal" bacteria, stimulated growth, and effected a body weight increase by 10 to 15 %.

Later, we studied the effects of bacterial overgrowth in the small intestine in germfree and gnotobiotic rats with a surgically produced blind-loop of the small intestine, associated with defined intestinal bacteria. We came to the conclusion that the malabsorption syndrome in these animals with bacterial overgrowth in the small intestine resulted from several causes, the effects on the gut wall morphology and gut motility and the effects on bile acid metabolism presumably being the most important ones. Later, other investigators found that bacterial deconjugation of bile acid was a major cause of the malabsorption syndrome in humans.

More recently, our efforts have been focussed on intestinal microbial transformations of cholesterol, plant sterols, bile acids, steroid hormones and fatty acids. Several of the intestinal microorganisms that metabolize these substances are strictly anaerobic, very fastidious and yet unnamed.

The double bond between carbon 5 and carbon 6 of cholesterol and plant sterols can be hydrogenated by intestinal bacteria to yield coprostanol. We isolated an unnamed anaerobic _Eubacterium_ species carrying out this reaction. The microorganism uses the double bond of sterols as an hydrogen acceptor and does not grow unless high levels of a suitable unsaturated sterol are available in the culture medium.

Intestinal microorganisms also dehydroxylate bile acids. Thus, deoxycholate is formed from cholate and Prof. T. Midtvedt was first to isolate a 7-dehydroxylating intestinal microorganism. Dehydroxylation of chenodeoxycholate yields the highly hepatotoxic 3-monohydroxylated lithocholate.

We have recently isolated a very fastidious microorganism that, in the rat, dehydroxylates certain bile acids, the so-called muricholates, to yield hyodeoxycholate, the major fecal bile acid in conventional rats. This microorganism does not grow on any of the commercially available culture media unless the medium is supplemented with that specific bile acid.

Bile acids, as well as steroid hormones, can be sulfated prior to excretion in the bile or the urine. We have shown this sulfation to be a sex-dependent process, which is especially active in female rats and male mice. The sulfated compounds, i.e. lithocholate sulfate, are less efficiently reabsorbed from the intestine, and this contributes to their elimination from the body. However, sulfatase producing intestinal microorganisms, by desulfating these substances, promote their entero-hepatic recycling. For example, taurolithocholate is slowly excreted in feces and urine by germfree and conventional rats. In contrast, taurolithocholate-3-sulfate is very rapidly eliminated by germfree rats or gnotobiotic rats associated with a microflora that does not desulfate bile acids.

Similarly, sulfated steroid hormones are poorly reabsorbed, unless hydrolyzed by desulfating microorganisms. A number of such desulfating strains have been isolated in our laboratory.

Certain intestinal microorganisms hydrogenate double bonds in long-chain unsaturated fatty acids. We have isolated several intestinal microorganisms that carry out these reactions. For example, the anaerobe _Eubacterium lentum_ first isomerizes the 12-13 _cis_ double bond of linoleic acid to an 11-12 _trans_ double bond, and then hydrogenates the 9-10 _cis_ double bond to yield the mono-unsaturated transvaccenic acid.

Prof. P. De Somer has always had a keen interest in the apparently multifold activities of the intestinal microbial world. The intestinal microbiota and the biochemical reactions they catalyze will be further discussed by the lecturers, D.C. Savage, T. Midtvedt and P. Goldman.

DWAYN C. SAVAGE studied microbiology at the University of California in Berkeley. From 1965 till 1967 he was Research Associate and Visiting Fellow at Rockefeller University in New York where he continued his research under Professor René Dubos on the development and characteristics of the intestinal microflora of conventional mice. From 1967 till 1972 he was at the University of Texas in Austin, and in 1973 he became Professor of Microbiology at the University of Illinois in Urbana. His research work is focussed on the colonization and the adherence of intestinal microorganisms to epithelial (i.e. gut wall) surfaces.

TORE MIDTVEDT studied medicine and microbiology in Oslo. In the second half of the 1960's he came to work at the Karolinska Institute in Stockholm, where he joined the germ-free animal unit headed by Prof. Bengt Gustafsson, a pioneer of germ-free research and the secretary of the Nobel Committee. While working with Gustafsson, Tore Midtvedt successfully isolated the first intestinal microorganism converting, by 7-dehydroxylation, the bile acids cholic acid and chenodeoxycholic acid into deoxycholic and lithocholic acid, respectively. Tore Midtvedt returned to Norway until 1982, where, following Prof. Gustafsson's retirement, he was called back to Stockholm to become Professor and Director of the Gnotobiotic Animal Research Center. Prof. Midtvedt has been especially interested in the metabolism of endogenous substances by the intestinal microflora, which is also the subject of his presentation.

PETER GOLDMAN studied Engineering Physics at Cornell University and at Harvard University where he graduated as a master in physics in 1963. He then decided to start medical studies at Johns Hopkins University where he graduated as M.D. 4 years later. After several research fellowships in the laboratories of distinguished scientists at N.I.H. in Bethesda he became Senior Investigator at the National Institute of Arthritis and Metabolic Diseases. From 1972 till 1975 he was Professor of Clinical Pharmacology at Harvard University. He was Director of the Division of Clinical Pharmacology from 1975 till 1982, when he became Professor of Health Sciences at the Department of Nutrition and Director of the Food and Drug Laboratory. Since 1984 he is Acting Chairman of the Department of Nutrition, and since 1986 also Acting Chairman of the Department of Pharmacology at Harvard University. His major research interests cover pharmacological problems in the areas of clinical pharmacology, toxicology and nutrition. We are grateful to Prof. Goldman for accepting our invitation to lecture on the therapeutic implications of drug metabolism by the intestinal microflora.

D.C. Savage

THE INDIGENOUS GASTROINTESTINAL MICROBIOTA

D.C. SAVAGE
Department of Microbiology, University of Illinois, 131 Burrill Hall, 407 South Goodwin, Urbana, Illinois 61801, U.S.A.

1. INTRODUCTION

Although representatives of only few species have been studied in detail, little doubt now exists that birds, adult mammals including man, and animals of many other species have microbiotas (microfloras) indigenous to their alimentary canals (1,2). These biotas develop soon after birth, consist of prokaryotic and eukaryotic microorganisms of many genera and species, and are essential for the health and welfare of their animal hosts. Mammals of many species (i.e. ungulates) and insects of certain species (i.e. termites) are known to derive as adults essential elements of their nutrition from products of their alimentary microbiotas (3). Even when their essential nutrients need not be microbial products, animals of many and perhaps even all species physiologically interact in various ways with their biotas (3-5). Indeed, their capacity to survive in nature depends upon the biota. Thus, mammals and animals of most if not all species can be regarded as creatures consisting of eukaryotic animal cells in symbiosis with eukaryotic and prokaryotic microbial cells. In other words, animals exist under natural conditions as complex interactive organic units consisting of both animal and microbial cells.

These concepts reflect findings from the research of many investigators. Notable among the findings were those from studies of the effects of the intestinal microflora on the physiology and health of animals from the work of pioneers in the area such as Prof. Piet De Somer (6). Such early findings gave impetus to modern research on the mechanisms of the microbial influences, the composition and regulation of the microbiota and the biochemistry and genetics of microorganisms isolated from the biotas.

Professor De Somer's research directly influenced my earliest work in the field, and has continued to affect my experimental approaches. His important observations on the toxicity of penicillin for guinea pigs (6) influenced me in my early research to examine the toxicity of that and other antibacterial drugs on laboratory rodents. Those experiments were coupled with examinations of the effects of the antibiotics on the microbiota in animals. None of the drugs tested, including penicillin, proved to be toxic for the rodents. Nevertheless, as Professor De Somer's findings predicted, the drugs profoundly influenced the intestinal tract of the animals, enlarging the cecum in a fashion much the same as that seen in germfree rodents (7). Moreover, the drugs radically altered the population levels of the major taxonomic classes of bacteria present in the gastrointestinal canal. When lactating mice were given antibiotics, the development of the microflora in their suckling offspring also radically altered. In addition, the cecums enlarged in the baby mice when they were about 10 days to two weeks of age (8), the period during which the cecums of germfree mice also begin to enlarge. The antibacterial drugs proved to be particularly potent in disrupting the communities of indigenous microorganisms that colonize epithelial surfaces in the tract (7,8, and below). In mice given penicillin, lactobacilli that normally colonize the gastric epithelium disappeared allowing yeasts to grow abnormally on the surface (9). Extrapolations from these studies continue to be research subjects in my laboratory and can be directly traced to the pioneering efforts of Professor De Somer.

In this paper, I present a personal view of the current state of knowledge of the gastrointestinal microbiota from the perspective of a microbiologist interested in how the biota is regulated in location, composition and function. Emphasis is upon the biotas of man and mammals of a few other species having in their foreguts (stomachs) and midguts (small intestines) biotas of relatively uncomplicated composition and function as compared to the complex biotas in their hindguts (cecum and colon). The presentation is completed with a discussion of some findings from research aimed at generating knowledge on how the biota is regulated (1,2).

2. GASTROINTESTINAL MICROBIAL ECOLOGY

The gastrointestinal tract and its associated microbiota in a mammal of any given species constitutes an open ecosystem analogous to a stream flowing through a field with occasional rather stagnant eddies (1,2). In a normal system, indigenous (autochthonous) microorganisms, the native inhabitants, occupy specific niches in particular habitats in the system. In other words, the organisms live in specific ways while localized in particular areas of the tract. The habitats can be differentiated both vertically and horizontally. Different major regions of the tract (i.e., stomach, small bowel, large bowel), and even differentiated areas within those regions, can be distinctive habitats. More importantly, in animals of certain species, specific microbial habitats can be distinguished in some regions in the lumen of the tract, on the epithelial surface, and even in the crypts of Lieberkuhn in the bowel wall mucosae (10,11).

Microbial climax communities found in the lumen in gastrointestinal tracts are often associated with particles of digesta (10,11). In a given area or subarea, the community in the lumen, and especially on surfaces of particles of digesta, can differ from that on the epithelial surface in the relative proportions of the populations of the various species. Likewise, those communities can differ in composition from cryptal communities. Indigenous lumenal communities exist in major areas where the rate of passage of digesta does not exceed the rates at which microorganisms can multiply (for example, the cecum). Epithelial communities can exist in any area. Cryptal communities have been reported so far to be present only in the lower small intestine and cecums and colons of monogastric animals of certain species.

Indigenous microorganisms of an amazing variety occupy these many habitats, often at enormous population levels (1,2,10,11). Because the ecosystem is open, however, the habitats also may contain transient (allochthonous) microorganisms. Such organisms may enter with the food and other ingesta of the animal, from habitats above the tract (oral cavity, respiratory tract), from high gastrointestinal habitats into lower ones, and even from low intestinal habitats into higher ones in coprophagous animals such as rodents. Allochthonous organisms may temporarily colonize, and multiply in, habitats vacated by their autochthonous residents in disturbed systems. The ecosystems may be disturbed by, for example, malnutrition (3) or antimicrobial drugs (6-9). Normally, however, the habitats are colonized in adult animals by climax communities composed of indigenous microorganisms of particular species, each of which occupies a specific niche in the area. Microbial strains found in any particular habitat may be differentiated into autochthonous or allochthonous types in experiments in which the former are demonstrated to satisfy certain criteria for autochthony (1,2).

Microbial climax communities in the gastrointestinal ecosystem form as a result of successions that begin at birth of the animal (1,2). During succession, microbial strains colonize the system in sequential patterns that are characteristic of the animal, its diet and its environment. The successional and climax communities are regulated in composition and function by many complex forces, most of which are at best poorly understood. These issues are amplified below.

3. THE CLIMAX COMMUNITIES

3.1. Communities in the foregut

Microbial strains of many genera and species have been isolated from the contents of the stomachs of mammals of several monogastric species (1, 2). Most of the strains isolated enter with the food and other ingesta including feces. The stomachs of most monogastric animals undoubtedly can empty at an overall rate more rapid than that at which microorganisms can multiply. Therefore, microbial cells in the lumen can pass out of the stomach with its content, and should be regarded as transients in the area. However, certain microbial species able to associate with gastric epithelial surfaces may be autochthonous to habitats in the area. Lactic acid bacteria (for example, *Lactobacillus* species) and yeasts of the genus *Candida* are known to colonize epithelial surfaces in the stomachs of mammals of several species, including possibly also man (1,2,10,11). Thus, the autochthonous gastric microflora of those monogastric mammals is composed of aciduric bacteria and yeasts that associate with specific epithelial surfaces.

3.2. Communities in the midgut

The small intestines of monogastric mammals also yield in culture strains of many microbial species (1,2,10,11). Most of the organisms are probably transients in the area, especially in the upper two-thirds of the bowel where peristalsis moves lumenal content at a rate more rapid than that at which microorganisms can multiply. In the lower one-third, however, where the digesta moves sluggishly, strains of microorganisms found in the large bowel may be cultured, occasionally at high population levels. Such organisms may be indigenous to the region. Conversely, they may be transients from the cecum crossing the ileocecal valve into the area.

In monogastric mammals of some species, microorganisms have been found which are adhering to, or associating with, the mucous gel covering the epithelium of the small intestine (10,11). The organisms adherent to the membranes of the epithelial cells are found in the lower two-thirds of the intestines of rats and mice and may be indigenous inhabitants. They are filamentous prokaryotes with Gram-positive ultrastructure. Their population levels are quite high per unit area of the epithelial surface (12). Microbial cells have also been seen in the mucous gel overlying the epithelium in the small intestines of laboratory rodents (13), humans (11) and monogastric mammals of a few other species (1,2,10,11). These microorganisms are principally bacteria of mixed species and have not been demonstrated as of yet to be indigenous to habitats in the area. Nevertheless, as in the stomach, the indigenous microbiota in the small intestine probably consists of microorganisms that associate with the epithelial surface.

3.3. Communities in hindgut

The lumens of the cecums and colons of adult monogastric mammals of all species studied, including man, contain enormous populations of microorganisms (over 100 billion microbial cells per gram of content) (1,2,10). In these regions, the digesta moves slowly allowing ample time for microbial multiplication. The populations are composed primarily of Gram-positive and Gram-negative bacteria that are unable to multiply in atmospheres containing oxygen. Indeed, many of the species are intolerant of oxygen and are killed by exposure to oxygen or to culture media or diluting fluids with positive oxygen-reduction potentials. Over 99 % of the total microbial population obligately gains its energy through anaerobic processes. In adult animals, facultative bacteria (those that are able to use both aerobic and anaerobic processes to generate energy for growth), such as Escherichia coli, are minority members in the ecosystem usually outnumbered by the anaerobes by as much as 1000 to 1. The microfloras may differ somewhat in generic and species composition, but are generally similar in overall composition and in the types of microbial products synthesized.

Many of the microbial cells found in the lumens of the cecums and colons of monogastric mammals adhere in communities (sometimes called "consortia") to surfaces of digesta, such as plant materials (10,11). Morever, as in the stomach and small bowel, microorganisms also associate with epithelial surfaces in the cecum and colon (10,11). Some microorganisms have been found colonizing the crypts of Lieberkuhn in some animal species. Thus, in the hindgut, the indigenous microbiota consists of both lumenal, epithelial and cryptal communities of considerable complexity in composition.

3.4. Succession of the communities

Up to the time of birth, the gastrointestinal tract is sterile in the healthy fetus (1,2). During normal birth, however, the baby is contaminated with microorganisms from the vagina and external genitalia of the mother and other environmental sources to which it is exposed. Many of these microorganisms are unable to colonize habitats in the neonatal tract and disappear from it soon after birth. Others produce the offspring eventually forming during succession the climax communities in the adult.

The successional patterns have been best studied for laboratory rodents (1,2). In suckling mice and probably also rats, lactic acid bacteria colonize the tracts to population levels near those found in adults within one or two days after birth. Shortly after birth as well, populations of facultative anaerobes, such as Escherichia coli and Streptococcus faecalis, appear and rise to high population levels. Once the animals begin to sample solid food, within 10 to 14 days after birth, strictly anaerobic bacteria can be found in the intestine. The population levels of these bacteria increase quickly and are at adult climax levels by the time the animals are weaned. As the population levels of strict anaerobes increase, the levels of facultatives (i.e. E.coli and S.faecalis) decline concomitantly. At weaning time (3 weeks after birth), the climax populations of the facultatives are normally low, and remain at those low levels unless the ecosystem is disturbed in some drastic way such as by antimicrobial drugs (7-9) or starvation (1,2). In laboratory rodents and lagomorphs at least, the succession of the flora is completed only after the animals have stopped taking mother's milk. Certain microorganisms that colonize epithelial surfaces in the stomach and small intestine in rodents and possibly most of the flora in rabbits colonize the tract only after the animals are weaned (1,2).

Successions in mammals other than laboratory rodents have been examined less systematically than those in rodents, although several attempts have been made to study the microflora of various species including humans at sequential times after birth (14). Such studies, especially those with human infants, are complicated by many factors that limit the type and number of samples that can be taken. In addition, the studies with humans are complicated by dietary practices of different populations. Thus, the successions in man and other large mammals are less well understood in detail than those of laboratory rodents. Nevertheless, enough detail is known to support the hypothesis that microorganisms colonize the tracts of such animals in a sequential fashion suggestive of successions, and that the process is significantly influenced when the diet is shifted from milk to solid food (1,2,14).

3.5. Regulation of the communities

The population levels and species of microorganisms in the climax communities in the gastrointestinal ecosystem, the successions of those communities, and the biochemical and genetic activities of the microorganisms in them, are regulated by multifactorial processes (1,2,15). Some of the forces (allogenic factors) in the processes are exeted by the animal host and its diet and environment, while some (autogenic factors) are exerted by the microorganisms in the communities themselves. Microorganisms can exert direct influences to exclude other organisms from their habitats and niches, and can effect changes in functions of the host that regulate the biota and thus exert indirect influences on its composition, biochemistry and geographic distribution. Thus, autogenic and allogenic factors interact in exceedingly complex ways in effecting their regulatory functions.

A detailed discussion of all the factors involved in regulating the biota is beyond the scope of this chapter. Details on the factors vary from one animal species to another and even from area to area in the alimentary tract (1,2,15). For example, the host's diet and the capacity of indigenous microorganisms to associate with epithelial surfaces are powerful forces regulating the biota. The host's diet profoundly effects the composition and function of the foregut biota in animals with such biotas (1,2,15). By contrast, the extensive biotas found in the hindguts (cecums and colons) of monogastric animals are affected by the host's diet in only subtle and complex ways by comparison with the foregut biotas (1,2,15-18). In the hindguts, the biotas appear to be buffered against changes in the host's diet by materials produced by the host and found in relative abundance in the area [e.g., mucus, antibody and enzymatic proteins, and components of exfoliating epithelial cells (3)], while those in the stomach may be almost solely dependent for nutrients on components of the host's diet. Therefore, to be understood in detail, the forces regulating the communities in various habitats in the tract must be studied separately for each community and each animal species. In our laboratory, we are focussing on the capacity of microorganisms to colonize epithelial surfaces. Some findings from one of our projects are detailed below as an illustration of one approach in the analysis of the factors involved in regulating the microbiota.

Except possibly for surfaces lined with stratified squamous epithelium (in the esophagus and part of the stomach in mammals of some species), gastrointestinal "surfaces" are dynamic three-dimensional compartments when viewed from the microbial point of view (10,11). These compartments are composed of mucous gels flowing on glycocalyces overlying epithelial cell membranes that are convoluted into microvilli. The mucous gel, composed of complex glyco- and lipoproteins forming a three-dimensional hydrated matrix, can be more or less viscous depending upon its state of hydration. The membranes themselves and the glycocalyx are presumably in a dynamic fluid state with continuous movement and transposition of their macromolecular constituents. The secretive and absorptive functions of the epithelial cells and movement engendered by epithelial replacement, villous motility and peristalsis add to the complexity of the "compartments". Microorganisms that colonize the compartments have mechanisms, however, for overcoming and exploiting that macromolecular complexity. One such mechanism is the capacity of bacteria of some taxonomic groups to move by motility mechanisms, allowing them to respond to chemotactic gradients attracting them into viscous mucous gels (19,20). We are studying the motility of one such microorganism that colonizes a mucous gel habitat.

In adult laboratory rodents, communities of indigenous microorganisms can be found colonizing epithelial surfaces in all areas of the tract (11). One such community can be found in the mucous gel on the epithelium of the cecum and proximal colon. Most of the bacteria forming the community are species of oxygen-intolerant anaerobic bacteria, although strains of _E.coli_ and _S.faecalis_ can be found there as well. In normal adults, the populations of the latter species rarely exceed about 1 to 10 million bacteria per gram of whole organ with contents intact, while the populations of anaerobes can exceed 100 billion cells per gram of material.

The bacterial populations probably do not adhere to the underlying epithelial membranes; the microbial layers in the mucous gel are easily washed from the epithelial surface (11). Most of the cells in the populations are motile (20). Motility is such an obvious feature of the microbial community (R.W. Schaedler, personal communication), that several years ago we were motivated to study whether the capacity of the bacteria to move was essential from them to colonize their habitat. To effect the study, we used methods for isolating from the community some strains of rapidly motile strictly anaerobic bacteria. One of the strains isolated was a Gram-negative, slightly curved, rod-shaped bacterium with a polar to subpolar fascicle of flagella. From 25 to 35 flagella could be seen in the fascicle on each cell. [The magnitude of the commitment of the bacterium to motility can be estimated from the fact that over 5 % of the cellular protein is flagellin, the molecular subunit of the flagella (21)]. The cells were capable of rapid and sustained motility in culture media maintained free of oxygen at low oxidation-reduction potential. As soon as they were exposed to oxygen, however, they immediately stopped moving. Moreover, after such exposure, they no longer reproduced. In attempts to classify the organism, we found that it matched no known genus or species of anaerobic bacteria and therefore named it _Roseburia cecicola_ in honor of Theodore Rosebury (20).

R.cecicola appeared to us to be an ideal tool for studies of flagellar motility as a bacterial mechanism for colonizing mucous gels. It had been isolated from scrapings of mucous gel from the cecal epithelium of a mouse, was highly motile and invested a substantial percentage of its machinery for synthesizing flagellin. Therefore, we selected it for our further studies.

In earlier efforts to learn how intestinal microorganisms of species other than _R.cecicola_ colonize the gastrointestinal tract, we had associated germfree mice with specific microbial species and then examined the tracts of the animals for the microbial cells. Such examinations were accomplished by microscopy and attempts to culture the bacteria from specific sites. We decided to adopt a similar approach with _R.cecicola_, and, therefore tested whether it would colonize the tracts of germfree mice, and if so, then where it localized in the tract (20). The general experience of most investigators has been that oxygen-intolerant anaerobes will not colonize the tracts of germfree mice (monoassociated animals) until they are previously associated with one or more strains of more aerotolerant species. Nevertheless, R.cecicola colonized to high population levels the tracts of the mice without other bacteria present (20). Moreover, they localized predominately in the cecum, the area from which they had been originally isolated. A bonus from the study was the observation that the bacteria passed from mother to infant, a phenomenon that must also occur in conventional animals.

Success in those experiments allowed us to use germfree mice to test whether motility is essential for R.cecicola to colonize its mucous habitat (20). To prepare for the experiments, we treated the wild-type strain with the mutagen nitrosoguanidine and screened for motility mutants. Some of the mutants were immotile; some were not as actively motile as the parent. Some of the immotile strains were selected along with the parental strain for monoassociating germfree mice. After 3 to 5 weeks, the cecums of the animals were cultured for R.cecicola by a procedure that allowed the microbial population levels to be estimated, and the motility of the cells to be determined. To our astonishment, essentially all of the colonies from the animals associated with the parental strain as well as the nonmotile mutant strains were nonmotile.

After considering the design of the experiments, we drafted the hypothesis that R.cecicola requires its motility to colonize its habitat only when bacteria of other species able to colonize that habitat are present in the animal. To test that hypothesis, the experiments described above were repeated. This time, however, in addition to germfree mice, animals associated with microorganisms of other genera, some of which were known from previous experience to colonize the mucous gel on the cecal epithelium, were also associated with the parental or a nonmotile mutant strain of R.cecicola. Again after 5 weeks, most of the colonies obtained from the monoassociated animals contained nonmotile cells. By contrast, most of the colonies from the animals associated with the motile or the nonmotile R.cecicola and other microorganisms contained motile cells, even those from the animals originally associated with the nonmotile mutant. Therefore, in the fact of competition from other species, R.cecicola's motility is important for it to be able to colonize its habitat in the mucous gel (20).

As mentioned earlier, mucous gels consist of glycoproteins and glycolipids organized into complex hydrated supramolecular matrices. Especially when dehydrated, such structures could be expected to restrict bacterial motility. Moreover, the macromolecules in the matrix could serve to bind bacterial cells and their flagella, thereby immobilizing them. As also described, however, bacteria such as R.cecicola often reach enormous population levels in mucous gels. These considerations stimulated us to follow molecular methods to study how flagellar motility functions for R.cecicola in its native mucous habitat. Our overall goal was to learn how surface domains of macromolecules of the flagella and on bacterial surface interact with macromolecules in the mucous as the bacterial cells move into and about in the gel.

To begin this study, we isolated, purified and analyzed the amino acid content of the flagellar protein (flagellin) from the wild-type strain of R.cecicola (21). (Even though R.cecicola is an oxygen-intolerant anaerobe, the flagellin proved to be unremarkable in amino acid composition by comparison with E.coli, B.subtilis and some other species able to grow aerobically). Thereafter, the amino acid sequence of the amino terminal region of the flagellin was determined, of which two regions served as patterns for the synthesis of pools of oligonucleotides of complementary sequence and limited degeneracy (Martin JH, Savage DC, Abstracts of the Annual Meeting of the American Society of Microbiology, p 185, 1986) (Fig. 1).

									10			
Met	Val	Val	Gln	Gln	Asn	Met	Thr	Ala	Met	Asn	Ala	Asn
AUG	GUN	GUN	CAP	CAP	AAQ	AUG	ACN	GCN	AUG	AAQ	GCN	AAQ
				$_{3'}$GTQ	TTP	TAC	TGN	CGN	TAC	$TT_{5'}$		

						20					
Arg	Gln	Leu	Gly	Ile	Thr	Thr	Gly	Ala	Gln	Ala	Lys
CGN	CAP	CUN	GGN	AUQ	ACN	ACN	GGN	GCN	CAP	GCN	AAP
AGP		UUP		AUA					$_{3'}$GTQ	CGN	TTQ

				30							
Thr	Asn	Glu	Lys	Leu	Asn	Ala	Gly	Try	Ala	Ile	Ala
ACN	AAQ	GAP	AAP	CUN	AAQ	GCN	GGN	UAQ	GCN	AUQ	GCN
TGN	TTP	$CT_{5'}$		UUP						AUA	

FIGURE 1. Amino terminal amino acid sequence (first of each set of 3 lines) of R.cecicola flagellin, the inferred nucleotide sequence (second of each set of 3 lines) and synthesized oligonucleotide sequences (dodecanucleotide on the third line of first set of 3 lines and heptadecanucleotide illustrated on the third line of the second and third set of 3 lines); N, any nucleotide; P, purine; Q, pyrimidine.

The oligonucleotides were end-labeled with ^{32}P and applied to Southern blots of agarose gels in which had been electrophoresed R.cecicola chromosomal DNA digested with the restriction endonucleases Eco RI, Pst I, Hind III or Mbo I. The oligonucleotide probes hybridized to single fragments produced by each endonuclease. The fragments homologous to the probes were inserted into their respective restriction sites in either pBR322 or pBR328. The recombinant plasmids were then transformed into E.coli HB101. Strains of E.coli containing inserts of DNA from R.cecicola were selected by insertional inactivation. DNAs from the strains were screened in Southern blots with the oligonucleotide probes for the flagellin gene. Strains containing that gene were screened for expression of the flagellin with antibodies specific for the R.cecicola protein (Martin JH, Savage DC. Abstracts of the XIV International Congress of Microbiology, p 242, 1986). The gene has been subcloned and sequenced (Martin JH, Savage DC, in preparation, 1986).

The sequence of the gene will be instrumental in studying how the flagella of R.cecicola function in the mucous gel. The amino acid sequence of the entire flagellin protein has been inferred from the genetic sequence. We are now using the amino acid sequence as a guide for inserting site-specific mutations into the DNA sequence. Strains of R.cecicola into which such mutations have been transferred by marker exchange will be studied for their capacity to colonize and move about in mucous gels. Such studies should make possible analysis of how the bacterial cells and their flagella interact with macromolecules in the mucous gel.

4. SUMMARY

Mammals and animals of most if not all species exist under natural conditions as complex interactive organic units consisting of both animal and microbial cells. The microbial cells in vast majority are found in microbiotas indigenous to the gastrointestinal tracts. The microbiotas of adult monogastric mammals including man consist of microorganisms of many genera and species (predominantly species of strictly anaerobic bacteria) organized into climax communities that can occur in the lumen, on the epithelial surface or in the crypts of Lieberkuhn in any region of the tract. These communities form during succession in neonatal animals. Some are not completely organized until after the animals are weaned. The communities are regulated in microbial composition, localization and function by many complex interactive factors. One such factor, bacterial flagellar motility, is under study at the molecular level for its function in communities of strictly anaerobic bacteria found in epithelial mucous gels.

REFERENCES

1. Savage DC. Ann. Rev. Microbiol. 31:107-133, 1977.
2. Savage DC. in The Germ-Free Animal in Biomedical Research (Coates ME, Gustafsson BE, Eds). Laboratory Animal Ltd., London, pp 119-140, 1984.
3. Savage DC. Ann. Rev. Nutr. 6:155-178, 1986.
4. Wostmann BS. Ann. Rev. Nutr. 1:257-279, 1981.
5. Savage DC. in Bacterial Adhesion (Savage DC, Fletcher M, Eds). Plenum Press, New York, pp 437-463, 1985.
6. De Somer P. Zbl. Bakteriol. 170:338-340, 1957-1958.
7. Savage DC, Dubos R. J. Exp. Med. 128:97-110, 1968.
8. Savage DC, McAllister JS. Infect. Immun. 3:342-349, 1971.
9. Savage DC. J. Bacteriol. 98:1278-1283, 1969.
10. Savage DC. in Human Intestinal Microflora in Health and Disease (Hentges DJ, Ed). Academic Press, New York, pp 55-78, 1983.
11. Savage DC. Int. Rev. Cytol. 82:305-334, 1983.
12. Garland CD, Lee A, Dickson MR. Microbial Ecol. 8:181-190, 1982.
13. Rozee KR, Cooper D, Lam K, Costeron JW. Appl. Environ. Microbiol. 43:1451-1463, 1982.
14. Cooperstock MS, Zedd AJ. in Human Intestinal Microflora in Health and Disease (Hentges DJ, Ed). Academic Press, New York, pp 79-100, 1983.
15. Tannock GW. in Human Intestinal Microflora in Health and Disease (Hentges DJ, Ed). Academic Press, New York, pp 517-540, 1983.
16. Lhuillery C, Demarne Y, Ducluzeau R, Clara A. Infect. Immun. 36:621-627, 1982.
17. Freter R, Brickner H, Botney M, Cleven D, Aranki A. Infect. Immun. 39:676-685, 1983.
18. Dubos F, Pelissier JP, Andrieux C, Ducluzeau R, Raibaud P. Appl. Environ. Microbiol. 50:1258-1261, 1985.
19. Freter R, Allweiss B, O'Brien PCM, Halstead SA, Macsai MS. Infect. Immun. 34:241-249, 1981.
20. Stanton TB, Savage DC. J. Gen. Microbiol. 130:173-183, 1984.
21. Martin JH, Savage DC. J. Gen. Microbiol. 131:2075-2078, 1985.

T. Midtvedt

METABOLISM OF ENDOGENOUS SUBSTANCES

T. MIDTVEDT
Department of Medical Microbial Ecology, Karolinska Institutet, Stockholm, Sweden

1. INTRODUCTION

In Tennysons poem, Ulysses is saying : "I am part of all that I have met". I met Professor P. De Somer at a scientific meeting more than 20 years ago and after that meeting, Dr. De Somer and his research became an important part of my scientific life. My chapter is focussed on the metabolism of endogenous substances, one of De Somer's major research topics.

The term "endogenous substances", although quite simple, deserves some comments. It can be defined as substances derived from the body and presented into the gastrointestinal tract. Then, the question of whether the gastrointestinal tract is belonging to the "milieu interieur" or "milieu exterieur" has to be discussed. In modern ecology, man and his microbes are one, or several, ecological systems, and instead of saying, as Claude Bernard did, that "the fixity of the milieu interieur is the essential condition of free life", we now feel that "the fixity of the "milieu total" is the essential condition of free life". De Somer devoted many years of his life to study essential factors in the "milieu total". Germfree animals are invaluable tools for such studies, as they can be conventionalized by association with microbes capable of performing the parameter under study. On the other hand, conventional animals may, when treated with antibiotics, loose part of their microflora, and hence behave as germfree animals with respect to one or several parameters.

In germfree research, the terms MAC and GAC are often used (1). MAC (Microflora-Associated Characteristic) refers to any anatomical structure, physiological property or biochemical function in the macroorganism which is influenced by microorganisms. When the latter are absent, as in germfree animals, newborns, and, sometimes, in individuals receiving antibiotics, the parameter concerned can be defined as GAC (Germfree Animal Characteristic).

Germfree animal research was established in Leuven more than two decades ago, and during the past, Prof. De Somer and his colleagues have studied several GAC/MAC parameters, some of them will be described later.

In investigating the microbial metabolism of endogenously (as well as exogenously) derived substances, the following four questions can be addressed :

(1) The action(s) of the microbe(s) on the substance.
(2) The action(s) of the substance on the microbe(s).
(3) The interaction(s) of other substances/microbes with the actions sub (1) and (2).
(4) The resultant effect(s) on the host. The latter effect should be analyzed from various viewpoints : place (local or at a distance), form (direct or indirect), mechanism (both biochemical and immunological), and the possible consequence (physiological or pathophysiological).

In all mammalian species, including man, the number of endogenous substances reaching the gastrointestinal tract is quite high, and includes compounds which will be excreted as well as compounds which enter the enterohepatic circulation. The liver, pancreas and other glands lining the gastrointestinal tract are the main sources for such compounds.

2. HEPATICALLY DERIVED ENDOGENOUS SUBSTANCES

Normal mammalian bile contains bile acids, bile pigments, cholesterol, steroid hormones, lipids, proteins, various carbohydrates, etc. Cholesterol, bile acids and bile pigments will be further addressed.

2.1. Cholesterol

In mammals, cholesterol is an important part of the cell membrane; it also acts as a basic source for compounds involved in a broad variety of functions, such as steroid hormones, bile acids, etc. The total body cholesterol pool comprises endogenous cholesterol as well as cholesterol from dietary sources. Although several tissues are capable of synthesizing cholesterol *de novo*, only the liver and the small intestine synthesize cholesterol at high rates. Both exogenous and endogenous cholesterol are eliminated from the body almost exclusively by the feces, and the two major elimination pathways are (*1*) biliary excretion of cholesterol and (*2*) hepatic conversion of cholesterol to bile acids, followed by biliary excretion of the bile acids. The intestinal microflora plays an important role in the metabolism of cholesterol by interacting with the bile acids and cholesterol itself.

In the feces from man and mammals, the major cholesterol derivative is coprostanol. Unmetabolized cholesterol may be subject of entero-hepatic circulation, whereas coprostanol is excreted in the feces. The reduction of cholesterol to coprostanol occurs mainly in the large intestine, and the isolation of a pure bacterial strain capable of performing this reaction was first published in 1973 by Eyssen and his coworkers (2). The organism responsible for this conversion appeared to be a strictly anaerobic, non-sporing, unnamed *Eubacterium* species which required a culture medium with high amounts of cholesterol or another suitable sterol for growth. Apart from cholesterol, this *Eubacterium* also reduced stigmasterol, campesterol, sitosterol and allocholesterol, as well as the 3-keto group of cholestanone and coprostanone. Similar cholesterol-reducing strains of *Eubacterium* have also been isolated from human and baboon feces (3,4). Two major pathways have been postulated for the conversion of cholesterol to coprostanol, and the mechanisms involved have been studied in Leuven (5) and elsewhere (6,7). Coprostanol production is greatly influenced by the diet. In gnotobiotic rats associated with a coprostanol-producing microflora, replacement of starch by lactose caused a 100-fold decrease in the number of coprostanol-producing microorganisms, and a subsequent reduction in coprostanol production by more than 70 % (8).

In 1959, Danielsson & Gustafsson (9) found higher serum cholesterol levels in germfree than conventional rats fed on the same diet. Several factors such as *de novo* synthesis, oxidative catabolism, conversion to bile acids, and others, have been investigated (10), to account for this phenomenon. Since germfree animals lack an intestinal "microbial excretion-knife" for cholesterol, they are trying to compensate by reducing the *de novo* synthesis. Depending on the amount of dietary cholesterol, the animals may succeed or fail in maintaining a stable serum concentration of cholesterol.

This microbial excretion-knife may be targeted upon by antimicrobial drugs. Some antibiotics reduce the intestinal microbial conversion of cholesterol to coprostanol (1,10,11). Such drugs may also influence cholesterol metabolism independently of their action upon microbes; i.e. either indirectly by an increased binding to bile acids (12,13) or by other mechanisms. Pertinent in this regard is a statement made by Prof. De Somer's research group in 1971 : "sulfaguanidine, a poorly absorbed sulfonamide, specifically inhibits the absorption of cholesterol in rats and mice, and promotes fecal excretion of cholesterol without affecting the excretion of bile salts. Although the mode of action of this drug remains uncertain, our investigations demonstrate that the effect on cholesterol is independent of its antimicrobial activity since it also occurs under germfree conditions. Sulfaguanidine does not prevent formation of micelles, does not displace cholesterol from preformed micelles, and does not inhibit esterification of cholesterol by pancreatic esterase *in vitro*. Several closely related analogues of sulfaguanidine fail to inhibit cholesterol absorption" (14). Even in 1984, "the way by which sulfaguanidine inhibits cholesterol absorption in rats and mice is still a matter of uncertainty" (10); and the authors continue "... from all this it becomes clear how important gnotobiotic animals are in this kind of investigation. The elimination of the microflora on one hand and the observation of its impact under precise and defined conditions on the other hand are indispensable steps towards a better understanding of the overall metabolism of cholesterol".

2.2. Bile acids

In all mammals, bile acids are derived from cholesterol in the liver. Cholic acid and chenodeoxycholic acid are common, but many other primary bile acids may be found (15,16). The primary bile acids are conjugated, usually with taurine or glycine, and excreted into the bile. Recently, it has been shown that bile acids may be present as sulfate or glucuronide conjugates as well. Sulfation occurs by esterification, mainly at the OH-group at C-3 (rat, man) or at C-7 (mice). In the intestinal tract, the conjugated primary bile acids are attacked by microbial enzymes and converted to a variety of metabolites. The so-called secondary bile acids thus formed may then either be excreted in the feces, or reabsorbed and sometimes further metabolized by hepatic enzymes to so-called tertiary bile acids before re-excretion in the bile. These (primary, secondary or tertiary) bile acids, when presenting in the intestine, are subject of a number of microbial transformations such as deconjugation, desulfatation or deglucuronidation, dehydroxylation and various oxidation-reduction reactions at the hydroxyl groups. The hydroxyl groups are particularly vulnerable to a microbial attack. Both dehydrogenases and dehydroxylases have been found in the species *Eubacterium*, and their substrate specificity is intriguing. In the last years, most interest has been focussed on the hydroxyl groups at C-6 and C-7, and most reactions found to take place *in vivo* can now be duplicated *in vitro*, and *vice versa* (17,18). In general, the metabolites formed are less water-soluble, less active in forming micelles, and sometimes also more toxic to the host (if absorbed).

It is easy to understand why gnotobiotic animals have been widely used to study the metabolism of bile acids. They exclude interference from the microflora or allow to assess the specific interference of certain defined microbes. Taken together, "all this complies with the hypothesis that the intestinal microflora limits the intestinal bile acid pool by causing less efficient reabsorption and faster excretion" (15).

Of particular interest is bacterial desulfation. The presence of sulfated bile acids in human bile was ascertained about 15 years ago. In mice, Parmentier *et al.* identified cholic acid-7-sulfate (19), a previously unidentified substance, accounting for up to 50 % of the total bile acids. It was later found that sulfation is an important sex-linked event in bile acid metabolism in both germfree and conventional rats (20). It also occurs in patients with cholestasis, where the urine is a major route of bile acid excretion (21). Recently, a clostridial strain capable of desulfating 3-alpha-sulfates of 5-beta-bile acids and 3-beta-sulfates of 5-alpha-bile acids, but not 3-alpha-sulfates of 5-alpha-bile acids, was described (22,23), and, even more recently, a clostridial strain capable of hydrolyzing both 5-alpha- and 5-beta-bile acid 3-alpha-sulfates was isolated from rat feces (24). Working with gnotobiotic animals, the Leuven group has demonstrated that these reactions take place in intestinal segments from where a certain degree of reabsorption still is possible, thus inferring that microbial desulfation is an important factor in the overall elimination of bile acids (25). Recent investigations indicate that these microbes may also influence the enterohepatic circulation and excretion of estrone sulfate (26), thus further extending their biological importance.

It is clear that antibiotics may interfere with the fecal excretion of bile acids and cholesterol in animals. Other studies indicate that this may also be the case in man (27-29). However, to what extent an altered microbial intestinal metabolism may interfere with the long-term homeostasis of bile acids and cholesterol, remains to be investigated.

2.3. Bile pigments

Bile pigments, consisting almost exclusively of bilirubin, are the end products of the catabolism of hemoglobin and some heme-containing enzymes. Bilirubin is conjugated to glucuronate in the liver, and the conjugates (most often mono- and di-glucuronides) of bilirubin are excreted in the bile. In the intestine, the bilirubin conjugates are transformed by microbial enzymes to a variety of metabolites, including urobilinogen, which is then excreted in the feces. Urobilinogen represents the final end product of hemoglobin catabolism in man and conventional animals.

This microbial transformation of bilirubin may in several ways interfere with its enterohepatic circulation. Bilirubin glucuronides are poorly absorbed, whereas free bilirubin is absorbed from both the small and large intestine. Some urobilinogen is also absorbed. In contrast with bilirubin, urobilinogen does not seem to need conjugation to be excreted by the liver in the bile.

A minor part of intestinal β-glucuronidase may be derived from endogenous sources (30), but the major part is of microbial origin. Microbial β-glucuronidase is associated with the common intestinal microorganisms, although it is not clear which species are most active *in vivo* (31). Microbial conversion of bilirubin to urobilinogen appears to be a rare property of intestinal microorganisms and has so far been associated with one species only, that is, the anaerobic, gram-positive, spore-forming rod, *Clostridium ramosum* (32,33).

Antibiotics may interfere with the intestinal metabolism of bilirubin. Recently, we have evaluated the effects of 10 antibiotics, given orally to healthy subjects, on fecal excretion of urobilinogen (34). Bacitracin, vancomycin, clindamycin, erythromycin and ampicillin resulted in a pronounced suppression of fecal urobilinogen excretion, whereas doxycycline, metronidazole, nalidixic acid, ofloxacin and cotrimoxazole had no significant effect. Ampicillin, clindamycin and metronidazole were also investigated for their effects on fecal excretion of conjugated bilirubin. Clindamycin and metronidazole significantly increased the output of conjugated bilirubin.

2.4. General remarks concerning intestinal metabolism of hepatically derived endogenous substances

A common feature of many of the reactions described above is that they are carried out by only few species. Biochemically, they can be looked upon more as transport systems for hydrogen ions rather than energy-yielding processes. Depending on the oxidation/reduction potential, the various substances may act as donors/acceptors of hydrogen ions. In ecological terms, the intestinal flora seems to consist of two groups, i.e. r-selected and K-selected species (see below). The cholesterol-, bile acid- and bilirubin-transforming microbes belong to the latter group. Their presence reflects an ecosystem in a steady state.

3. PANCREATICALLY DERIVED ENDOGENOUS SUBSTANCES

Pancreatic juice may contain a variety of substances, such as enzymes or proenzymes, other proteins, mucin, etc. Trypsin will be commented upon as a model substance.

Metabolism of trypsin. Trypsin is excreted by the pancreas as a precursor molecule, trypsinogen, which is activated in the small intestine mainly by intestinal brush border enzymes. Fecal tryptic activity is the net sum of a series of complex processes involving the secretion of trypsinogen and trypsin inactivators from pancreas, activation of trypsinogen to trypsin by brush border enzymes, and the presence in the intestine of microbially or dietarily derived compounds that inactivate/degrade the trypsin as well as its inactivators. Germfree rats excrete high amounts of fecal tryptic activity whereas their conventional counterparts do not (35,36). Obviously, intestinal microorganisms are responsible for the inactivation of trypsin. The identity of these microorganisms is not known however.

The mean fecal tryptic activity in healthy adults on a western diet is approximately 100 mg per kg feces (37). Perorally administered antibiotics do not cause significant alterations in fecal tryptic activity. Some data indicate increased levels of tryptic activity in patients with Crohn's disease (38-40), and there might exist other pathophysiological conditions in which fecal tryptic activity is increased. Furthermore, individual variations in the intestinal levels of tryptic activity might play an important role in the utilization of proteins by the host and its intestinal microflora (41). However, the mere fact that the microbe(s) responsible for the intestinal inactivation of trypsin is yet unknown, makes further considerations along these lines rather speculative.

4. METABOLISM OF INTESTINALLY DERIVED ENDOGENOUS SUBSTANCES

The cells lining the gastrointestinal tract produce various substances which are extruded into the intestinal lumen either by secretion (i.e. mucin) or by turn-over of epithelial cells. In man, the loss of intestinal epithelial cells has been calculated to be as high as 200 grams per day. These cells are digested by the intestinal microorganisms and the resultant products are subsequently absorbed.

Breakdown of mucin. Endogenous mucin, derived from Goblet cells and mucus glands throughout the gastrointestinal tract, may be considered as representative for endogenous substances derived from other sources than the liver and the pancreas. Mucin has several important physiological and pathophysiological roles. It acts as a lubricant, and also as a barrier for the intestinal microclimate, and might also serve as a source of energy for the intestinal flora. There is a growing evidence that mucin may be relevant to the pathophysiology of some intestinal diseases, such as ulcerative colitis, Crohn's disease, gastric and duodenal ulcerations, colon adenocarcinomas, intestinal strangulation obstruction as well as mucus obstruction in cystic fibrosis (42).

Germfree rats excrete high amounts of mucin in their feces, whereas conventional rats, and healthy humans, do not. In conventional animals and humans mucin is degraded in the colon, probably by strains of Peptostreptococcus, Bacteroides, Ruminococcus and Bifidobacterium (42). This microbial degradation is markedly affected by antibiotics both in man and animals. In healthy individuals receiving oral antibiotics, the most pronounced effects were seen after administration of bacitracin, clindamycin or vancomycin (43). In these subjects the mucin excretion pattern was similar to that found in germfree rats. Altered patterns were also found in some individuals receiving ampicillin, doxycycline, erythromycin or metronidazole, whereas no alterations were noted in subjects receiving ofloxacin or cotrimoxazole. Similar findings have been reported for rats receiving antibiotics (1).

5. GENERAL CONSIDERATIONS

Studies on the metabolism of endogenous (and exogenous) substances only give an estimation of the consequences. These consequences occur locally or distantly and can be measured either directly or indirectly. This may be exemplified as follows.

The degradation of mucins yields energy, which may be of direct benefit to the microbes taking part in this process, and which may, indirectly, stimulate the synthesis of various compounds, such as short chain fatty acids, which, in turn, are either utilized by other bacteria, or absorbed and utilized by the host, or excreted in the feces and then lost.

Also, it is well established that short-chain fatty acids may suppress the growth of some pathogens like Salmonella. Previously, it was believed that intestinal production of short-chain fatty acids gave rise to diarrhea; now, we assume that they act against diarrhea by promoting water and electrolyte absorption from the colon. When absorbed, the short-chain fatty acids may directly supply energy to the host, and indirectly regulate carbohydrate metabolism by stimulating the release of insulin from the pancreas. They are also suspected to be involved in several pathological conditions, such as ulcerative colitis, Crohn's disease and colon adenocarcinomas.

Another example may be the bile acids. Many of the bile acids formed in the intestine may have a regulatory effect on the microbial flora, suppressing the growth of some species and stimulating the growth of others. They may also have an effect on the breakdown and turn-over of the intestinal cellular lining. If absorbed, they may interact by a complex feedback mechanism with cholesterol metabolism, and, from a pathophysiological viewpoint, bile acids may, of course, be involved in gall-stone formation.

The terms "r- and K-selected" were mentioned above, and they may deserve some comments. They originated from studies in macroecology, but may be adapted to intestinal microecology. The so-called r-selected species are characterized by a high resource allocation to reproduction and a high adaptability to resource utilization, a high migratory tendency and large variations in population size; these tendencies are in contrast to the tendencies found in K-selected species.

When applying these terms to newborn babies acquiring an intestinal microflora, ex-germfree animals undergoing conventionalization, or some individuals receiving antibiotics, the establishment, or re-establishment, of an intestinal flora leads to the following. Initially, r-selected species with many qualitative and quantitative variations are found, the ecological system is not in equilibrium, and various MAC functions may not be established. Later, the so-called K-selected species are favoured, and, as they are trying to utilize all metabolic niches, they establish all, or nearly all, MACs. In fact, several MACs represent final stages of metabolism, and the establishment of a MAC can be considered as filling a metabolic niche. Consequently the ecological system is now in a more steady state, which is to the benefit of the "milieu total".

In studies of this "milieu total", many aspects have to be investigated and the results often have to be re-evaluated. May I draw your attention to an investigation made by Professor De Somer and his group more than 30 years ago (44). They gave penicillin, in very small amounts, to guinea pigs, and observed a marked mortality. Approximately 60 per cent of the animals died after receiving 5000 IE penicillin intraperitoneally. They noted that the animals suffered from a marked diarrhea. At autopsy, an hemorrhagic enterocolitis was found, and bacteriological cultures yielded high numbers of coliforms. They stated that "the cause of death must be attributed to this altered intestinal flora, and is probably due to resorption of the toxins of the gram-negative bacteria from the intestine". Except for the gram-negatives, this statement would fit a description of a Cl.difficile enterocolitis. Prof. De Somer and his colleagues continued their work, and after examining several more antibiotics for toxicity in guinea pigs, questioned whether gram-negatives were the sole explanation (45). In discussing their findings in guinea pigs and humans receiving antibiotics, they stated : "Finally, it is possible that the antibiotics have a greater influence on the animal organism than we know at this moment ... Perhaps all these factors together with the toxicity of antibiotics for guinea pigs could be the different forms through which the same basic phenomenon manifests itself". This reflects the present interpretation of antibiotic-associated colitis, which is most often attributed to toxin-producing Cl.difficile.

Reiterating De Somer's words, I may conclude that studies of the metabolism of endogenous substances and analysis of the consequences are in fact studies of different forms by which the same basic phenomenons manifest themselves.

REFERENCES

1. Midtvedt T, Björneklett A, Carlstedt-Duke B, Gustafsson BE, Höverstad T, Lingaas E, Norin KE, Saxerholt H, Steinbakk M. in Germfree Research : Microflora Control and Its Application to the Biomedical Sciences. Progress in Clinical and Biological Research (Wostmann BS, Ed). Alan R. Liss Inc, New York, vol 181, pp 241-244, 1985.
2. Eyssen HJ, Parmentier GG, Compernolle FC, De Pauw G, Piessens-Denef M. Eur. J. Biochem. 36:411-421, 1973.
3. Brinkley AW, Gottesman AR, Mott GE. Appl. Environ. Microbiol. 43:86-89, 1982.
4. Sadzikowski MR, Sperry JF, Wilkins TD. Appl. Environ. Microbiol. 34: 355-362, 1977.
5. Parmentier G, Eyssen H. Biochim. Biophys. Acta 348:279-284, 1974.
6. Björkhem I, Gustafsson J-Å. Eur. J. Biochem. 21:428-432, 1971.
7. Skrede S, Björkhem I, Buchmann MS, Midtvedt T. J. Biol. Chem. 260:77-81, 1985.
8. Eyssen H, De Pauw G, Parmentier G. J. Nutr. 104:605-612, 1974.
9. Danielsson H, Gustafsson B. Arch. Biochem. Biophys. 83:482-485, 1959.

10. Van Eldere J, Eyssen H. in The Germ-free Animal in Biomedical Research. Laboratory Animal Handbooks 9 (Coates ME, Gustafsson BE, Eds). Laboratory Animals Ltd, London, pp 317-332, 1984.
11. Midtvedt T, Carlstedt-Duke B, Höverstad T, Lingaas E, Norin E, Saxerholt H, Steinbakk M. Eur. J. Clin. Invest. 16:11-17, 1986.
12. Eyssen H, Evrard E, Van den Bosch J. Life Sci. 5:1729-1734, 1966.
13. Eyssen H, Evrard E, Vanderhaeghe H. J. Lab. Clin. Med. 68:753-768, 1966.
14. Eyssen H, Vanderhaeghe H, De Somer P. Fed. Proc. 30:1803-1807, 1971.
15. Eyssen H, Van Eldere J. in The Germ-free Animal in Biomedical Research. Laboratory Animal Handbooks 9 (Coates ME, Gustafsson BE, Eds). Laboratory Animals Ltd, London, pp 291-316, 1984.
16. Haslewood GAD (Ed). Bile Salts. Methuen & Co Ltd, London, pp 1-116, 1967.
17. Eyssen H, De Pauw G, Stragier J, Verhulst A. Appl. Environ. Microbiol. 45:141-147, 1983.
18. Eyssen H, De Pauw G, Van Eldere J. in Germ-free Research : Microflora Control and Its Application to the Biomedical Sciences (Wostmann BS, Ed). Alan R Liss Inc, New York, pp 103-106, 1985.
19. Parmentier G, Mertens J, Eyssen H. in Advances in Bile Acid Research. III. Bile Acid Meeting, Freiburg, June 13-15 (Matern S, Hackenschmidt J, Back P, Gerok W, Eds). F.K. Schattauer Verlag, Stuttgart, New York, pp 139-144, 1975.
20. Eyssen H, Smets L, Parmentier G, Janssen G. Life Sci. 21:707-712, 1977.
21. Van Berge Henegouwen GP, Brandt K-H, Eyssen H, Parmentier G. Gut 17: 861-869, 1976.
22. Huijghebaert SM, Mertens JA, Eyssen HJ. Appl. Environ. Microbiol. 43: 185-192, 1982.
23. Huijghebaert SM, Eyssen HJ. Appl. Environ. Microbiol. 44:1030-1034, 1982.
24. Van Eldere J, Parmentier G, Robben J, Eyssen H. J. Steroid Biochem. 26:in press, 1987.
25. Robben J, Parmentier G, Eyssen H. Appl. Environ. Microbiol. 51:32-38, 1986.
26. Eyssen H, Van Eldere J, Parmentier G, Huijghebaert, S, Mertens J. J. Steroid Biochem. 22:547-554, 1985.
27. Samuel P, Holtzman CM, Meilman E, Sekowski I. Circulation Res. 33: 393-402, 1973.
28. Canzi E, Ceccarelli A, Ferrari A, Fesce E, Pacini N. Current Microbiol. 12:1-4, 1985.
29. Midtvedt T, Andreasson AK, Carlstedt-Duke B, Höverstad T, Lingaas E, Midtvedt AC, Norin KE, Saxerholt H, Steinbakk M. in Recent Advances in Chemotherapy. Animicrobial Section. Proc. 14th Int. Congress of Chemotherapy, Kyoto (Ishigami J, Ed). University of Tokyo Press, Tokyo, pp 2609-2610, 1985.
30. Röd TO, Midtvedt T. Acta Path. Microbiol. Scand. Sect. B 85:271-276, 1977.
31. Gadelle D, Raibaud P, Sacquet E. Appl. Environ. Microbiol. 49:682-685, 1985.
32. Gustafsson BE, Swenander-Lanke L. J. Exp. Med. 112:975-981, 1960.
33. Midtvedt T, Gustafsson BE. Acta Pathol. Microbiol. Scand. Sect B 89: 57-60, 1981.
34. Saxerholt H, Carlstedt-Duke B, Höverstad T, Lingaas E, Norin KE, Steinbakk M, Midtvedt T. Scand. J. Gastroenterol. 21:991-996, 1986.

35. Norin KE, Gustafsson BE, Midtvedt T. Lab. Anim. 20:67-69, 1986.
36. Norin KE, Midtvedt T, Gustafsson BE. Lab. Anim. 20:234-237, 1986.
37. Norin KE, Carlstedt-Duke B, Lingaas E, Höverstad T, Saxerholt H, Steinbakk M, Midtvedt T. To be published.
38. Bergstrand LO, Gustafsson BE, Homström B, Norin KE. Acta Chir. Scand. 147:707-709, 1981.
39. Bergstrand LO, Gustafsson BE, Holmström B, Norin KE. Acta Chir. Scand. 147:711-715, 1981.
40. Van der Merwe JP, Mol GJJ. Digestion 24:1-4, 1982.
41. Macfarlane GT, Allison C. FEMS Microbiol. Ecol. 38:19-24, 1986.
42. Carlstedt-Duke B, Midtvedt T, Nord CE, Gustafsson BE. Acta Path. Microbiol. Scand. Sect. B 94:293-300, 1986.
43. Carlstedt-Duke B, Höverstad T, Lingaas E, Norin KE, Saxerholt H, Steinbakk M, Midtvedt T. Eur. J. Clin. Microbiol. 5:634-638, 1986.
44. De Somer P, Van De Voorde H, Eyssen H, van Dijck P. Antibiot. Chemother. 5:463-469, 1955.
45. Eyssen H, De Somer P, van Dijck P. Antibiot. Chemother. 7:55-64, 1957.

P. Goldman

SEEKING EXPLANATIONS FOR PHARMACOLOGICAL AND TOXICOLOGICAL PHENOMENA IN THE METABOLIC REACTIONS OF THE INTESTINAL FLORA

P. GOLDMAN
Department of Pharmacology (Harvard Medical School) and Department of Nutrition (Harvard School of Public Health), Harvard University, 665 Huntington Avenue, Boston, Massachusetts 02115, U.S.A.

1. INTRODUCTION

It is an honor to attend this Symposium organized to pay respects to the late Professor Piet De Somer. The range of the subject matter under discussion is testimony to the breadth of Prof. P. De Somer's scientific interests just as the Rega Institute is an Embodiment of his recognition that scientific progress requires the commitment of tangible resources. Clearly Prof. De Somer was a scientist who recognized the need to explore practical means to foster a productive relationship between scientific progress and its practical applications.

The work I will describe relates only to one of Prof. De Somer's interests in microbiology, namely whether characteristics of the intestinal flora might help to explain enigmatic phenomena in biology and medicine. I, too, have pursued this scientific goal and tried to elucidate the biochemical reactions carried out by the intestinal flora in hopes that these reactions might help to explain the toxic and beneficial actions of some xenobiotic compounds.

Of course, it is now well recognized that both the pharmacological and toxicological effects of many compounds can be explained by reactive metabolites that are generated as the result of the body's detoxication reactions. The cytochrome P450 system has received particular attention because its reactions, along with those carried out by various conjugating enzymes, have been shown to explain the action of a number of drugs as well as the action of a number of classic chemical carcinogens.

Less widely appreciated in this regard are the involvement of the metabolic reactions of the intestinal flora in explaining toxic phenomena. Professor De Somer and his group were among the first contributers to this field when they demonstrated that toxins produced by intestinal bacteria were responsible for the exquisite sensitivity of the guinea pig to penicillin (1).

How does one determine whether a reaction carried out in the whole animal can be attributed to the flora ? What kinds of reactions are carried out by the flora ? How can one determine whether these reactions have pharmacological or toxicological implications ? I will try to provide answers to these questions and to illustrate such answers with examples from our own studies.

2. METHODS OF STUDYING THE METABOLIC ACTIVITY OF THE FLORA

Germfree animals are particularly useful in answering the question of whether a metabolic reaction that occurs in the whole animal can be attributed to the flora. When a compound administered to a conventional rat forms a metabolite that is not found when the same compound is administered to a germfree rat, it strongly suggests that the flora participates in the metabolic pathway. Caution must be exercised, however, before concluding that such metabolic differences are necessarily a direct effect of the flora. Germfree and conventional animals differ significantly in anatomy and physiology. Some of these differences, for example, the size of the cecum or basal metabolism rate (2), may alter the absorption and distribution of compounds and, as a result, limit the access of such compounds to the mammalian enzymes that metabolize them. The result may be a quantitative difference in the metabolism of a compound between germfree and conventional animals, that is not necessarily a direct consequence of bacterial metabolism.

The flora is often said to be implicated in a metabolic pathway when the formation of a metabolite is inhibited by the treatment of a conventional animal with antibiotics. Although such a finding may be considered presumptive evidence for the flora's participation in a metabolic pathway, such evidence may also have other interpretations. An example of this was the demonstration by Professor De Somer's group that antibiotics may have pharmacological actions other than those related to the suppression of the flora. By showing that the aminoglycoside neomycin may bind cholesterol within the gut, Professor De Somer and his colleagues (3) provided a mechanism by which antibiotics might suppress a compound's metabolism simply by retarding its absorption, and therefore its access to sites of mammalian enzyme activity. If an antibiotic can affect a compound's pharmacokinetics, as well as suppress the flora, then the effect of antibiotics in diminishing the formation of a metabolite in the conventional animal is not so simple to interpret. Thus, Prof. De Somer was one of the first to appreciate the value of germfree and gnotobiotic animals in defining the biological implications of the flora.

When a reaction present in the conventional animal is missing in the germfree animal, the involvement of the flora might be strongly suspected. Suspicion moves closer to certainty if the missing reaction can be demonstrated in cultures of the fecal or colonic flora of the conventional animal. Of course, even this simple rule could be violated because of uncertainties concerning the cultivation of the flora _in vitro_. As implied by the results of Professor De Somer's experiments with penicillin in the guinea pig, and as shown very elegantly by Professor Savage's contribution to this Symposium, the flora is a complex ecosystem, comprising as many as several hundred different bacterial species (4). Each species exists within its own niche under constraints imposed by such factors as the nutrients supplied either by the gut or by other bacteria. Such constraints will no longer apply when the intestinal bacteria are cultivated _in vitro_, and, as a result, bacteria that are ordinarily of minor metabolic importance may tend to proliferate to the point where they assume a dominant role. Thus, cultures of bacteria derived from the flora may have metabolic activities that differ from those of the normal flora, such changes resulting either from changes in bacterial composition or from altered metabolic reactions that occur in an altered ecosystem.

Our answer to these difficulties is to attempt to cultivate the flora under conditions most suitable for the growth of anaerobic bacteria. In this way we hope to give as much assurance as possible that the bacteria isolated from either the feces or cecal contents are representive of the normal flora. Thus the use of complex media, along with the rigorous exclusion of oxygen (5), makes it possible to cultivate essentially all of the 10^{11} bacteria per gram present in feces. Although the reactions found in such cultures will not necessarily replicate those carried out in the bacteria's normal environment, such methods probably offer the best _in vi-_

tro method for assessing the metabolic capability of the normal flora. Therefore, we must assume that the flora is responsible for a metabolic reaction found in a conventional rat when the reaction is absent in the germfree rat and can be demonstrated in either pure or mixed anaerobic cultures of the predominant flora (6).

At first we simply applied this algorithm to various compounds in order to show what reactions in the metabolism of the whole animal might be attributed to the flora, in a sense to amplify the work of others (7) who had shown that metabolic reactions found to occur in vivo might be carried out in the flora as well as in mammalian tissues. It was quite clear from such studies that the flora has a propensity for certain kinds of reactions and that often the reactions carried out by intestinal bacteria contrast with those carried out elsewhere in the body. Thus, for example, reactions carried out by the cytochrome P-450 system, in liver and other tissues, are oxidative and actually depend on the availability of molecular oxygen. The microorganisms of the flora, on the other hand, are adapted to an oxygen-free environment and therefore must depend on molecules other than oxygen as a final electron acceptor; such bacteria will tend to reduce many xenobiotic compounds. Thus, while the liver tends to oxidize xenobiotic compounds, the flora tends to reduce them. Another contrast is that the liver tends to carry out reactions for the conjugation of xenobiotics or their metabolites, the result being metabolites conjugated to such compounds as glucoronic acid or sulfate. The flora, on the other hand, has enzymes that hydrolyze these conjugates. From a chemical standpoint, therefore, the flora tends to reverse the metabolic reactions carried out by mammalian tissues (6).

Because of the enterohepatic circulation, it is often possible that the same compound may be subjected to the contrasting, or complementary, reactions of the liver and the flora. Sometimes transformations at the two sites nullify each other. For example, a compound that is conjugated to glucornic acid in the liver and excreted in the bile may be restored to its original form by the deconjugation reactions of the flora. On the other hand, sometimes mammalian tissues and the flora may make successive transformations of a compound the result being a very diverse metabolic profile. The pesticide propachlor, for example, yields only three major metabolites in the germfree rat, but may result in as many as eleven metabolites in the conventional rat. In this case, the diversity of metabolites apparently results from what might be described as a pingpong metabolism between liver and flora (8).

A dramatic illustration of the consequences of metabolic reactions by the flora was provided by Laqueur and his coworkers, who used gnotobiotic rats to show that reactions of the flora were obligatory for the toxicity of the plant glycoside, cycasin. Laqueur's group confirmed that cycasin is hepatotoxic and carcinogenic for conventional rats, but also demonstrated that it is harmless for germfree rats. On the other hand, they showed that cycasin's aglycone, methylazoxymethanol, which is liberated from cycasin by the flora, is toxic for both conventional and germfree rats. Thus, their evidence indicated that cycasin's toxicity results from the action of its aglycone when that aglycone is liberated by certain strains of the flora. This hypothesis was confirmed by showing that cycasin becomes toxic to germfree rats after they are associated with a strain of bacteria that is capable of liberating the aglycone (9). Similar evidence indicates that the liberation of cyanide from another glycoside, amygdalin, is also mediated by the activity of the flora. Thus, conventional rats experience both an elevated serum cyanide concentration and the symptoms of amygdalin toxicity when amygdalin is administered. On the other hand, when germfree rats are exposed to the same dose of amygdalin, toxic symptoms are not observed and there is no elevation of the serum cyanide (10).

I would like next to consider two additional drugs whose metabolism by the flora helps to understand their modes of action.

3. SULFASALAZINE

Sulfasalazine is a drug that has been shown to be effective both for the treatment of mild ulcerative colitis and as prophylaxis against its recurrence (11). The drug has an interesting history, having been conceived in the late 1930's as a treatment for rheumatoid arthritis. The rationale for the design of sulfasalazine was based on the view, current at that time, that rheumatoid arthritis had both an infectious and an inflammatory component. It seemed rational, therefore, to design a drug that combined in one molecule the two desired therapeutic properties. Thus, a molecule was sought that combined the anti-inflammatory action of salicylate with the anti-bacterial activity of the newly discovered sulfa drugs. The chemical synthesis to accomplish this goal was achieved by diazotizing salicylate to the N-4 position of a sulfa drug. Sulfasalazine, in which salicylate is diazotized to sulfapyridine (Fig. 1), was one of a series of similar compounds that was synthesized at AB Pharmacia in Sweden.

Sulfasalazine was not impressive in the therapy of rheumatoid arthritis, but fortunately the physician responsible for developing this drug, Dr. Nana Svartz, also had an interest in ulcerative colitis. At that time, ulcerative colitis also was considered to have an infectious component, and so, while other physicians experimented with more traditional sulfa

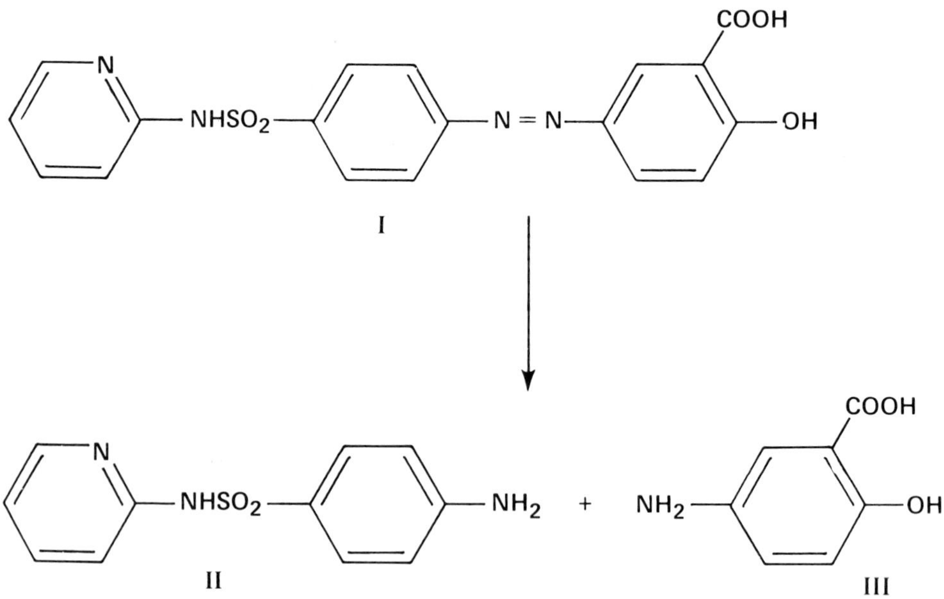

FIGURE 1. Sulfasalazine (I) is reduced by the intestinal microflora to form sulfapyridine (II) and 5-aminosalicylate (III).

drugs, Dr. Svartz treated her patients with the sulfa derivative in which she had a special interest. She soon reported that sulfasalazine was successful in the treatment of acute ulcerative colitis, a finding that has now been amply confirmed by several systematic clinical trials.

It was known for many years that the azo bond of sulfasalazine was reduced during its metabolism in both human beings and rats. In the early 1970's however, when comparing the metabolism of sulfasalazine in germfree and conventional rats, Dr. Mark Peppercorn and I found that the azo bond of sulfasalazine was cleaved only in conventional rats. Thus, when sulfasalazine was administered to germfree rats, their excreta contained no sulfasalazine metabolites, only the parent drug. The flora was further implicated in azo bond cleavage when it was found that all anaerobic bacteria isolated from the intestinal flora were capable of reducing sulfasalazine. Furthermore, reduction of sulfasalazine could also be demonstrated in ex-germfree rats that were selectively associated with such bacteria (12). Thus, it seemed indisputable that the intestinal microflora was obligatory for the cleavage of sulfasalazine's azo bond.

The reduction of the azo bond by the flora in the lower bowel proved to be of particular interest, however, because of the distribution of sulfapyridine and 5-aminosalicylate that resulted. 5-Aminosalicylate was found almost exclusively in the feces, whereas the other portion of the sulfasalazine molecule, sulfapyridine, appeared almost exclusively in the urine. This distribution of metabolites was found when sulfasalazine was administered orally to both rats and human subjects (13). In the rats, however, it was also possible to administer sulfasalazine parenterally, and when this was done the metabolites were found to be distributed as they were after oral administration. Metabolites have the same distribution after both oral and parenteral administration because sulfasalazine is readily absorbed high in the gastrointestinal tract and thus readily gains access to the circulation regardless of its route of administration. Once in the circulation the drug apparently gains access to the colon as a result of excretion into the bile. When azo bond reduction occurs in the colon, sulfapyridine, being uncharged, readily crosses the colonic barrier and enters the systemic circulation to appear later in the urine. In contrast, 5-aminosalicylate, having charge on both the amino and the carboxyl group at the pH of the colon, tends to remain within the colon and, as a result, is eliminated in the feces. It therefore seemed reasonable to postulate that the therapeutic benefit of sulfasalazine for ulcerative colitis results from its capacity to deliver 5-aminosalicylate to the colon (14).

This hypothesis for the mechanism of action of sulfasalazine is now supported by several clinical trials which show that rectal installation of 5-aminosalicylate is effective in treating ulcerative colitis of the distal colon. How 5-aminosalicylate actually works, however, is not clear. One possibility is that, like salicylate itself, 5-aminosalicylate is a prostaglandin inhibitor. Support for this idea comes from the observation that prostaglandins are found to be increased in the feces during the active phase of ulcerative colitis. It seems plausible, therefore, that 5-aminosalicylate, acts to suppress inflammation in the gut wall by suppressing the synthesis of prostaglandins. Whatever the mechanism, however, it now seems quite clear that the rectal administration of 5-aminosalicylate can provide much of the therapeutic effectiveness of orally administered sulfasalazine.

This observation on the effectiveness of 5-aminosalicylate is of particular importance because sulfasalazine frequently causes such adverse drug reactions as rashes and hematological abnormalities, and therefore may have to be withdrawn from many patients with ulcerative colitis. Furthermore, some patients with ulcerative colitis, and still more with Crohn's disease, receive no discernible benefit from sulfasalazine. Because many of the adverse reactions to sulfasalazine are similar to those caused by sulfapyridine, it seemed logical to hope that a more effective and safer drug might be found which used another approach to releasing 5-aminosalicylate within the colon. Thus, a search has been launched to discover new ways of delivering 5-aminosalicylate to the intestinal tract (15).

One approach has been to design a drug that uses a more innocuous carrier molecule for 5-aminosalicylate. Thus it has been suggested to use such amino compounds as 4-aminobenzoate or its glycine conjugate to replace sulfapyridine. This approach has been extended to the design of a drug, olsalazine, that consists of 2 molecules of 5-ASA diazotized to each other, thus in a sense providing a double dose of the active ingredient (16). Of course all such drugs require bacterial reduction to release 5-aminosalicylate at the appropriate site. Dependence on the action of the intestinal bacteria can be avoided, however, by incorporating 5-aminosalicylate into a pharmaceutical preparation that releases the drug slowly as it traverses the gastrointestinal tract (17). If the release of 5-aminosalicylate is controlled by pharmaceutical means, then the delivery of 5-aminosalicylate does not depend on the flora; furthermore, release controlled by pharmaceutical means rather than by the flora enables the drug to be delivered to areas of the bowel where the flora is sparse.

Selective release of 5-aminosalicylate, which does not depend on the flora, might be of particular value in treating inflammatory bowel disease present in the small intestine (Crohn's disease) where there are normally few intestinal bacteria. Indeed, the proposal that sulfasalazine acts through the flora-mediated release of 5-aminosalicylate is supported by the clinical observation that Crohn's disease of the colon is more responsive to sulfasalazine than Crohn's disease of the small intestine. In other words, there is circumstantial evidence to indicate that the activity of sulfasalazine depends on the density of the flora.

The search for drugs that are superior to sulfasalazine has therefore been stimulated by the observation that the therapeutic benefit of sulfasalazine may depend on its activation by the flora. There is now considerable hope that the effectiveness of sulfasalazine may be improved by drugs that are better designed for the dependable release of 5-aminosalicylate in the appropriate areas of the gastrointestinal tract (15).

4. METRONIDAZOLE

Metronidazole is a 5-nitroimidazole that has found an increasing number of therapeutic uses since it was introduced more than 20 years ago as the first effective systemic chemotherapy for trichomonal vaginitis. Later it was found also to be effective in the treatment of infections with the parasites *Entamoeba histolytica* and *Giardia lamblia*. In the meantime it was reported that a patient treated with metronidazole for trichomoniasis had a remarkable remission in an intercurrent ulcerative gingivitis. This observation stimulated the investigation of the effectiveness of metronidazole in treating an increasingly recognized class of clinical infections, those due to anaerobic bacteria. Slowly at first, but then with increasing frequency, there appeared in the literature systematic studies documenting the remarkable effectiveness of metronidazole in the treatment and prophylaxis of anaerobic bacterial infections (18).

Metronidazole has a number of other biological properties. Thus it is mutagenic for bacteria, including the histidine auxotrophs of *Salmonella typhimurium* (the Ames tester strains); like other nitroimidazoles it is able to sensitize anaerobic cells to radiation and to exert a cytotoxic effect which is synergistic with that due to alkylating drugs. Reduction of the nitro groups appears to play a role in explaining all these properties. Attention has turned, therefore, to the chemistry of nitro group reduction as a means of understanding how the nitroimidazoles work (19).

The available evidence suggests that reduction of metronidazole and other 5-nitroimidazoles occurs through a series of such intermediates as those contained in brackets in Fig. 2. When reduced, either by isolated enzymes or by bacteria, the imidazole ring of metronidazole is fragmented; the result is the formation of a wide variety of metabolites (20). Two of these bacterial metabolites of metronidazole, acetamide and N-(2-hydroxyethyl)oxamic acid have been characterized and are also shown in Fig. 2. These 2 metabolites are complementary in the sense that together they contain all of metronidazole's carbon atoms as well as all of its nitrogen atoms except that which was originally in the nitro group. In spite of the

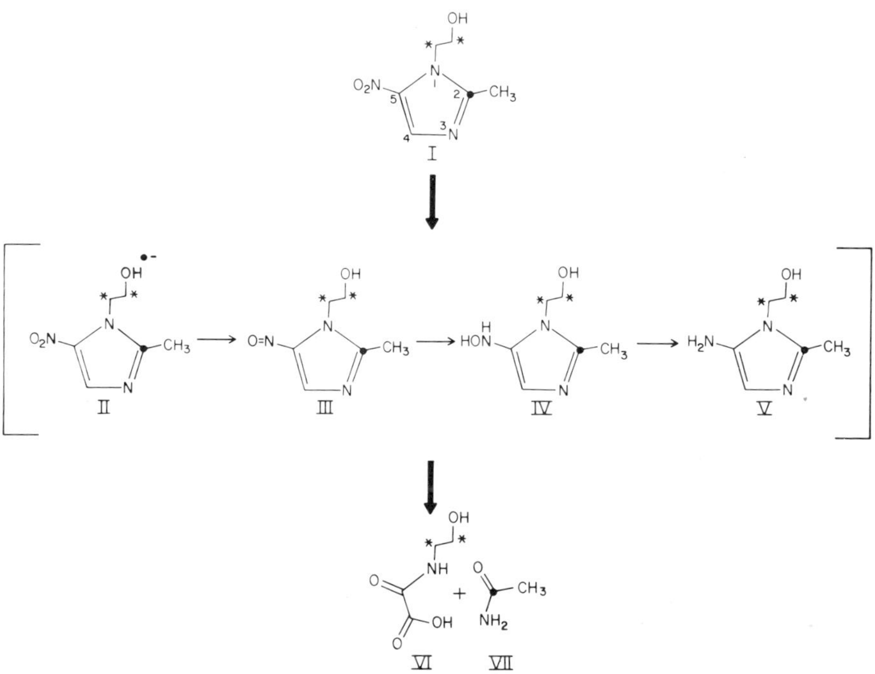

FIGURE 2. Reduction of metronidazole (I) to form N-(2-hydroxyethyl)oxamic acid (VI) and acetamide (VII). Possible intermediates, in the brackets, include the radical anion (II), and the nitroso (III), hydroxylamino (IV), and amino (V) functionalities of metronidazole. Solid circles indicate the fate of the radiolabel in [2-^{14}C]metronidazole, and asterisks the fate of radiolabel in [1',2'-$^{14}C_2$]metronidazole, the two forms of radiolabeled metronidazole used in the experiments described in this paper. Reproduced with permission of the Cold Spring Harbor Laboratory.

fact that these two metabolites appear to be complementary, the yields of acetamide (approximately 10 %) and of N-(2-hydroxyethyl)oxamic acid (approximately 2 %) indicate that their formation is neither stoichiometric with each other, nor with the disappearance of metronidazole (21).

It is not exactly clear how the formation of acetamide and N-(2-hydroxyethyl)oxamic relate to the reduction of metronidazole. Suffice it to say that these 2 stable metabolites can be isolated after metronidazole is subjected to reducing conditions, such as those necessary for its biological properties. In a sense, then, such metabolites might be considered to be markers of the prior existence of other molecules (perhaps those contained in brackets in Fig. 2, which are more difficult to isolate and characterize) that are the actual mediators of metronidazole's biological activity.

It was therefore of interest to find that small amounts of acetamide and N-(2-hydroxyethyl)oxamic acid are formed from metronidazole in the conventional rat (21). That such metabolites are not formed either in the germfree rat, or in tissue culture preparations incubated anaerobically, suggest that the reaction sequence proposed in Fig. 2 must be limited to the flora.

How can we infer the reactions that occur in the flora and relate them to the formation of the biologically active form of metronidazole ? To approach this question we offered the following hypothesis. We assumed that metronidazole is reduced to some biologically active labile intermediate, M^*, which then has one of several fates : it may interact with a bacterial target to cause a lethal event or it may interact, for example, with water or other molecules to yield a stable metabolite such as acetamide, which we singled out for further study (Fig. 3). It must be recognized, however, that we cannot yet assign a structure to the hypothetical active form of metronidazole, M^*. The role of M^* is portrayed schematically in Fig. 3; its structure might be one of those depicted in brackets in Fig. 2, whose existence is based only on limited circumstantial evidence.

Nevertheless, the existence of a labile intermediate such as M^* is consistent with several biological observations. One of these is the observation that *E.coli*, which is one of the bacteria shown to form acetamide from metronidazole, increases the potency of metronidazole's bactericidal effect on *B.fragilis* (22). Such an observation is consistent with the suggestion that M^* forms in *E.coli* and is stable enough to diffuse out of this bacteria and, before it decomposes, to interact with the more sensitive *B.fragilis* present in the same culture. Thus, *E.coli* enhances the effect of metronidazole on *B.fragilis* by causing it to be exposed to more M^* than it generates by itself.

The existence of M^* is also supported by the kinetic relationship found experimentally between the bactericidal effect of metronidazole and the formation of acetamide. If metronidazole is converted to M^* at a constant rate, which depends only on the redox potential of the bacterial culture, and if the further breakdown of M^* to acetamide depends only on the concentration of water, then the rate at which acetamide accumulates in the culture medium will be proportional to the rate of exposure of the culture to M^*. If the bactericidal effect of M^* is first order, then the log of bacterial survival should be related linearly to the accumulation of acetamide. Indeed this relationship was found for *B.fragilis* and *Clostridium perfringens*, which are sensitive to metronidazole, as well as for the wild type *E.coli*, which is not very sensitive to metronidazole (22, 23).

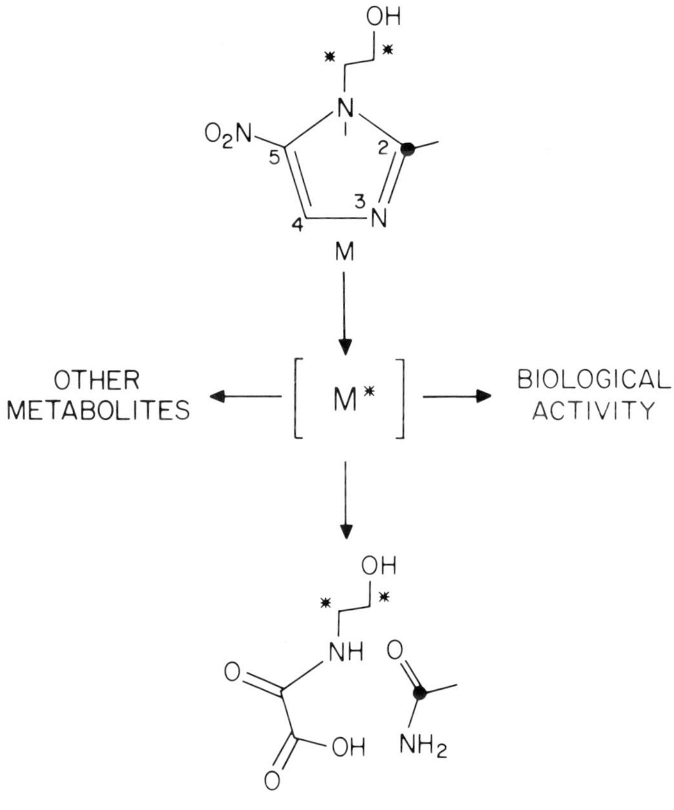

FIGURE 3. A schematic diagram of the fate of metronidazole. Metronidazole is reduced by nitroreductase to form its hypothetic bactericidal form M*. M* can then interact either with the bacterial target to give manifestations of biological activity or it may decompose to yield acetamide and N-(2-hydroxyethyl)oxamic acid or other metabolites.

The proposed model suggests that there will be at least two ways in which bacteria might achieve resistance to metronidazole. On the one hand, they might have diminished nitroreductase activity and hence a diminished ability to carry out the "lethal synthesis" of M*. On the other hand, they might be resistant because the target for M* is less sensitive. The finding that diminished susceptibility to metronidazole is associated with diminished nitroreductase activity in both a strain of _Bacteroides fragilis_ and a strain of _Trichomonas vaginalis_ (24,25) is consistent with the proposed model. The proposed model, however, makes a further prediction about susceptible and less susceptible strains of the same microorganism. If it is assumed that two strains of the same species of microorganism have targets for M* with the same sensitivity, then the model predicts that both sensitive and less sensitive strains would show the same relationship between acetamide accumulation and bactericidal effect. In other words, the 2 strains differ in their susceptibility to metronidazole because of differences in their nitroreductases and hence differences in the rates at which they form M^*. Once formed, however, the relative fate of M^* should be the same in both strains, and therefore both strains should display the same relative relationship between lethal hits by M^* and the breakdown of M^* to form acetamide. This prediction was verified for both *Bacteroides fragilis* and *Trichomonas vaginalis* when it was found that the strain which was relatively resistant to metronidazole nevertheless showed the same relationship between acetamide accumulation and metronidazole mediated microbiocidal action as its more metronidazole-sensitive counterpart (24,25).

No bacterium has been described that appears to be resistant to metronidazole because it has a target that is resistant to M^*. On the other hand, bacteria might be found that are more sensitive to metronidazole because they had a target that was more sensitive to M^*. Since metronidazole is a bacterial mutagen it seemed likely that the target for M^* would be DNA. We reasoned, therefore, that greater susceptibility to metronidazole might be found among bacterial mutants that had defects in DNA repair. In screening such mutants we found that *Escherichia coli* mutants with increased sensitivity to ultraviolet light, i.e. with defects in either *uvr*, *recA*, or *recB* genes, exhibit an increased sensitivity to metronidazole in comparison with the wild type (Table 1). When the sensitivity to metronidazole of these mutants is examined and the results expressed in terms of the relationship between bactericidal activity and acetamide formation (Fig. 4), the prediction of the model is affirmed, namely that such mutants exhibit greater susceptibility in terms of acetamide accumulation than the wild type. Furthermore, since other forms of repair mutants have no increased sensitivity to metronidazole, these results strongly support the idea that M^* damages DNA in the same way as ultraviolet light does.

TABLE 1. A comparison of the minimum inhibitory concentration (MIC) of metronidazole for the wild type and various DNA repair mutants of *E.coli*.

E.coli strain[a] (phenotype)	Metronidazole MIC (μg/ml)
AB1157 (Uvr^+ Rec^+)	> 200
P90C (Uvr^+ Rec^+)	> 200
AB1885 ($UvrB^-$ Rec^+)[b]	100
SR58 ($RecA^-$ $UvrB^-$)[b]	13
EE128 (Uvr^+ $RecA^-$)[b]	25
EE348 ($UvrB^-$ Rec^+)[c]	> 200
EE349 ($UvrB^-$ Rec^+)[c]	> 200
SM172 ($UvrB^-$ Rec^+)[c]	> 200

[a] A more complete description of these strains will be found in 22.
[b] Point mutants.
[c] Deletion mutants in which the deletion also removes one form of nitroreductase activity.

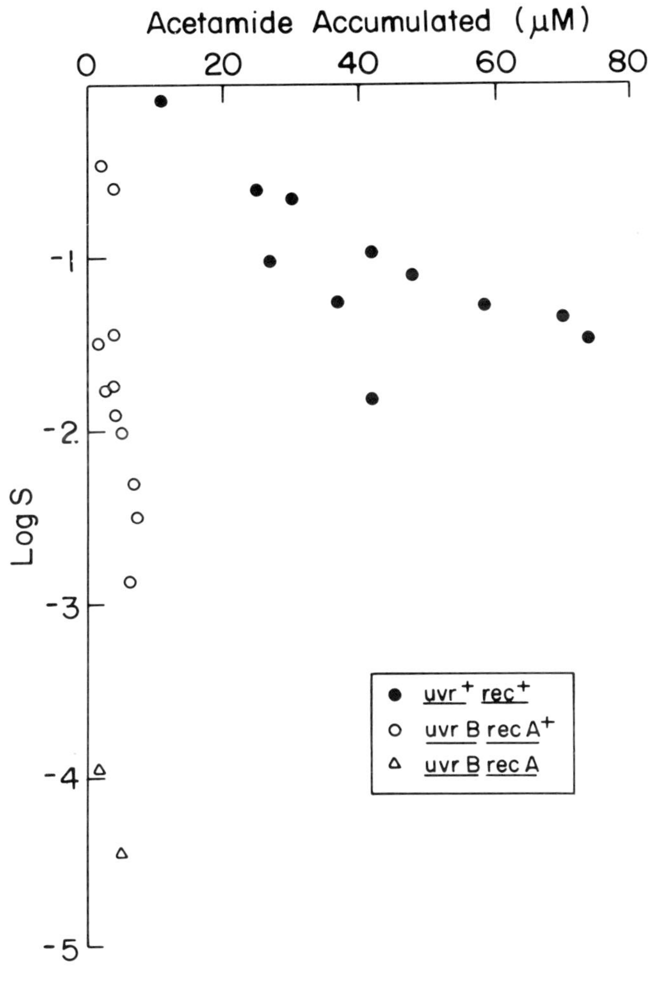

FIGURE 4. The relation betweeen acetamide accumulation and the log of surviving bacteria when metronidazole is incubated with either wild type E.coli (●) or various DNA repair mutants (Δ, O).

When we later examined the sensitivity to metronidazole of mutants with deletion mutations (rather than point mutations) of the same genes, we were dismayed to find that metronidazole's MIC for these mutants was indistinguishable from that for the wild type (Table 1). This apparent inconsistency was resolved, however, when it was found that the deletion in the DNA repair genes extended to include a deletion in one of the genes for nitroreductase activity. The failure of the deletion mutants to exhibit sensitivity to metronidazole could therefore be attributed to their lack of nitroreductase and as a consequence their inability to generate M^*. Such mutants therefore had compensating defects. When examined in terms of bacterial survival vs acetamide production, however, these deletion mutants behaved in the same way as the point mutants shown in Fig. 4, or in other words, as the model would predict for bacteria whose target is more sensitive to M^* (23).

Most recently we have been trying to characterize the reactive form of metronidazole and related 5-nitroimidazoles. We began this aspect of the project by attempting to synthesize compounds such as those in brackets in Fig. 2. Confirming previous work, we were able to synthesize the amino functionality of metronidazole and to show that it had no bactericidal activity for bacteria sensitive to metronidazole (26). We were also able to get evidence that the amino derivative of metronidazole forms when metronidazole is reduced by bacteria in culture. Thus, when ^{14}C-metronidazole was incubated with metronidazole-susceptible bacteria in the presence of a pool of this amino functionality (compound V in Fig. 2), we were able to trap radiolabel in the amino functionality. It therefore appears that this amino functionality forms when metronidazole is reduced by bacteria, but that this 6-electron reduction product is not "M^*", the bactericidal form of the drug (26).

We later tried to synthesize the hydroxylamino derivative of metronidazole (Compound IV in Fig. 2), but were not successful. We pursued this problem further, however, using dimetridazole, a 5-nitroimidazole closely related to metronidazole which has a methyl substituent at N-1 rather than the hydroxyethyl substituent of metronidazole (Fig. 5). Dimetridazole has virtually the same activity as metronidazole for most microorganisms and is indistinguishable from metronidazole in our test systems. It was easier for us to use dimetridazole in the study to be described because the substituent at its N-1 position would not be acetylated when acetic anhydride was present. Therefore we could synthesize chemically the hydroxylamino functionality of dimetridazole (the compound comparable to compound IV in Fig. 2) and stabilize it by acetylation to yield the diacetyl derivative shown as compound 1 in Fig. 5 (27).

This diacetyl derivative interested us because it could be regarded as a drug which might yield the hydroxylamino functionality by a route other than that provided by the action of nitroreductase on dimetridazole. Thus compound I, or compound II (Fig. 5), offered the possibility of generating the hydroxylamino functionality of a 5-nitroimidazole by a deacylation reaction rather than simply by reducing the nitro group (Fig. 5). In a sense, then, either compound I or compound II might be a new antibiotic that was selectively bactericidal for microorganisms capable of deaceylating one of these compounds just as the 5-nitroimidazoles themselves are selective for microorganisms capable of carrying out nitro group reduction. Unfortunately, however, we have not yet been successful in finding the necessary deaceylase activity and so have not been able to generate the hydroxylamino functionality in order to test the concept of reaching the active form of a 5-nitroimidazole by another route.

POSSIBLE PATHWAYS OF LETHAL SYNTHESIS

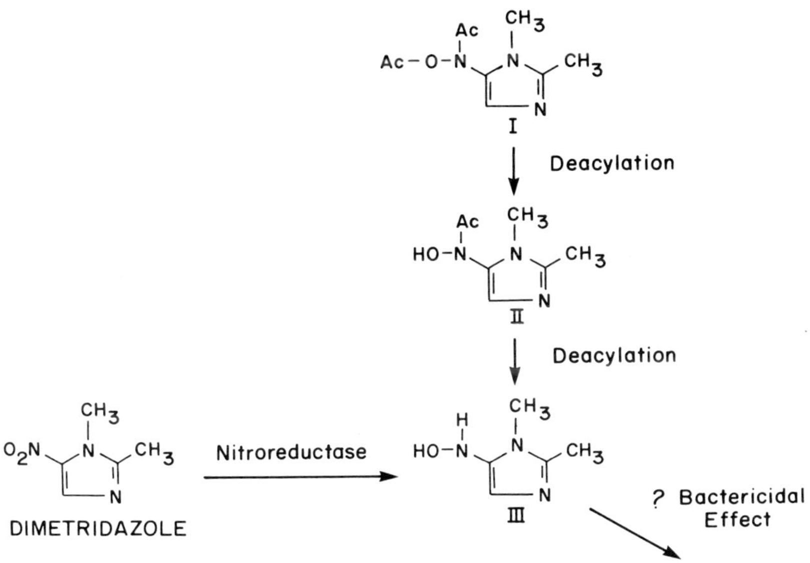

FIGURE 5. Two pathways to the potentially bactericidal hydroxylamino functionality of dimetridazole (III). On the one hand, compound III may arise directly from dimetridazole by nitroreductase activity. On the other hand, one may synthesize the diacetyl derivative of the hydroxylamino functionality of dimetridazole (I), which can be deacylated by acetylesterase to form the monoacetyl derivative (II) (27). The complete deacylation of either compound I or II therefore offers another pathway for the lethal synthesis of compound III.

5. USING GNOTOBIOTIC RATS FOR A HOST-MEDIATED ASSAY

The studies with the nitroimidazoles remind us of the usefulness of cells or bacteria for detecting toxic metabolites that are too labile to isolate and characterize chemically. The occurrence of methemoglobinemia in rats in response to nitrobenzene is a good example of a host-mediated assay that has been used to implicate a reaction for which the intestinal bacteria are responsible. Thus, the reduction of nitrobenzene, which occurs in conventional but not in germfree rats, parallels the formation of methemoglobin, which also occurs in conventional but not in germfree rats (28). In a sense, then, the formation of methemoglobin serves as an indication that the metabolism of the flora has some biological significance. A similar approach has been used to show that the flora is the mediator of the effect of 2,4-dinitrofluorene on the host's unscheduled DNA synthesis (29).

Another biological system to detect the chemical mediators of toxicity is the reversion to histidine independence of the Ames Salmonella typhimurium histidine auxotrophs. This assay for mutagenicity is of interest because of an apparent correlation between a compund's mutagenicity in this system and its animal carcinogenicity. Such a correlation is improved by including in the bacterial mutagenesis assay the elements of a liver microsomal (cytochrome P450) enzyme system. In other words, the bacterial detection system is more fully exploited by ensuring that it is coupled to a major source of mammalian metabolic activation. Clearly such value is enhanced by a host-mediated assay system which deploys the bacterial detection system within the animal's body (30) and therefore exploits activation of the compound that may occur at any site in the body.

As a modification of this traditional short term host-mediated assay (30), we have associated the Ames histidine auxotrophs (strain TA1538) with otherwise germfree rats and found that these bacteria colonize the gastrointestinal tract and appear in the feces at concentrations between 10^8 and 10^9 bacteria per gram. The host-bacteria association remains stable in terms of the number of bacteria at these sites during experiments that last as long as 7 months. At sacrifice, the salmonella are found at concentrations greater than 10^7 per gram in the contents of the forestomach and at concentrations greater than 10^8 per gram in the contents of the lower bowel. Ordinarily the concentration of His^+ revertants remains less than 100 per gram in the feces, but we found that the concentration usually rises promptly in response to the administration of carcinogens. Indeed this response can be detected when the carcinogens are administered in a daily dose comparable to those causing tumors on more prolonged administration. That this response was specific was suggested by the finding that the His^+ response was not provoked by the administration of structurally related compuonds that are not considered to be carcinogens (31).

At this point we felt that the long-term host-mediated assay system might be worth pursuing as a potential model of colon carcinogenesis. A number of observations were encouraging. We found, for example, that the concentration of His^+ revertants in the colons of sacrificed rats correlated very closely with that in the feces, indicating that the revertant concentration determined in the feces reflected that in the lower bowel. (Of course, this correlation did not establish that the revertants had their origin in the colon. His^+ revertants might have originated in the upper gastrointestinal tract and merely traversed the gastrointestinal tract to reach the lower bowel.) The finding that the His^+ response in the feces subsides within a few days after terminating the administration of the carcinogen also seemed fortunate because it suggested that test animals, which are both expensive and technically difficult to maintain, might be used repeatedly. On the other hand, we were discouraged by the finding that the His^+ response in the feces was quite variable both from animal to animal and from day to day (32). Furthermore, the height and duration of the His^+ responses showed long term changes when the same animal was dosed repeadly with the carcinogen 2-nitrofluorene. Indeed, this change in the response was only one of several phenomena which indicated that the properties of the tester bacteria changed progressively during their association with the animal (32). Nevertheless, when animals were used promptly and for only one administration of a carcinogen, the variability of the His^+ response was controlled sufficiently to demonstrate a clear dose-response relationship for orally administered 2-nitrofluorene (33).

It was not possible to examine the effect of a complete flora on the association of the histidine auxotrophs and the gnotobiotic rats because when the animals were removed from their isolators the auxotrophs were no longer detected in the feces, and presumably were eliminated from the gastrointestinal tract. It appeared, therefore, that a normal flora, the components of which are selected in the course of evolution for their adaptation to the gastrointestinal tract, "crowd out" the less well adapted Ames tester bacteria. Nevertheless, the concentration of strain TA1538 in the colon was retained if a limited number of strains selected from the normal flora were added to the gnotobiotic association (33). We selected 3 strains for this associatin, each of which did not interfere with the enumeration of the tester strains and themselves could be enumerated in the feces by the use of selective media.

The addition of *Bacteroides vulgatus* to the association resulted in the blunting of the His^+ response to 2-nitrofluorene, a phenomenon for which there are at least 2 explanations. On the one hand, the blunted response might have been a manifestation of the capacity of *B.vulgatus* to reduce 2-nitrofluorene to its non-mutagenic metabolite 2-aminofluorene. On the other hand, it might have been simply that the *B.vulgatus* displaced the salmonella from the site at which they were mutagenized. Unfortunately we could not establish where mutagenesis was taking place within in the gastrointestinal tract. That the size of the revertant response in the feces of individual animals was found to correlate inversely with the concentration of *B.vulgatus* in their stomachs (33) when they were sacrificed, suggested, however, that the revertant response in the feces reflected merely the mutagenic activity that occurred in the stomach. Such an explanation is in accord with the observation that the concentration of revertants in the feces generally did not become elevated until approximately two days after the carcinogen was ingested. Thus, although histidine auxotrophs are highly concentrated in the colon, and remain so in the presence of selected bacteria of the normal flora, the mutagenesis that we observe was very likely to have occurred in the stomach. The uncertainty about the site of mutagenesis, and the vexing variability of the His^+ response, which was even greater in several other Ames tester strains than in strain TA1538 (33), discouraged us from further attempts to use these histidine auxotrophs to detect labile metabolites within the gastrointestinal tract. We report such experiments, however, because they illustrate what may become a useful approach in the future as a means to gain evidence that implicates reactions of the intestinal flora as a cause of a compound's biological activity.

ACKNOWLEDGEMENTS

The experimental work from the author's laboratory was supported mainly by U.S. Public Health Service Grant CA34957 from the National Cancer Institute (DHHS).

REFERENCES

1. De Somer P, Van De Voorde H, Eyssen H, Van Dijck P. Antibiot. Chemother. 5:463-468, 1955.
2. Wostman BS. World Rev. Nutr. Dietetics 22:40-92, 1975.
3. Eyssen H, Vanderhaeghe H, De Somer P. Fed. Proc. 30:1803-1807, 1971.
4. Moore WEC, Holdeman LV. Cancer Res. 35:3418-3420, 1975.
5. Holdeman LV, Cato EP, Moore WEC (Eds) Anaerobe Laboratory Manual. Anaerobe Laboratory, Blacksburg, 1977.
6. Goldman P. Ann. Rev. Pharmacol. Toxicol. 18:523-539, 1978.
7. Scheline RR. Pharmacol. Rev. 25:451-532, 1973.
8. Bakke JE, Gustafsson JA, Gustafsson BE. Science 210:433-435, 1980.

9. Laqueur GL, Spatz M. Gann 17:189-204, 1975.
10. Carter JH, McLafferty MA, Goldman P. Biochem. Pharmacol. 29:301-304, 1980.
11. Goldman P, Peppercorn MA. New Engl. J. Med. 293:20-23, 1975.
12. Peppercorn MA, Goldman P. J. Pharmacol. Exp. Ther. 181:555-562, 1972.
13. Peppercorn MA, Goldman P. Gastroenterology 65:240-245, 1973.
14. Goldman P, Peppercorn MA. Gastroenterology 65:166-169, 1973.
15. Goldman P. Gastroenterology 83:1138-1141, 1982.
16. Selby WS, Barr GD, Ireland A, Mason CH, Jewell DP. Brit. Med. J. 291: 1373-1375, 1985.
17. Rasmussen SN, Bondesen S, Hvidberg EF, Hansen SH, Binder V, Halskov S, Flachs H. Gastroenterology 83:1062-1070, 1982.
18. Goldman P. New Engl. J. Med. 303:1212-1218, 1980.
19. Goldman P. J. Antimicrob. Chemother. 10, Suppl. A:23-33, 1982.
20. Chrystal EJT, Koch RL, Goldman P. Mol. Pharmacol. 18:105-111, 1980.
21. Koch RL, Chrystal EJT, Beaulieu BB Jr, Goldman P. Biochem. Pharmacol. 28:3611-3615, 1979.
22. Chrystal EJT, Koch RL, McLafferty MA, Goldman P. Antimicrob. Agents Chemother. 18:566-573, 1980.
23. Yeung T-C, Beaulieu BB, McLafferty MA, Goldman P. Antimicrob. Agents Chemother. 25:65-70, 1984.
24. Beaulieu BB Jr, McLafferty MA, Koch RL, Goldman P. Antimicrob. Agents Chemother. 20:410-414, 1981.
25. McLafferty MA, Koch RL, Goldman P. Antimicrob. Agents Chemother. 21: 131-134, 1982.
26. Ehlhardt WJ, Beaulieu BB Jr, Goldman P. Biochem. Pharmacol. 36:259-264, 1987.
27. Ehlhardt WJ, Beaulieu BB Jr, Goldman P. Biochem. Pharmacol. 36:in press, 1987.
28. Reddy BG, Pohl LR, Krishna G. Biochem. Pharmacol. 25:1110-1112, 1976.
29. Mirsalis JC, Hamm TE, Sherrill JM, Butterworth BE. Nature 295:322-323, 1982.
30. Legator MS, Malling HV. _in_ Chemical Mutagens, Principles and Methods for their Detection. Plenum Press, New York, vol 2, pp 569-589.
31. Wheeler LA, Carter JH, Soderberg FB, Goldman P. Proc. Natl. Acad. Sci. USA 72:4607-4611, 1975.
32. Carter JH, Wheeler LA, Ingelfinger JA, Soderberg FB, Goldman P. Mutation Res. 41:209-216, 1976.
33. Goldman P, Carter JH, Wheeler LA. Cancer 45:1068-1072, 1980.

SECTION III : HUMORAL FACTORS IN IMMUNITY

SECTION CHAIRMAN : PROF. A. BILLIAU

A. Billiau

INTRODUCTION TO SECTION III : HUMORAL FACTORS IN IMMUNITY : CYTOKINES, INTERFERONS AND INTERLEUKINS

A. BILLIAU
Rega Institute, Katholieke Universiteit Leuven, B-3000 Leuven, Belgium

A subject that has become very popular in cell biology, immunology and medical practice is that of the cytokines. Cytokines represent a group of substances that have much in common with peptide hormones. The distinction with hormones lies in the fact that cytokines carry short-range messages from cell to cell, while hormones more often carry long-range signals from one organ to another. Interferon can be considered as the prototype of cytokines. Professor P. De Somer has fostered a keen interest in interferon research ever since interferon was discovered in 1957. His interest was mainly based on the belief that interferon would once become a successful antiviral drug in medicine. If sometimes he was sceptic, it was simply because his personal way of probing his belief prompted him to interrogate his fellow scientists on their beliefs or disbeliefs in interferon.

Interferons are in fact good prophylactic but bad therapeutic means in the control of virus infections, and so are the inducers of interferon. Already in 1966 Prof. P. De Somer felt that interferon would have immunomodulatory properties. He showed that inducers of interferon could cause an inhibition and/or enhancement of antibody formation. Up to the present day we have continued to pursue this line of research. Thus, De Somer was very much involved when, in 1982, we found that IFN-β_1 could be induced by the 22K protein which was subsequently identified as interleukin-1.

Together with Dr. J. Content of the Pasteur Institute, we found in 1984 that interleukin-1 not only induced IFN-β_1 but also another molecule that meanwhile had been dubbed IFN-β_2, but that in Belgium we obstinately call 26K. During the last weeks of his life, De Somer was very much intrigued by this finding and repeatedly admonished us that this had to be a very important molecule. When we succeeded in making a rather specific antibody to the 26K protein he personally went to see several clinicians to collect serum samples from patients with immune disorders that he wanted to test for the presence of 26K. In the autumn of 1985, shortly after De Somer's death, we found evidence that the 26K protein was in fact a potent growth factor for hybridoma and plasmacytoma cells.

Another example of De Somer's interest in basic problems was his unabated quest for an explanation of the Shwartzman phenomenon. This skin reaction which seems to have been forgotten by modern immunology text books never ceased to intrigue him. He felt that somehow interferon was involved in this reaction. This was the reason why he injected his own skin with fibroblast interferon in 1975 and showed that it caused an inflammatory response. Skin reactions ever fascinated him, and he wanted us to inject each potentially new cytokine that we had isolated into the skin of animals and, occasionally, man as well.

We are still continuing this line of investigation. Inflammatory reactions represent an old and very complex problem, but complexity has now become accessible to analysis in that the cytokines that may be involved can be obtained in pure form. Even more important is the fact that we now possess monoclonal antibodies to these individual cytokines, and this will allow us to define their specific role in the inflammation reaction.

After having sketched Professor De Somer's involvement in cytokine research, I would like to introduce the three speakers in this section : S. Baron, J.T. Vilcek and C.A. Dinarello.

SAMUEL BARON is Professor and Chairman of the Department of Microbiology at the University of Texas Medical Branch in Galveston. Like De Somer he has devoted almost his entire career to the study of interferon, starting to work in this area under the late A. Isaacs, the discoverer of interferon. Professor Baron puts his research on interferon in the more general framework of non-immune specific defense mechanisms. From 1955 till 1975 Prof. S. Baron was at the National Institute of Allergy and Infectious Diseases and I still praise myself happy that I could be his coworker at this Institute in 1967-68. Dr. Baron has been and still is an all-round authority in the field of interferon and a long-time friend of Prof. De Somer and the Rega Institute.

JAN T. VILCEK is Professor of Microbiology at New York University School of Medicine where he is also Head of the "Biological Response Modifiers Unit". Prof. Vilcek has a long standing reputation of excellence in the field of interferon. Born in Czechoslovakia, Jan Vilcek studied at the Comenius University in Bratislava and also did his first work on interferon in the laboratory of Prof. L. Borecky in Bratislava. His affiliation with New York University dates from 1965. Prof. Vilcek was one of Prof. De Somer's best friends whom he went to see whenever he visited New York. Also, the work of Dr. Vilcek has always shown a striking parallelism with the work done in the Rega Institute. We have often been in competition with each other, i.e. in the production of fibroblast interferon or, more recently, in the study of the antiviral effects of interleukin-1 and tumor necrosis factor.

CHARLES C. DINARELLO is Associate Professor of Medicine and Pediatrics at Tufts University in Boston. Fever is the subject that has been close to his heart since he started his scientific career in 1968. Having Dr. Dinarello as participant in our Symposium is another tribute to Prof. De Somer, who has always been interested in the fever response. A simple increase in body temperature as a mechanism to inhibit bacteria or viruses, has a special appeal because of its simplicity and elegance. However, the mechanism by which fever is brought about is rather complex. The key cytokine in the fever response is interleukin-1 (IL-1), previously called endogenous pyrogen. Professor Dinarello's work has contributed immensely to our understanding of the biology of this cytokine. Through our own interest in this molecule we have become friends. What I admire most in Charles Dinarello is his ability to combine his commitment in the clinic with a tremendously productive laboratory activity. Describing the career of Charles Dinarello would be tantamount to describing the history of IL-1. This task I will leave to himself.

MECHANISMS OF NATURAL RESISTANCE

S. BARON
Department of Microbiology, University of Texas, Medical Branch, Galveston, Texas 77550, U.S.A.

1. INTRODUCTION

The most dramatic changes in human society and the quality of our lives have been associated with the discovery of new ideas and principles. There are many obvious examples, such as the discovery that the earth is round, that the earth is not the center of the universe, that the sun is fueled by nuclear fusion, that beliefs are testable by a scientific method, that evolution determines existing life forms, that electromagnetic waves occur, that many diseases are contagious, and that the template for life is encoded in genes. Without these fundamental concepts the modern miracles of space travel, energy generation, world-wide communication, and health might not exist. This manuscript will explore the development of the concepts that explain a modest miracle, the survival of people and animals despite constant immersion in a sea of microorganisms that outnumber us many million times. Prof. Piet De Somer and his colleagues, Prof. A. Billiau and Prof. E. De Clercq, were among a small band of scientists who championed a major change in the paradigms by which the body defends itself against this incessant onslaught.

2. PROFESSOR PIET DE SOMER

In his lifetime, Prof. P. De Somer was involved in the establishment of several major principles. Early in his career, Prof. De Somer participated in the establishment of the importance and medical use of antibiotics. I did not know him at that time but later he and I did participate in the unravelling of the nonspecific host defenses and interferon mechanisms. He and his colleagues contributed major discoveries : (1) the induction of the complex interferon system by relatively simple polymers (1,2), (2) the fundamental mechanisms by which interferon alters the cells to resist infection (3-6), (3) the unexpected finding that interferon inducers could act against bacteria and protozoa parasites (2); (4) the role of interferon and other natural defenses in the pathogenesis of viral infections (4); (5) a new mechanism by which interferon inhibits the multiplication of the retroviruses that are related to human leukemia and AIDS viruses (7); (6) the large scale production of interferons beta and gamma (8,9), their purification (8); (7) their toxicities (10) and clinical trials (11); (8) the fact that interleukin-1 induces interferon beta (12); and (9) the structure and configuration of the interferon molecule (5). I will attempt to relate the story of how Prof. De Somer, his colleagues and several of us around the world shared in the development of a new insight in natural host defenses against viruses and some other microorganisms, and how some of these new concepts have been applied to successfully control some life-threatening viral infections and some forms of cancer. Have we really lost Professor De Somer with such a legacy of ideas ?

S. Baron

3. CONTAGION

One of the seminal concepts that was discovered in antiquity is contagion, i.e., many diseases can spread from an infected individual to others (for example, plagues and diarrheal disease). Without further understanding of the mechanisms by which the disease was spread, mankind built on this concept to save more lives, by preventing direct and indirect contact with the products of infected individuals, than all of the other medical advances in history. The discovery of contagion led to the discovery of its cause.

4. HUMAN SURVIVAL IN AN ENVIRONMENT OF MICROBES

Perhaps the next major discovery was that "microscopic animals" (microbes) were responsible for this contagion. This paradigm ultimately led to the development of vaccines and antibiotics to further control contagious diseases. However, this discovery and their multitude number generated the great paradox that humans and animals survive while outnumbered many millionfold by microorganisms, many of which are highly virulent.

5. IMMUNE DEFENSES AND AVIRULENT MICROBES

Study of the mechanisms behind our paradoxical survival led to three major understandings. The first is that the body can mount effective immune defenses (antibody and immune cells) against invasion by microorganisms. Second, that there is selective evolutionary pressure on the microorganisms to develop less virulent offspring so that their host animals and humans will survive and continue to spread the microbes. Third, microorganisms, if they succeed in invading the body, most often succumb to the body's internal immune defense, resulting in host survival but with continued shedding of the microorganisms. At this time it was believed that the newly discovered immune defenses were mainly responsible for our survival. However, this was only half true.

6. IMMUNE DEFENSES

The first glimpse of these host defenses occurred in the 1700's. An exhilarating finding must have been the recognition that humans and animals that had recovered from an infection were usually resistant to reinfection by the same organism. Developing from this seminal observation was the discovery of the components of the immune responses (antibody and immune cells) to foreign organisms. This paradigm, that the immune response is the major host defense that permits us to survive invasion by microorganisms, was the dominant concept until this century when it was discovered that a series of nonspecific body defenses also played a major defensive role, especially during acute infections. The discovery, nature, and medical application of these nonspecific defenses, in which Prof. P. De Somer and several of us were involved, will be considered further.

7. NONSPECIFIC DEFENSES

As noted above, prior to the recognition of the nonspecific defenses, scientists believed that infection occurred only in nonimmune hosts and that in most cases the body's immune response (antibody and immune cells) was sufficient to overcome the invading viruses and restore the body to health. However, the recognition that other critical host defenses had been overlooked was hinted at by three key findings. The first finding was that the specific immune response occurred too late during viral infection, i.e., after the time that virus multiplication was brought under control during the acute infection (13). The second was that during recovery from viral infections the host develops a powerful resistance, not only to superinfection with the same virus, but an equally strong resistance to other viruses (14). Consequently, the specific immune defenses could not account for the heightened defenses against unrelated viruses. These missing nonspecific defenses were discovered as a group of early-appearing and potent nonspecific resistance mechanisms that were operative within hours of infection and prior to the development of the specific immune responses (13). We will consider the evidence for these nonspecific defenses with emphasis on one of the prototype defenses, interferon, which Professor De Somer, his colleagues, and a few other laboratories (including Alick Isaacs) pioneered (15). Thus, the established paradigm that recovery from the primary viral infection is due mainly to the specific immune response was supplanted by the modified paradigm that nonspecific defenses are the main defenses only during the early phase of viral infections and that the immune defenses become important during the later stage of infection. Also, the induced immunity, but not the nonspecific defenses, persist, and the induced immunity is manifested as specific resistance to reinfection. The evidence for this new concept will be considered below, as well as the application of this new knowledge to prevent and control viral infections and some forms of cancer.

8. OVERVIEW OF NONSPECIFIC DEFENSES

Most viral infections are self-limited by host defenses, which enable the host to accommodate successfully to large accumulations of viruses that would have been lethal if presented initially (Fig. 1). Some nonspecific defenses such as anatomic factors, nonspecific inhibitors in body fluids, and phagocytosis exist independently of infection; other nonspecific defenses (fever, inflammation, and interferon) are produced by the host in response to infection (13,16).

The intracellular site of virus replication and the ability of some viruses to spread by cell fusion protect them against such extracellular defenses as antibody, phagocytosis (Fig. 2), and many nonspecific inhibitors; however, because they replicate within the cell, viruses are vulnerable to intracellular alterations caused by other host responses to infection. Nonspecific responses that alter the intracellular environment and thereby inhibit virus multiplication include fever, inflammation, and interferon.

9. NONSPECIFIC DEFENSE MECHANISMS PREEXISTING INFECTION

9.1. Anatomic barriers

Anatomic barriers to viruses exist at the surfaces and within the body (16). At body surfaces, the dead cells of the epidermis and live cells that may lack viral receptors resist virus penetration and do not permit virus replication; however, this barrier is easily brached (e.g., for rabies virus by animal bites, for togaviruses by insect bites, and for wart virus by minor traumas). At mucosal surfaces, only the mucous layer stands between invading virus and live cells. The effectiveness of the mucous layer is due to its formation of a physical barrier, entrapment of foreign particles that are cleared by the mucous flow and its content of nonspecific inhibitors. When viruses infect cells at mucosal surfaces, these cells may sometimes serve as a barrier to infection of deeper cells because the surface cells are normally shed and because viruses are often adapted to multiply mainly in the surface cells with special metabolic and environmental characteristics. Also a large function of virus replicating in surface cells may be released in the lumen of the origin rather than into the

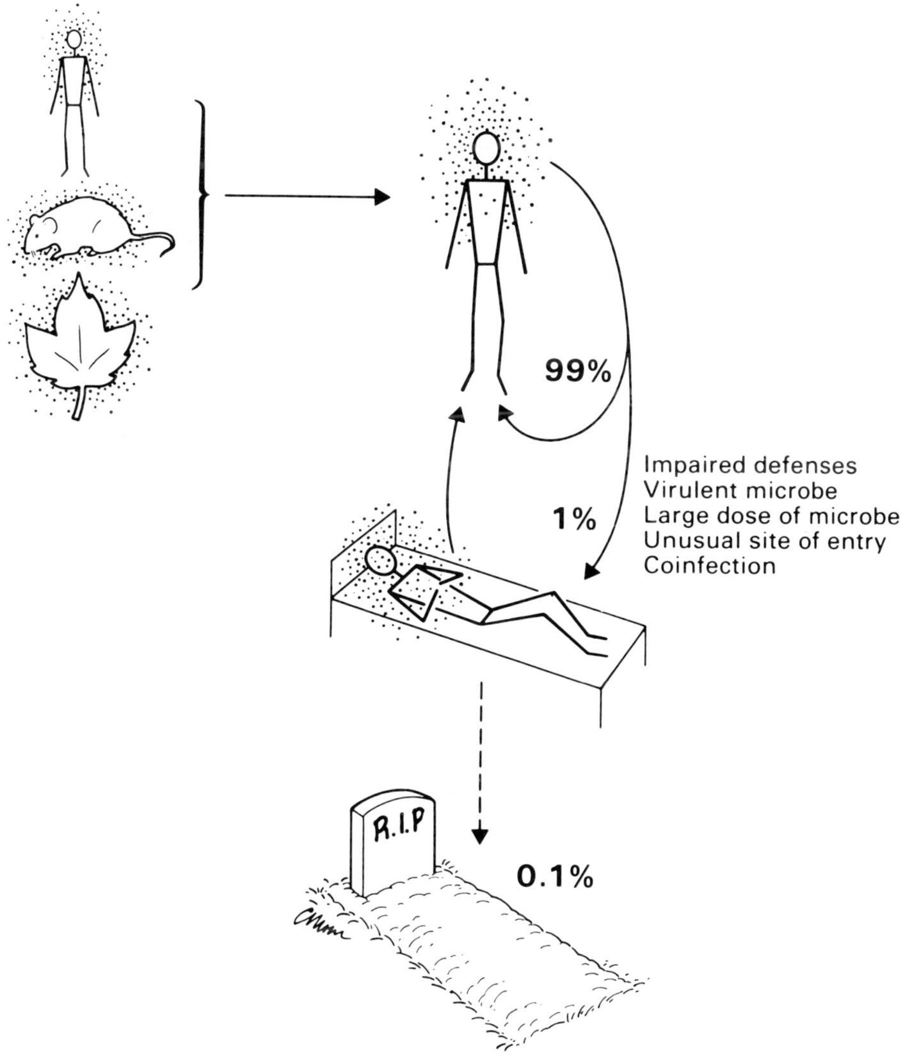

FIGURE 1. Disease or a lethality is an unusual outcome of most infections. We are exposed to microbes (˙.˙.˙) from other people, animals and plants. Illnes is not the usual outcome of most infections and mortality is even less common.

deeper tissue. This mucosal barrier is not absolute, however, since sufficient quantities of many viruses can infect by this route. In fact, most viruses use mucosal surfaces as the portal of entry and initial replication site.

Within the body, an anatomic barrier to virus spread is formed by the

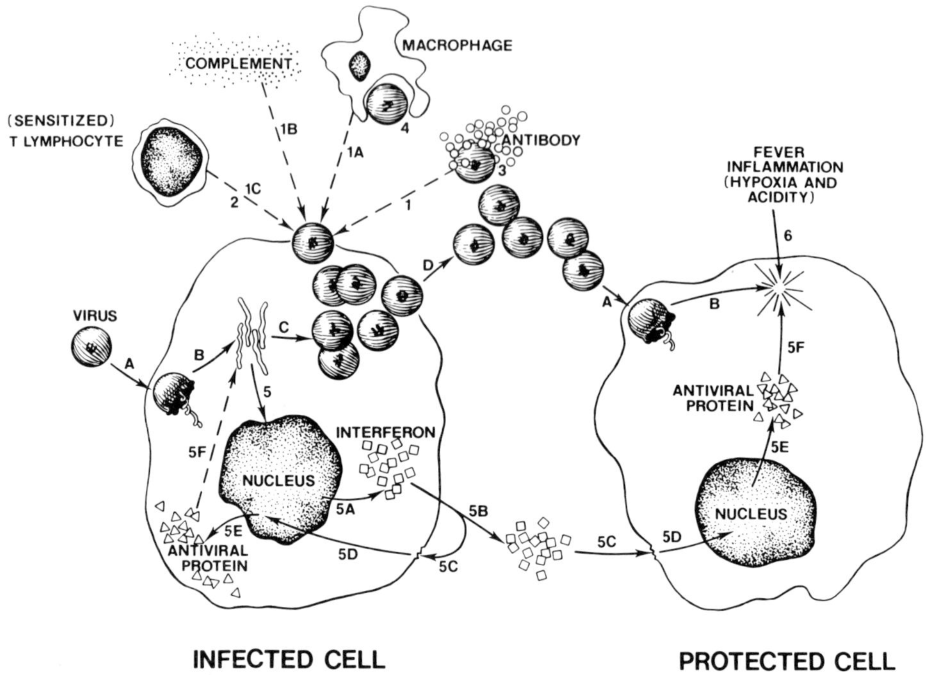

FIGURE 2. Host defenses against virus infection. Virus multiplication is shown in steps A-D. The extracellular defenses include antibody (3) and, to a lesser extent, the phagocytic macrophage (4) which may combine with and inactivate virus. Defenses against intracellularly replicating virus include cytolysis of virus-infected cells bearing virus surface antigens by synthesized T lymphocytes (2) acting directly or through their mediators: antibody plus macrophages (1A); antibody plus complement (1B); antibody plus T lymphocytes (1C); and antibody plus other effector leukocyte cells (not shown). Additional inhibition of intracellular viral replication (C) can be caused by the nonspecific defense mechanisms of interferon (5-5F), local acidity, local hypoxia (6), and interfering virus (not shown).

layer of endothelial cells that separates blood from tissues (for example, the bloodbrain barrier). Under normal conditions, these barriers have low permeability for viruses unless the virus penetrates by replicating in the endothelial cells or in circulating leukocytes. These internal barriers may explain in part the high level of viremia required to infect organs such as the brain, placenta, and lungs.

9.2. Nonspecific inhibitors

A number of viral inhibitors occur naturally in most body fluids and tissues. They vary both chemically (lipids, polysaccharides, proteins, lipoproteins, and glycoproteins) and in degree and range of viral inhibition (17). Some inhibitors are related to the viral receptors of the cell surface, but most are of unknown origin. Almost all protect by preventing virus attachment to cells. Most of these inhibitors block one or a few viruses; however, a recently described contact-blocking viral inhibitor (CVI) has a broad antiviral spectrum. CVI seems to act also by blocking virus attachment. Other important properties are a molecular weight of approximately 1,000 daltons, distribution in most internal and surface fluids, and ability to protect mice against several different viruses. (A special case of viral inhibition is the inactivation of many viruses by acid, bile salts, and enzymes in the gastrointestinal tract.) Although the effectiveness of the various inhibitors has not been fully established *in vivo*, their importance as host defenses is suggested by their antiviral activity in cell culture and *in vivo*, and by the direct correlation between the degree of virulence of some viruses and the degree of resistance to certain inhibitors. The presence of these inhibitors may explain the relatively high doses of virus required to initiate infection *in vivo* as compared with cell cultures.

9.3. Phagocytosis

The limited information available suggests that phagocytosis is less effective against viral infections than against bacterial infections; however, few of the factors that control uptake of virions or infected cells by phagocytes (chemical and electrical affinities) and their digestion by lysosomal enzymes have been systematically studied (16). The observed result of this process is that different viruses are affected differently by the various phagocytic cells. Specifically, some viruses are not engulfed while others are engulfed but may or may not be inactivated. In fact, some viruses may even multiply in the phagocytes. Macrophages seem to be more effective against viruses than are granulocytes, and large viruses seem to be more affected than small viruses. An important mechanism of protection by phagocytes is the marked reduction of viremia by macrophages but only when the conditions are optimal. For example, the degree of virulence in several strains of herpes simplex virus correlates with their ability to survive or even multiply within macrophages. Thus, depending on the situation, macrophages in their phagocytic function may reduce virus, help spread the infection or, more commonly, have little or no effect. Macrophages in their nonphagocytic functions may be permissive or nonpermissive for multiplication of certain viruses and thereby affect the outcome of these infections.

10. DEFENSES MECHANISMS EVOKED BY INFECTION

10.1. Fever

Viral replication is strongly influenced by temperature (13,16,18). Even a modest increase can cause strong inhibition; a temperature rise from 37°C to 38°C may decrease drastically the yield of many viruses. This phenomenon has been observed in cell culture as well as in many experimental and natural infections. The artificial induction of fever in mice infected with viruses reduces mortality rates as compared with mortality rates in control animals maintained at normal temperature. On the other hand, artificial lowering of temperature during the infection may increase mortality, as shown in suckling mice infected with coxsackie viruses and taken away from the warmth of their mother's nest.

Several observations strongly suggest the same phenomenon may occur during human viral infections. Retrospective studies have shown that the incidence and severity of paralysis among children infected by polioviruses were significantly higher in patients treated with antipyretic drugs than in untreated children. Also consistent with these findings is the observation that virus strains that replicate best at fever temperature are usually virulent, whereas virus strains that replicate poorly at fever temperature are usually low in virulence and are therefore often used as live virus vaccines.

Temperatures as low as 33°C are normal at body surfaces; viruses that infect these sites and replicate optimally at these temperatures establish only local infections that do not spread to deeper tissues where body temperature is higher. For example, rhinoviruses that cause common colds replicate optimally at 33-34°C (a temperature found when swelling of the edematous mucosa interrupts air flow). An interesting question is whether a temperature increase may be important for recovery from common colds.

Probably, the same general considerations concerning temperature apply to other human viral infections such as measles, rubella, and mumps, although suitable and controlled studies have not yet been conducted. Nevertheless, available information suggests conservative use of antipyretic drugs in virus infections.

10.2. Inflammation

Several antiviral mechanisms are generated by the local inflammatory response to virus-induced cell damage (13,16,18). The major components of the inflammatory process are circulatory alterations, edema, and leukocyte accumulation. The resulting phenomena are elevated local temperature, reduced oxygen tension in the involved tissues, altered cell metabolism, and increased CO_2 and organic acids. All these alterations, which occur in a cascading and interrelated fashion, may drastically reduce replication of many viruses. For instance, altered energy metabolism of the infected and surrounding cells as well as the accumulating lymphocytes can generate local hyperthermia. When inflammation occurs at superficial sites, where temperature is normally lower, hyperthermia can be generated also by hyperemia during early stages of inflammation. As inflammation progresses, hyperemia becomes passive, thereby greatly reducing blood flow and decreasing oxygen tension. Two factors account for this decrease in oxygen tension : limited influx of erythrocytes and lower diffusion of oxygen through edemia fluid. In turn, the decreased oxygen tension causes less ATP production (and thus reduces energy available for viral synthesis) and increases anaerobic glycolysis, which increases the accumulation of CO_2 and organic acids in the tissues. These acid catabolites may decrease the local pH further to levels that have been shown to strongly inhibit replication of many viruses. Local acidity may also increase by accumulation and subsequent degradation of the leukocytes in the affected area; and possibly other less-defined factors may also play a significant role.

Thus, the local inflammation resulting from viral infection clearly activates several metabolic, physicochemical, and physiologic changes; acting individually or synergistically, these changes probably constitute an effective defense to virus multiplication. Although further study is needed in whole animals and humans to assess the role of local inflammation in the host's reponse to virus infection, its importance is supported by the finding that treatment with anti-inflammatory drugs (corticosteroids) often increases the severity of virus infection in animals.

10.3. Viral interference and interferon

Generally, infection by one virus renders host cells highly resistant to other, superinfecting viruses. Called viral interference, this intriguing phenomenon occurs frequently in cell cultures and in the whole animal (14). Although interference mostly occurs between unrelated viruses, sometimes it may occur only between homologous viruses and in these ways function as a defense against viral infections. Some types of interference are caused by competition among different viruses for critical replicative pathways (extracellular competition for cell surface receptors or intracellular competition for biosynthetic machinery). Similarly, interference may result from competition between defective viruses and infective viruses that may be produced concurrently (19). Another type of interference, which occurs most commonly in nature, is directed by the host cells themselves; these infected cells may respond to viral infection by producing a new protein, called interferon, which can react with uninfected cells to render them resistant to infection by a wide variety of viruses.

10.4. Interferon

Interferon was discovered thirty years ago by Isaacs and Lindenmann (19), who observed that fluids from virus-infected cell cultures contained a protein that could react with cells to render them resistant to infection by a wide variety of viruses. Since then much has been learned about the complex interferon system, including its applicability for treatment of disease. Although interferons were first recognized for their extraordinarily potent antiviral properties, it has now been established that they may profoundly affect other vital cellular and bodily functions, including cell metabolism and growth, immunity, and tumors. Thus the interferon system has been used as a model for studying membrane changes and molecular events involved in viral replication, gene expression, cell regulation, immune regulation, and hormone production and action. Clinically, trials of interferon and its inducers during viral infections and cancer have reached the stage of controlled clinical trials and licensing for specific uses (see below).

10.5. Overview of the interferon system

In the 30 years since interferon's discovery, the field of interferon research has progressed tremendously. We now know that the interferon of Isaacs and Lindenmann is really a family of molecules which can be divided into three species : IFN-α, IFN-β, and IFN-γ. The interferons differ in the agent which induces them and in the cell type which produces them (Fig. 3).

The interferon proteins have been purified and the interferon genes have been cloned in bacteria. A wide variety of interferon activities including antiviral effects, antiproliferative effects, immunomodulatory effects, and hormonal effects have been described and partially characterized. To induce these activities, interferon must activate cells to produce effector proteins (Fig. 4).

10.6. Biochemical effects - Production of new proteins

IFN-α and IFN-β induce the production of 12 new proteins in responding cells. IFN-γ induces these same proteins and an additional 12 new proteins. Some of these proteins have now been identified and some possible mechanisms of antiviral effects are shown in Fig. 4. However, the precise mechanisms by which these proteins express interferon's varied activities are still unknown.

10.7. Comparison of the interferon system with other host defenses against viruses

As discussed above, the known defenses against viruses may be divided into those which act against extracellular virus (neutralizing antibody and the phagocytic cells) and those which inhibit virus replication within cells [(1) the sensitized-lymphocyte, (2) cytolytic antibody plus complement, macrophages, or lymphocytes which can destroy infected cells, (3) fever, (4) hypoxia, (5) local acidity, and (6) interferon] (13,16).

The interferon system is the earliest-appearing of the known host defenses. It is operative within hours of infection. Fig. 5 illustrates the early production of interferon in comparison with antibody during experimental infection of man with influenza virus.

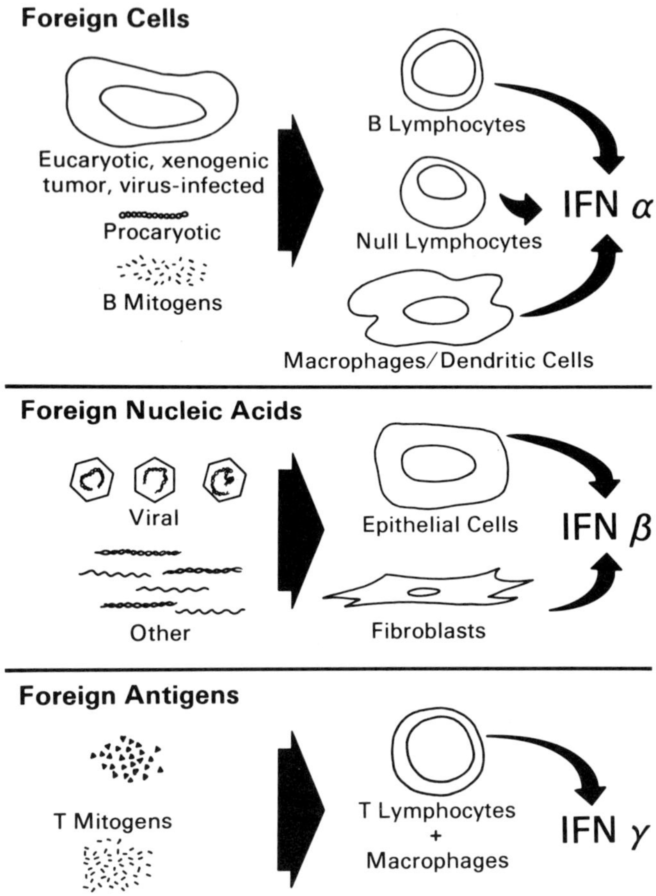

FIGURE 3. A general scheme for induction of human interferons α, β or γ respectively by foreign nucleic acids, foreign cells, or foreign antigens. In comparison, in the mouse system interferons α and β are usually produced together.

The available evidence strongly supports a causal relationship between the interferon system and natural recovery from many established viral infections (Table 1). The evidence also indicates that interferon is not the sole defense that influences recovery, and in certain virus infections it may not be the most important defense. This depends on both the effectiveness of the virus as a stimulator of interferon production and on the susceptibility of the virus to the antiviral action of interferon. The interferon system appears to play its defensive role during viral infection of body surfaces, solid tissues and also during virus spread through the vascular system to target organs (20).

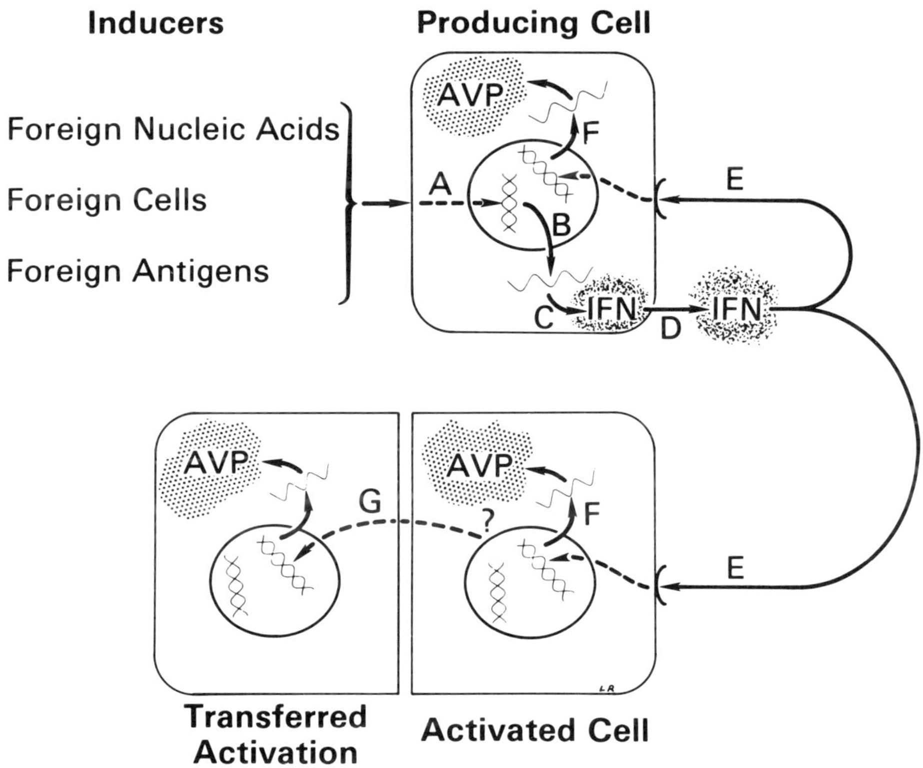

FIGURE 4. Induction, production and action of interferon (IFN) at the cellular level. IFN inducers react with cells to derepress the IFN gene(s) (A), leading to the production of IFN mRNA (B), that is translated to the IFN protein intracellularly (C) followed by its secretion extracellularly (D) where it reacts with the membrane receptors of cells (E). The IFN-stimulated cells derepress other genes (F) for effector proteins (AVP) that establish antiviral resistance and other cell changes. The activated cells also directly stimulate contacted cells (G) by a still unknown mechanism to produce AVP.

10.8. Antiprotozoal and antibacterial activities

There is also growing evidence that IFN may play a role in natural protection against certain bacterial and protozoal infections. This evidence centers around induction of interferon during many of these infections, protection by IFN against certain of these infections, and the identification of antimicrobial mechanisms activated by IFN. For example, administration of endogenous IFN is protective in mice infected with malarial organisms. The mechanism of this protection involves leukocyte activation in addition to effects on susceptible cells (see 21-23).

10.9. Large scale production of human interferon

Large amounts of all three human interferons are now being produced in systems using cultured human cells or bacterial cells containing human genes. These large amounts have made possible the clinical trials that

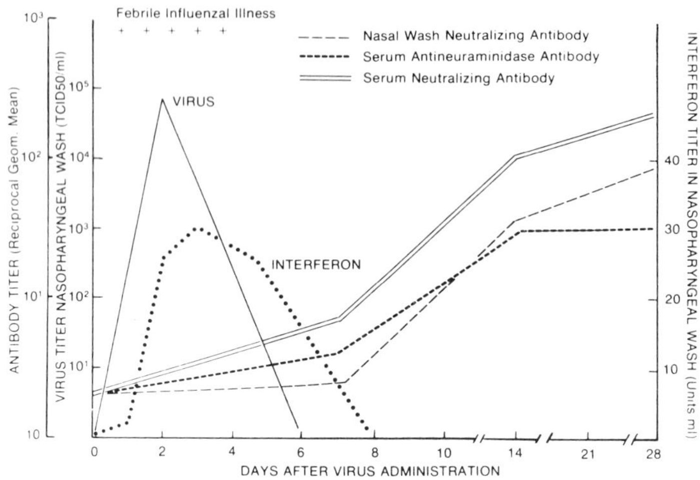

FIGURE 5. Production of virus, interferon, and antibody during experimental infection of humans with influenza wild-type virus (from a study by B. Murphy, E. Chalhub, S. Nusinoff, J. Kasel, C. Uhlendorf, S. Baron, and R. Chanock, National Institutes of Health).

TABLE 1. Evidence that interferons play a defensive role during viral infections.

Type of evidence	Example
Time of correlation	Interferon produced just before and during arrest of virus replication
Place of correlation	Interferon produced at the site of its action
Transfer	Exogenous interferon controls viral infections
Absence	Deletion of interferon in cell culture, eggs, and animals increases severity of viral infections
Sufficiency of quantity	Concentrations of interferon produced *in vivo* are sufficient to account for defensive action
Specificity of probes	Definitive characterization of interferon produced during infection
Reproducibility	Confirmation in many animal species and humans

have led to governmental approval for treatment of some human diseases (22,23).

10.10. Therapeutic application of interferon

As sufficient quantities of interferons have become available, interferons have been employed in clinical trials to determine their efficacy against malignancies and viral infections. It should be pointed out that, while interferon is effective against certain viral diseases and tumors, it is not yet as broadly effective in people as it is in experimental animal systems. In general, the clinical trials with IFNs have shown (1) a strong prophylaxis against rhinoviruses that cause the common cold and (2) effective therapy of HSV eye infections, laryngeal papilloma, warts, hepatitis B, varicella, hairy cell leukemia, multiple myeloma and renal cell carcinoma. Some of the clinical data are summarized below (see 21-23).

10.11. Interferon therapy for viral infections

IFN-α has been found to be effective in preventing influenza and rhinovirus infections.

A therapeutic effect of IFN-α or IFN-β has been seen in ocular infections such as herpes keratitis and adenovirus conjunctivitis (the latter not confirmed). IFN-α has also been shown to have a modest effect on infections caused by various members of the herpesvirus group (EBV, CMV, VZV, HSV-1) but the eradication of latent infections has not been demonstrated.

IFN-α treatment has been shown to decrease viremia during chronic hepatitis B virus infections; however, it is not clear whether recovery is affected.

Finally, IFN-α has been effective against several tumors which are believed to have a viral origin such as warts and condyloma acuminatum, juvenile laryngeal papillomatosis, and hairy cell leukemia (see below).

10.12. Interferon therapy for neoplastic disease

The antitumor potential of interferons has been shown in a number of studies against a variety of tumors. For some tumor types, interferon does appear to have efficacy. Unfortunately, the demonstrated therapeutic potential of the interferon treatment ranges widely from one tumor type to another. Also, for any given type of tumor, some patients may show a complete response (total regression of the tumor) while others may have a partial or no response.

Perhaps the best examples of interferon's potency are found in patients with hairy cell leukemia, warts, condyloma acuminatum or juvenile laryngeal papillomatosis. These patients have response rates which are quite high, approaching 90 %. While long-term studies with hairy cell leukemia and condyloma acuminatum are not complete, long-term studies of patients with juvenile laryngeal papillomatosis suggest that the interferon effect represents management of this condition and not cure, since recurrence of the tumor often occurs following withdrawal of interferon therapy. Effective management with interferon can be continued until puberty when spontaneous recovery usually occurs.

11. SUMMARY

The most dramatic changes in human society and the quality of our lives have been associated with the discovery of new ideas and principles. I have discussed the development of a modest miracle, the survival of people and animals despite constant immersion in a sea of microorganisms that outnumber us many million times and I have also considered the contributions of Prof. P. De Somer and his colleagues to our current paradigms. One of the major discoveries in antiquity was the recognition of contagion, i.e., many diseases can spread from an infected individual to others (for example, plagues and diarrheal diseases that have decimated mankind). The discovery of contagion led to the discovery of its cause, microbes (bacteria, fungi and viruses). However, the discovery of microbes and their multitude of numbers generated the great paradox that humans and animals survive while outnumbered many millionfold by microorganisms, many of which are highly virulent. An explanation for our survival came from the finding that if microbes succeed in invading the body, they usually succumb to the body's internal defenses. At this point, it was believed that these internal defenses were the newly discovered specific, immune responses; however, this was only half-true.

The recognition that other critical host defenses had been overlooked was hinted at by three key findings; (1) the specific immune response occurred too late during viral infection to account for the onset of recovery from the infection; (2) the heightening resistance to viruses (defensive response) that occurs during the early phases of virus infection does not share the specific properties of the immune responses; and (3) these missing host defenses were identified as a group of early-appearing and nonspecific resistance mechanisms that were operative within hours of infection and prior to the development of the specific immune responses. Prof. P. De Somer, his colleagues and a few other laboratories participated in the discovery of natural host defense mechanisms that changed medical thinking and resulted in an effective therapy of several forms of viral infection and cancer. The nonspecific defense mechanisms include, (1) anatomic barriers at the body surfaces and within the body; (2) soluble viral inhibitors in most body fluids and tissues; (3) phagocytosis of viruses; (4) the local inflammatory response to the virus infection, resulting in locally elevated temperature, reduced oxygen tension, altered cell metabolism and increased acidity, all of which may drastically reduce viral replication; (5) viral interference whereby one virus particle may preempt a cell and prevent other virus particles from multiplying therein; and (6) interferon which is produced by infected cells and which can protect surrounding cells. Prof. P. De Somer made major contributions to the understanding and medical application of the interferon system.

We now know that the interferons can be divided into three species, interferon alpha, interferon beta, and interferon gamma. To be protective, interferon must activate cells to produce effector proteins. The interferon system is the earliest appearing of the known host defences; it is operative within hours of the infection. The available evidence strongly supports a causal relationship between the interferon system and natural recovery from many established viral infections. Although interferons were first discovered for their extraordinarily potent antiviral properties, they may profoundly affect other vital cellular and body functions, including cell metabolism and growth, immunity, tumors, and the multiplication of some bacteria and protozoa. Large amounts of all three human interferons are now being produced in systems using cultured human cells or bacterial cells containing human genes. Using these large quantities of interferon, clinical trials have shown a strong prophylaxis against rhinoviruses that cause the common cold and beneficial effects in the treatment of herpes simplex virus eye infections, laryngeal papillomas, warts, hepatitis B, hairy cell leukemia, multiple myeloma and renal carcinoma. The results with other viral infections and tumors are encouraging and is under study. While the paradigm of how the body defends itself against microbial invasion has been definitely changed, the greater message may be that pioneers like Piet De Somer are essential for that change.

REFERENCES

1. De Clercq E, Stewart WE II, De Somer P. Infect. Immun. 6:344-347, 1972.
2. Eyckmans L, Billiau A, De Somer P. Biomedicine 19:187-192, 1973.
3. Billiau A, Heremans H, Ververken D, Van Damme J, Carton H, De Somer P. Arch. Virol. 68:19-25, 1981.
4. Heremans H, Billiau A, De Somer P. Infect. Immun. 30:513-522, 1980.
5. Stewart WE II, De Somer P, De Clercq E. Prep. Biochem. 4:383-393, 1974.
6. Cocito C, Schonne E, De Somer P. Life Sci. 4:1253-1262, 1965.
7. Billiau A, Heremans H, Allen PT, De Maeyer-Guignard J, De Somer P. Virology 73:719-721, 1976.
8. Heine JW, Van Damme J, De Ley M, Billiau A, De Somer P. J. Gen. Virol. 54:47-56, 1981.
9. Billiau A, Edy VG, Heremans H, Van Damme J, Desmyter J, Georgiades JA, De Somer P. Antimicrob. Agents Chemother. 12:11-15, 1977.
10. Billiau A, Heine JW, Van Damme J, Heremans H, De Somer P. Ann. N.Y. Acad. Sci. 350:374-375, 1980.
11. Billiau A, Bloemmen J, Bogaerts M, Claeys H, Van Damme J, De Ley M, De Somer P, Drochmans A, Heremans H, Kriel A, Schetz J, Tricot G, Vermylen C, Verwilghen R, Waer M. Eur. J. Cancer Clin. Oncol. 17:875-882, 1981.
12. Van Damme J, Opdenakker G, Billiau A, De Somer P, De Wit L, Poupart P, Content J. J. Gen. Virol. 66:693-700, 1985.
13. Baron S. _in_ Advances in Virus Research (Smith K, Lauffer M, Eds). Academic Press, New York, vol 10, pp 39-64, 1963.
14. Schlesinger R. _in_ The Viruses. (Burnet F, Stanley W, Eds). Academic Press, New York, vol 3, pp 157-189, 1959.
15. Isaacs A. _in_ Advances in Virus Research (Smith K, Lauffer M, Eds). Academic Press, New York, vol 10, pp 1-39, 1963.
16. Dianzani F, Baron S. _in_ Medical Microbiology (Baron S, Ed), pp 760-768, 1986.
17. Kumar S, McKerlie M, Albrecht T, Goldman A, Baron S. Proc. Soc. Exp. Biol. Med. 177:104-111, 1984.
18. Baron S. _in_ Modern Trends in Medical Virology (Health R, Waterson A, Eds). Butterworths, London, pp 77-110, 1967.
19. Isaacs A, Lindenmann J. Proc. R. Soc. Lond. (Biol.) 147:258-267, 1957.
20. Baron S, Buckler C, Friedman R, McCloskey R. J. Immunol. 96:17-24, 1966.
21. Baron S, Dianzani F (Eds). Texas Reports on Biology and Medicine, vol 35, 1977.
22. Baron S, Dianzani F, Stanton G (Eds). Texas Reports on Biology and Medicine, vol 41, 1982.
23. Baron S, Dianzani F. Stanton G, Fleischmann W II (Eds). Interferon: A Current Review to 1987. University of Texas Press, Austin, in press, 1987.

J. Vilček

REGULATION OF CELL GROWTH BY INTERFERON AND TUMOR NECROSIS FACTOR

J. VILČEK
Department of Microbiology, New York University Medical Center, New York, New York 10016, U.S.A.

"For every complex problem there is a simple solution -- and it is [usually] wrong."

H.L. Mencken

1. INTRODUCTION

Professor Piet De Somer was at once a romantic visionary and a down-to-earth realist. His approach to science was influenced by his early experiences in antibiotic research. He had admitted readily that his initial interest in interferon was based on the belief that interferons would mark the beginning of a new era in antiviral therapy, in the same way that penicillins marked the dawn of effective antibacterial therapies. However, Prof. De Somer must also be given credit for being among the first in the field to recognize that - at least in their natural form - interferons are only of limited value in antiviral therapy. Early clinical trials with fibroblast interferon, produced at the Rega Institute under the direction of Alfons Billiau, contributed to the recognition of many systemic effects of interferons related to the alteration of immune functions. These erstwhile "side-effects" of interferons today are the central themes of interferon research.

The major theme of this article concerns such "side effects", i.e. the actions of interferons and some other cytokines on the complex events of cell proliferation. Much of the paper will be devoted to the induction of interferon-beta (IFN-β) by tumor necrosis factor (TNF) and the role this event may play in the modulation and mediation of the mitogenic and antiviral activities of TNF in human fibroblasts. As will be explained, some of this work was inspired by similar studies with interleukin-1 (IL-1) done at the Rega Institute. Prof. De Somer was an active participant in this work at the Rega, but he was unfortunately deprived of learning about the latest unexpected series of amusing twistst and turns in this fascinating story.

2. INTERFERONS AS AUTOCRINE REGULATORS OF CELL GROWTH

Although interferon preparations were shown to inhibit cell growth by Paucker *et al.* (1) about 25 years ago, these findings were generally dismissed as actions likely due to contaminating impurities. Eventually it was recognized that purified interferon preparations indeed inhibited cell growth (2). A new era was initiated when the inhibitory effects of interferons were examined in synchronized cultures (3) and in cells stimulated by defined growth factors (4,5). Oleszak and Inglot (6) and Sreevalsan *et al.* (5) deserve credit for demonstrating the reciprocal antagonistic relationship of growth factors and interferons. Not only can interferons inhi-

TABLE 1. Interferons (IFNs) and growth factors (GFs) : period 1962-1985.

Demonstration	Authors	References
IFNs inhibit cell growth	Paucker *et al.*, 1962	1
	Gresser *et al.*, 1970	2
IFNs inhibit GF-induced cell growth	Sokawa *et al.*, 1977	3
	Inglot *et al.*, 1980	4
	Sreevalsan *et al.*, 1980	5
GFs inhibit antiviral and antiproliferative action of IFNs	Sreevalsan *et al.*, 1980	5
	Oleszak and Inglot, 1980	6
Autocrine IFN-ß regulates cell growth and function	Friedman-Einat *et al.*, 1982	7
	Creasey *et al.*, 1983	8
CSF-1 stimulates IFN release in monocytes	Moore *et al.*, 1984	9
IL-1 ("22kDa protein") induces IFN-$ß_1$	Van Damme *et al.*, 1983, 1985	10,11
IL-1 induces IFN-$ß_2$ ("26kDa protein")	Content *et al.*, 1985	12
Poly(I).poly(C) activates expression of the same genes as PDGF	Zullo *et al.*, 1985	13

bit growth factor activity, but various growth factors can interfere with the antiviral and antiproliferative action of interferons (Table 1).

The concept that interferons act as natural regulators of cell proliferation, originally advanced on theoretical grounds (14,15), received experimental support in the demonstration that spontaneous ("autocrine") interferon production can occur in a variety of cultured cells, either accompanying the process of terminal differentiation (7) or during saturation density-induced growth arrest (8). Analysis of the autocrine interferon produced in the course of these studies revealed that its activity was neutralized by antibodies specific for IFN-ß.

Induction of IFN-ß was also demonstrated in various types of cells exposed to growth and differentiation factors (7). For example, colony-stimulating factor-1 was shown to induce interferon in bone marrow cells (9) and this interferon was recently shown to be neutralized by antibodies to IFN-ß (16). [For the time being I am using the designation "IFN-ß" to denote any interferon whose antiviral activity is neutralized by anti-IFN-ß sera. I shall discuss the relationship between IFN-$ß_1$ and -$ß_2$ in the last section of this article (see Table 2).]

It is worthwhile to recount briefly the chain of discoveries leading to the recognition that IL-1 can act as an IFN-ß inducer. Prof. De Somer's group (17) described a "22kDa protein" -- present in supernatants of concanavalin A-stimulated cultures of mononuclear cells -- with antiviral activity in cultures of human skin fibroblasts and MG-63 osteosarcoma cells. The antiviral activity of this 22kDa protein was neutralized in the presence of antibodies to human IFN-ß, suggesting that it was either a novel type of IFN-ß or an inducer of IFN-ß. Eventually, Van Damme *et al.* (10) showed that the 22kDa protein was an inducer of IFN-ß. These findings were followed by two other surprising discoveries. Although the 22kDa protein was found to induce low levels of IFN-β_1 mRNA in MG-63 cells (18), Content et al. (12) demonstrated that it was a potent inducer of IFN-β_2 in the absence of any detectable induction of IFN-β_1 (whereby IFN-β_2 is referred to as "26kDa protein"). The next surprising finding was that the 22kDa protein from concanavalin A-stimulated leukocyte cultures proved to be identical to the independently identified major form of IL-1, termed IL-1β (11). This work of De Somer's group provided the first determination of the correct NH_2-terminal sequence of this form of IL-1.

In addition to a long list of other biological actions, IL-1 is known to be a growth factor for fibroblasts and some other types of cells (reviewed in ref. 19). Thus, the ability of IL-1 to induce IFN-β is consistent with the idea that induction of IFN-β by growth factors serves as a negative feedback control mechanism (7,8,13). A general model of interactions between various growth signals and endogenously produced (autocrine) β-interferons is shown in Fig. 1.

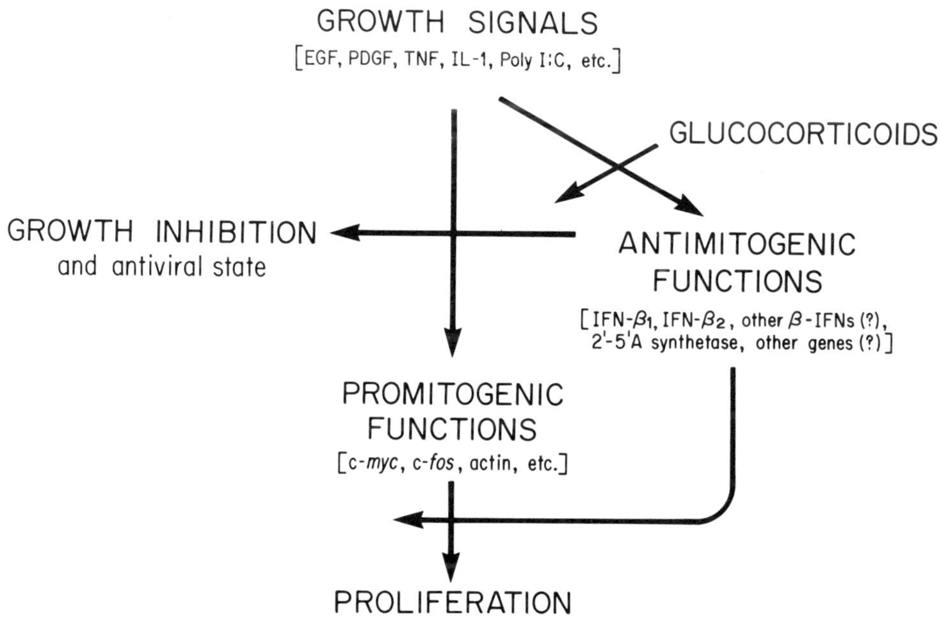

FIGURE 1. Model of antagonistic interactions of growth signals and autocrine interferons. The action of growth factors and other growth signals in generally accompanied by the activation of cellular proto-oncogenes and several other genes. Growth signals may also activate cellular genes for IFN-β_1 (e.g., poly I:C) or IFN-β_2 (e.g., IL-1, TNF, PDGF). Interferons (in part perhaps as a result of 2'-5'A synthetase induction by interferons) inhibit cell growth and thus may act as feedback regulators. This antiproliferative action of interferons may be due to an inhibitory effect on proto-oncogene synthesis and to other, yet to be identified mechanisms. Glucocorticoids (e.g., dexamethasone) generally inhibit interferon synthesis and may stimulate cell proliferation (20). It should be noted that the antagonistic relationship between growth factors and interferons is reciprocal, i.e., growth factors also can inhibit the antiproliferative and antiviral action of interferons (5,6,20,21).

3. TUMOR NECROSIS FACTOR AS A FIBROBLAST GROWTH FACTOR

As is apparent from its name, TNF was for a long time thought to be a specific inhibitor of tumor cell growth in the intact animal and in tissue culture. Although its cytotoxic/cytostatic action has been confirmed with the pure protein, it is now apparent that at least in some types of cultured cells TNF acts as a potent growth factor. Growth-stimulating activity of TNF was first seen in cultures of various human diploid fibroblasts which grew to a higher density when TNF was added to the growth media (22, 23, and W. Fiers, personal communication). A stimulation of DNA synthesis was also readily demonstrable in a standard [^{3}H]thymidine incorporation assay (23). In human FS-4 foreskin fibroblasts TNF exerts a significant growth factor-like activity at concentrations as low as 10^{-12} M. An optimal mitogenic effect is seen at concentrations around 10^{-10} M, i.e., concentrations similar to the Kd for TNF receptor binding.

Other results also support the view that the mitogenic action of TNF in fibroblasts resembles that of authentic growth factors. In promoting fibroblast growth TNF acts synergistically with serum or insulin (23). Under appropriate conditions TNF can also increase the responsiveness of cells to the mitogenic action of epidermal growth factor (EGF) and this effect correlates with an increased expression of EGF receptors in TNF-treated cells (24). TNF is also mitogenic in rat NRK fibroblasts and murine Balb/c-3T3 cells (V. Palombella and J. Vilček, unpublished data). Like many other growth factors, TNF can rapidly induced the expression of mRNAs for the proto-oncogenes c-*myc* and c-*fos* (J.-X. Lin and J. Vilček, unpublished data). The mitogenic action of TNF is apparently not limited to normal fibroblasts, because some transformed cell lines derived from tumor tissues also grow to higher densities in the presence of TNF. Lymphotoxin (termed by some investigators TNF-β because of its significant homology with TNF) also was shown to stimulate the growth of both normal and transformed cells (25).

Elucidation of the mechanism of the mitogenic action of TNF is somewhat complicated by our demonstration that TNF can induce the synthesis of both IL-1α and IL-1β in human fibroblasts (26). Like TNF, both forms of recombinant human IL-1 are mitogenic in human fibroblasts. In fact, the mitogenic action of recombinant human IL-1 was demonstrable at molar concentrations about 30-100 times lower than those at which TNF showed comparable activity (27). Nevertheless, available evidence suggests that the mitogenic action of TNF is not mediated by the generation of IL-1. More likely, binding of both TNF or IL-1 to their respective receptors can independently initiate the activation of intracellular pathways mediating the mitogenic response.

4. INTERFERON-β AS A FEEDBACK REGULATOR OF THE MITOGENIC ACTION OF TNF

In view of the reports that some growth factors can act as inducers of IFN-β (12,13,18), we examined the possibility that TNF too might stimulate the secretion of this interferon in human fibroblasts. Our initial approach was based on the assumption that stimulation of IFN-β synthesis might serve as part of a negative feedback loop and would eventually lead to a suppression of the mitogenic action of TNF. Therefore, we added antiserum to highly purified human IFN-β to FS-4 cultures stimulated with TNF and we examined the effects of this treatment on the rate of cell growth. In the first series of experiments antibody to IFN-β failed to affect cell growth in TNF-treated or control cells. Subsequently, we changed the experimental conditions. Rather than adding TNF to freshly seeded FS-4 cells, we used one week-old confluent and serum-starved cultures (28). When TNF was added to such cultures in serum-free medium or in medium containing a low concentration of fetal bovine serum (FBS), there was a modest stimulation of cell growth. However, the addition of anti-IFN-β serum under such conditions resulted in a clear enhancement of the growth-stimulating action of TNF (Fig. 2). This result showed indirectly that TNF induced the production of IFN-β in FS-4 cells and that generation of IFN-β was responsible for a suppression of the growth-stimulating action of TNF.

Of interest was the finding that amplification of the growth-stimulating action of TNF by anti-IFN-β was decreased or abolished with increasing FBS concentration in the medium (Fig. 2). At the same time, the mitogenic action of TNF (in the absence of anti-IFN-β) was increased and more sustained with increasing FBS concentrations. These findings, confirmed in another human foreskin cell line (28), suggest that the potentiation of the growth stimulating action of TNF by FBS could be due in part to the inhibitory action of FBS on the antiproliferative action of IFN-β. FBS, of course, is the source of growth factors (including PDGF, EGF, insulin and insulin-like growth factors) and, as already mentioned, growth factors were shown to have an antagonistic effect on the antiproliferative (5) as well as antiviral (6) activities of interferons.

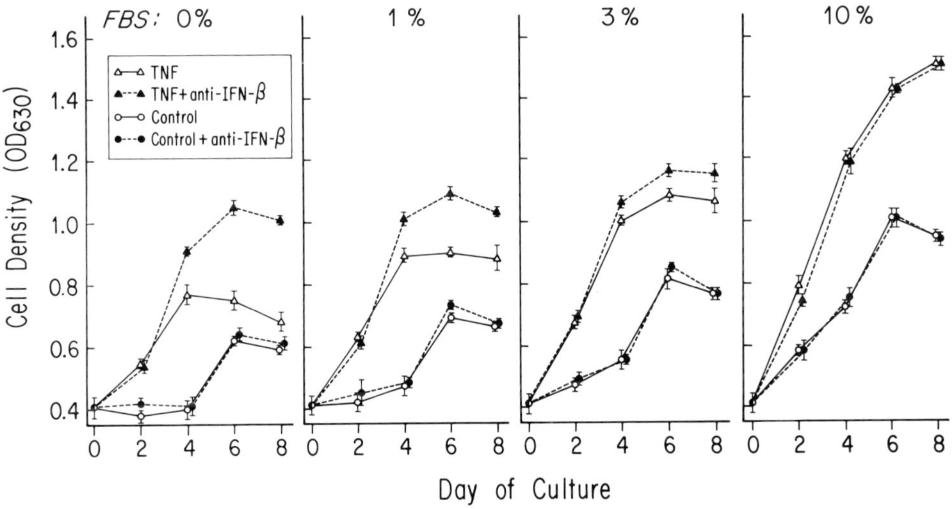

FIGURE 2. Growth-stimulating effect of TNF in FS-4 cells and its amplification by antiserum to IFN-β. Growth medium was removed from 7-day-old confluent cultures of FS-4 cells in 96-well plates and replaced with fresh medium containing 0 %, 1 %, 3 %, or 10 % FBS as indicated. Half of the cultures in each group were incubated with TNF (10 ng/ml). In addition, half of the TNF-treated and half of control cultures received calf antiserum to IFN-β (0.5 %); the other halves of the cultures were incubated with the same concentration of control calf serum. Day 0 refers to the onset of the treatments. Cultures were then incubated without medium change for 2, 4, 6, or 8 days. Cell density was determined by dye uptake. (From ref. 28, with permission).

Parenthetically, our laboratory described an inhibitory action of FBS (29) and of sera from several other animal species (30) on the antiviral action of interferon in chick cells almost twenty years ago. The inhibitor present in FBS was nondialyzable, precipitated by ammonium sulfate, resistant to heating at 60°C and resistant to trypsin (30). Some, but not all, of the inhibitory action of FBS could be mimicked by insulin (31). Whereas these prediluvial studies do not necessarily prove that we discovered the wheel, they support the idea that the inhibitory action of serum on interferon action is related to a similar inhibition seen with growth factors.

It is interesting to note that, despite its growth factor activity, TNF did not inhibit the antiviral action of IFN-β in FS-4 cells (20). In contrast, IL-1α, IL-1β, PDGF (21) and EGF (20) inhibited IFN-β action under the same conditions. These results may explain why some growth factors (e.g., IL-1 or PDGF) fail to induce an antiviral state despite their ability to induce IFN-β (see below).

5. INTERFERON-β AS A MEDIATOR OF THE ANTIVIRAL ACTION OF TNF

In the process of studying the cytotoxic and mitogenic activities of TNF in various types of cell culture, Kohase et al. (28) observed that TNF had antiviral activity in human diploid fibroblasts. Antiviral activity against encephalomyocarditis (EMC) virus in FS-4 foreskin fibroblasts (and in other foreskin cell lines) was demonstrable in confluent, aged cultures, but not in freshly seeded cultures or in cells that had undergone a recent medium change. TNF also protected FS-4 cells from the unrelated herpes simplex virus, suggesting an interferon-like mechanism. To determine whether TNF acted as an interferon inducer, we used a trick well known in the interferon trade. We examined the antiviral action of TNF in the presence of antibodies specific for IFN-α, -β or -γ. Anti-IFN-β sera from two different sources, but not the anti-IFN-α or -γ sera, abolished the antiviral activity; hence, it was clear that the antiviral effect of TNF in FS-4 cells was mediated by the induction of IFN-β.

Although at the time this work was being carried out it was not yet clear that TNF shared numerous biological activities with IL-1, it was already apparent that the characteristics of the antiviral action of TNF resemble the earlier demonstrated antiviral activity of IL-1 ("22kDa protein") as reported by Van Damme et al. (10,11). As had been the case with IL-1, the antiviral activity of TNF was not readily demonstrable under all conditions of culture, the degree of inhibition of virus replication was relatively modest, and the presence of IFN-β in the culture medium of human fibroblasts exposed to TNF was demonstrable in a highly sensitive bioassay only after concentration of the medium (28).

In order to confirm the similarities in the antiviral activities of TNF and IL-1, Masayoshi Kohase in our laboratory, in collaboration with the laboratory of Pravinkumar B. Sehgal, compared the induction of an antiviral state by recombinant human TNF, IL-1α and IL-1β in the FS-4 fibroblasts and in the MG-63 osteosarcoma cell line earlier employed by the Rega group for their studies of the 22kDa protein. In the MG-63 cells TNF and both forms of IL-1 induced protection against vesicular stomatitis virus (VSV) with very similar characteristics. Antiserum to human IFN-β abolished the ability of all three cytokine preparations to induce an antiviral state in MG-63 cells. However, both forms of IL-1 failed to induce antiviral protection in "aged" FS-4 cells, i.e., conditions under which TNF readily inhibited virus replication (20,21). Furthermore, both forms of IL-1, when added to cultures along with TNF, actually completely inhibited the induction of an antiviral state by TNF. As mentioned earlier, both forms of IL-1 also inhibited the antiviral action of exogenously added IFN-β in FS-4 cells whereas TNF did not have this activity and, in fact, could potentiate the antiviral action of exogenously added IFN-β. Thus, the inability of IL-1 to induce an antiviral state in FS-4 cells and the inhibitory action of IL-1 on the establishment of TNF-induced antiviral activity are likely due to the ability of IL-1 to interfere with the antiviral action of interferon. The mechanism of the inhibitory action of IL-1 on the antiviral activity of interferon is unknown but, as already stated, it is likely related to similar inhibitory activities seen with serum (30), EGF and PDGF (6).

Since the original demonstration of the antiviral action of TNF by Kohase _et al_. (28), other groups have reported inhibitory effects of TNF on virus replication. However, there is some disagreement about the mechanisms mediating the antiviral actions of TNF demonstrated in these experiments. In good agreement with our own data, Mestan _et al_. (32) found a virtually complete reversal of the antiviral action of TNF in the presence of antibodies to human IFN-β. In contrast, Wong and Goeddel (33) concluded that the antiviral action of TNF did not involve interferon as an intermediate. The reduction in virus yield observed by the latter authors might be attributable to a selective lytic effect of TNF on virus-infected cells, a phenomenon described earlier by Aderka _et al_. (34). Selective destruction of virus-infected cells by TNF would result in a decreased virus yield. It is possible that other, yet to be identified, mechanisms might also be involved in the antiviral activities of TNF. In any case, the demonstration that IL-1 and TNF can inhibit virus multiplication suggests that these proteins play a role in the host defense against viruses. This conclusion is further supported by the demonstration that viruses can act as inducers of TNF synthesis (33,35).

6. WHICH INTERFERON-β ?

As explained in the preceding two sections, our conclusion that TNF can induce IFN-β was based on indirect evidence obtained with specific antibodies to human IFN-β. In order to corroborate these results a series of Northern blot hybridization analyses were carried out in the laboratory of P.B. Sehgal at the Rockefeller University. In these studies we failed to obtain any evidence for the induction of IFN-$β_1$ mRNA in FS-4 fibroblasts exposed to TNF (28) or IL-1 (21). However, these studies demonstrated that TNF is a potent inducer of IFN-$β_2$ mRNA in FS-4 fibroblasts. Content _et al_. (12) had demonstrated that natural IL-1 (22kDa protein) induced IFN-$β_2$ mRNA in cultures of MG-63 cells or human fibroblasts. In full agreement with the work of Content _et al_., Kohase _et al_. (21) were able to show that recombinant human IL-1α and IL-1β both act as potent inducers of IFN-$β_2$ mRNA in FS-4 fibroblasts. In addition, Zilberstein _et al_. (36) showed that both TNF and IL-1 induce the secretion of the IFN-$β_2$ protein in cultures of human fibroblasts.

Despite the similarity in their designations, IFN-$β_1$ and -$β_2$ are very different from each other (Table 2). The name "IFN-$β_2$" was coined by Sehgal and Sagar (37) and Weissenbach _et al_. (15) to denote the translation product of a 1.3 kb mRNA produced in human fibroblasts stimulated with poly(I).poly(C) and/or cycloheximide. When injected into _Xenopus laevis_ oocytes this mRNA fraction gave rise to interferon activity neutralized by antisera to purified IFN-β, but distinct from the "authentic" IFN-$β_1$, which is a translation product of a 0.9 kb mRNA. Although the original publications of Weissenbach _et al_. (15) and Sehgal and Sagar (37) did not provide sequence information, both groups of authors believe that the molecules independently designated by them as IFN-β_2 are indeed identical. A cDNA corresponding to the IFN-β_2 mRNA was cloned by Weissenbach et al. (15), but the complete sequence of the IFN-β_2 cDNA and protein were published only very recently (36,38,39). Recent information indicates that IFN-β_2 is a secreted protein with an apparent molecular weight of 21 kDa that is derived from a gene located on human chromosome 7 (40). The IFN-β_2 gene contains at least 4 exons and the mature protein is about 180 amino acids long (36,38,39). Some authors concluded that the sequences of IFN-β_1 and -β_2 share no significant homology (38), whereas others believe that there is significant homology between some segments of the two sequences, perhaps explaining the similarities in antiviral activity and the cross-neutralization by specific polyclonal as well as monoclonal antibodies (13,15,39). It seems approciate to designate IFN-β_2 as a separate interferon family distinct from the IFN-α/β "superfamily" and IFN-γ (Table 3).

The degree of antiviral activity associated with IFN-β_2 has been a matter of some controversy (38). Recently, Zilberstein et al. (36) produced recombinant human IFN-β_2 by transfection of CHO cells with the IFN-β_2 cDNA. This material had antiviral activity in human cells which could be neutralized with various polyclonal sera and a monoclonal antiserum specific for human IFN-β. Antiviral activity of the material produced by Zilberstein et al. at the Weizmann Institute was also detected at the Rega Institute (J. Van Damme and A. Billiau: personal communication) and in my own laboratory. In my hands, not only was antiviral activity demonstrable against vesicular stomatitis and EMC viruses in two types of human fibroblasts, but this activity was also neutralized by bovine antiserum prepared in our laboratory against a highly purified human IFN-β preparation as well as by a commercially available monoclonal antibody with neutralizing activity against IFN-β_1. Thus, although the specific antiviral activity of IFN-β_2 is apparently lower than that of IFN-β_1 (36), the fact that IFN-β_2 has interferon activity has now been confirmed by several laboratories.

Is induction of IFN-β_2 responsible for the antiviral activity of TNF in human FS-4 cells ? It is fair to say that direct evidence is still lacking. The conclusion of Kohase et al. (28) that IFN-β_2, and not IFN-β_1, is the likely mediator of antiviral activity was based on the simple fact that induction of IFN-β_2 mRNA by TNF was readily demonstrable. In contrast, no evidence for an induction of IFN-β_1 mRNA could ever be obtained.

TABLE 2. Properties of human IFN-β_2 ("26kDa protein").

Feature	References
Antiviral activity of recombinant IFN-β_2 is neutralized by polyclonal and monoclonal antisera to IFN-β_1	36
Primary structure shows only small degree of homology (or no significant homology at all ?) with IFN-β_1	15,36,38,39
Intron-containing structural gene located on chromosome 7 (vs. chromosome 9 for IFN-β_1)	40
Induced by IL-1, TNF, PDGF, serum, poly(I).poly(C), cycloheximide, IFN-β_1, cell "aging"	12,15,21,28,36
Identical with hybridoma growth factor (HGF) and B cell differentiation factor (BSF-2)	41

TABLE 3. Characteristics of the human interferons.

	IFN-α/β (type I IFN)	IFN-γ (type II or immune IFN)	IFN-β_2
Subfamilies	IFN-α_I (≥ 14 subtypes) IFN-α_{II} IFN-β	-	-
Location of structural gene(s)	chromosome 9	chromosome 12	chromosome 7
Introns in genes	none	3	≥ 3
Location of receptor gene	chromosome 21	chromosome 6	chromosome 21(?)
Major biological functions	antiviral, induction of class I HLA, regulation of cell growth and differentiation	immunoregulation (monocyte/macrophase functions), induction of class II HLA	immunoregulation (B cells), regulation of cell growth and differentiation(?)

Since the IFN-β_2 protein itself is also produced in TNF-stimulated human fibroblasts (36), and since most investigators now appear to agree that IFN-β_2 has some antiviral activity, the conclusion that it is likely responsible for the TNF-induced antiviral protection is not too far-fetched.

One potential argument against the role of IFN-β_2 is our inability to produce an antiviral state in FS-4 cells with IL-1, despite a potent induction of IFN-β_2 mRNA (20,21). We believe that this apparent discrepancy can be explained by the inhibitory effect of IL-1 on the antiviral activity of interferon demonstrated by Kohase *et al*. (20,21, and unpublished data). As already mentioned earlier in this article, the ability of IL-1 to inhibit interferon action in FS-4 cells would likely override the ability of the endogenously produced IFN-β_2 to inhibit virus replication.

Is autocrine IFN-β_2 likely to be part of the negative feedback loop of growth factor activity, as suggested by Kohase *et al*. (28) ? I should point out that our approach to this question was very much influenced by the work of M. Revel, A. Kimchi and their colleagues of the Weizmann Institute, who have long been advocating a role for beta interferon in growth arrest and terminal differentiation (7,15,16). Direct evidence that IFN-β_2 can exert an antiproliferative effect in human fibroblasts (or other cells) has not yet been published. In fact, as a result of very recent findings at the Rega Institute, IFN-β_2 is now known to possess potent growth factor activity for B cell hybridomas and plasmacytomas (41). However, there are examples of molecules that can act as growth inhibitors in some cells and growth factors in others. The first such example that comes to mind is, of course, TNF. Another example is transforming growth factor-beta. On balance, I still believe that IFN-β_2 is likely to act as growth inhibitor in TNF-stimulated fibroblasts. Admittedly, this view would be strengthened by a clear demonstration of antiproliferative activity of purified recombinant IFN-β_2 in fibroblast cultures. It has not been ruled out that TNF can induce biologically relevant quantities of IFN-β_1 that escape detection by Northern blot hybridization but are sufficient to produce an antiviral or antiproliferative effect. Admittedly, TNF might be inducing other, yet to be identified, species of IFN-β (16,37), that also could have antiproliferative activity. Given the rapid progress in the elucidation of functions associated with IFN-β_2 in the last month and weeks, answers to at least some of these open questions may be known very shortly.

7. AFTERTHOUGHT

I first visited Leuven in 1962. At that time the Rega Institute was already emerging as a major center of interferon research. In the intervening quarter of a century Prof. De Somer and his colleagues in Leuven often pursued similar scientific goals as we did on a smaller scale in New York. There were times when we disagreed about the interpretation of our experiments. One question on which our points of view have differed recently is whether certain activities may or may not be mediated by IFN-β_2 (also known as 26kDa protein/hybridoma growth factor/BSF-2). Nonetheless we never allowed competition or minor differences in opinion stand in the way of complete scientific candor and we would always go out of our ways to give each other credit whenever credit is due. I consider this to be one of the important legacies of the De Somer era. Together with my old and new friends at the Rega Institute we shall see to it that this practice continues.

ACKNOWLEDGMENTS

I gratefully acknowledge the contribution of many individuals to the work described in this article. Participants from my laboratory included Masayoshi Kohase, Vito Palombella, Junming Le, Dorothy DeStefano and Masafumi Tsujimoto. Essential parts of our work were done in collaboration with Pravinkumar B. Sehgal and Lester T. May of The Rockefeller-University. I also thank Teruhisa Noguchi and his colleagues at the Suntory Institute for Biomedical Research in Osaka, Japan for the supply of precious reagents and support, and Cristy Minton and Karen Shipp for preparation of the manuscript.

REFERENCES

1. Paucker K, Cantell K, Henle W. Virology 17:324-334, 1962.
2. Gresser I, Brouty-Boyé D, Thomas M-T, Macieira-Coelho A. Proc. Natl. Acad. Sci. USA 66:1052-1058, 1970.
3. Sokawa Y, Watanabe Y, Kawade Y. Nature 268:236-238.
4. Inglot AD, Oleszak E, Kisielow B. Arch. Virol. 63:291-296, 1980.
5. Sreevalsan T, Rozengurt E, Taylor-Papadimitriou J, Burchell J. J. Cell. Physiol. 104:1-9, 1980.
6. Oleszak E, Inglot AD. J. Interferon Res. 1:37-48, 1980.
7. Friedman-Einat M, Revel M, Kimchi A. Mol. Cell. Biol. 2:1472-1480, 1982.
8. Creasey AA, Eppstein DA, Marsh YV, Khan Z, Merigan T.C. Mol. Cell. Biol. 3:780-786, 1983.
9. Moore RN, Larsen HS, Horohov DW, Rouse BT. Science 223:178-181, 1984.
10. Van Damme J, Billiau A, De Ley M, De Somer P. J. Gen. Virol. 64:1819-1822, 1983.
11. Van Damme J, De Ley M, Opdenakker G, Billiau A, De Somer P. Nature 314:266-268, 1985.

12. Content J, De Wit L, Poupart P, Opdenakker G, Van Damme J, Billiau A. Eur. J. Biochem. 152:253-257, 1985.
13. Zullo JN, Cochran BH, Huang AS, Stiles CD. Cell 43:793-800, 1985.
14. Gresser I. Cell. Immunol. 34:406-415, 1977.
15. Weissenbach J, Chernajovsky Y, Zeevi M, Shulman L, Soreq H, Nir U, Wallach D, Perricaudet M, Tiollais P, Revel M. Proc. Natl. Acad. Sci. USA 77:7152-7156, 1980.
16. Resnitzky D, Yarden A, Zipori D, Kimchi A. Cell 46:31-40, 1986.
17. Van Damme J, De Ley M, Claeys H, Billiau A, Vermylen C, De Somer P. Eur. J. Immunol. 11:937-942, 1981.
18. Van Damme J, Opdenakker G, Billiau A, De Somer P, De Wit L, Poupart P, Content J. J. Gen. Virol. 66:693-700, 1985.
19. Oppenheim JJ, Kovacs EJ, Matsushima K, Durum SK. Immunol. Today 7:46-50, 1986.
20. Kohase M, Tsujimoto M, May LT, Sehgal PB, Vilček J. in The Biology of the Interferon System 1986 (Cantell K, Schellekens H, Eds). Martinus Nijhoff, in press, 1987.
21. Kohase M, May LT, Tamm I, Vilček J, Sehgal PB. Cell. Mol. Biol., in press, 1987.
22. Sugarman BJ, Aggarwal BB, Hass PE, Figari IS, Palladino MA Jr, Shepard HM. Science 230:943-945, 1985.
23. Vilček J, Palombella VJ, Henriksen-DeStefano D, Swenson C, Feinman R, Hirai M, Tsujimoto M. J. Exp. Med. 163:632-643, 1986.
24. Palombella VJ, Yamashiro DJ, Maxfield FR, Decker SJ, Vilček J. J. Biol. Chem., in press, 1987.
25. Lee SH, Aggarwal BB, Rinderknecht E, Assisi F, Chiu H. J. Immunol. 133:1083-1086, 1984.
26. Le J, Weinstein D, Gubler U, Vilček J. J. Immunol., in press, 1987.
27. Le J, Vilček J. Lab. Invest., in press, 1987.
28. Kohase M, Henriksen-DeStefano D, May LT, Vilček J, Sehgal PB. Cell 45:659-666, 1986.
29. Vilček J, Lowy DR. Arch. Ges. Virusforsch. 21:253-264, 1967.
30. Rossman TG, Vilček J. Arch. Ges. Virusforsch. 31:18-27, 1970.
31. Vilček J, Rossman TF, Friedman-Kien AE. in L'Interferon. Colloques de l'Institut National de la Santé et de la Recherche Médicale (Paris), pp 243-250, 1970.
32. Mestan J, Digel W, Mittnacht S, Hillen H, Blohm D, Moller A, Jacobsen H, Kirchner H. Nature 323:816-818, 1986.
33. Wong GHW, Goeddel DV. Nature 323:819-822, 1986.
34. Aderka D, Novick D, Hahn T, Fischer DG, Wallach D. Cell. Immunol. 92: 218-225, 1985.
35. Aderka D, Holtman H, Toker L, Hahn T, Wallach D. J. Immunol. 136: 2938-2942, 1986.
36. Zilberstein A, Ruggieri R, Korn JN, Revel M. EMBO J. 5:2529-2537, 1986.
37. Sehgal PB, Sagar AD. Nature 288:95-97, 1980.
38. Haegeman G, Content J, Volckaert G, Derynck R, Tavernier J, Fiers W. Eur. J. Biochem. 159:625-632, 1986.
39. May LT, Helfgott DC, Sehgal PB. Proc. Natl. Acad. Sci. USA 83:8957-8961, 1986.
40. Sehgal PB, Zilberstein A, Ruggieri M-R, May LT, Ferguson-Smith A, Slate DL, Revel M, Ruddle FH. Proc. Natl. Acad. Sci. USA 83:5219-5222, 1986.
41. Van Damme J, Cayphas S, Opdenakker C, Billiau A, Van Snick J. Eur. J. Immunol., in press, 1987.

C.A. Dinarello

THE ROLE OF INTERLEUKIN-1 IN MEDIATING HOST DEFENSE MECHANISMS

C.A. DINARELLO
Department of Medicine, Division of Geographic Medicine and Infectious Diseases, Tufts University School of Medicine and New England Medical Center Hospitals, 750 Washington Street, Boston, Massachusetts 02111, U.S.A.

1. INTRODUCTION

Throughout the animal kingdom, one observes the elaborate measures and strategies which various species have developed in order to fight life-threatening microbial invasion. Clearly, before the antibiotic era, the outcome of this struggle between host and microbe was determined by the ability of each to develop evasion action. Mammals, in particular, have developed a large number of varied responses wich both signal the onset of microbial invasion and trigger host defense mechanisms. Some of the defenses include behavioral changes. For example, animals will often stop eating, reduce activity, conserve energy expenditure, and increase sleep-time when injured or infected. Despite these and substantial metabolic and hematological responses, the host is still highly vulnerable to infection. Since recorded time, efforts have been targeted at augmentation of various defense mechanisms in an attempt to shift the balance in favor of the host. Prof. P. De Somer's contributions towards shifting this balance in the direction of the host scans years of great achievements. More recently, the Rega Institute has carried on this tradition by uncovering fundamental mechanisms by which the host accomplishes its defense. One of Dr. De Somer's principal concepts for the Rega Institute was as a center to study host defense mechanisms. This has been unquestionably fulfilled by its leading role in the field of interferons, interleukins and other mediators of host defense. In this paper, one of the interleukins, interleukin-1 (IL-1), will be discussed as one of the body's key mediators of responses to microbial invasion, inflammation, immunological reaction and injury. Evidence will be presented which implicates IL-1 as one of the first and most prominent molecules the body synthesizes in response to infection and injury. Its biological effects are manifested in nearly every tissue and organ system. In addition, an attempt will be made to integrate the various biological activities of IL-1 into a pattern of augmenting both early and late components of host defenses.

2. PHYSICAL NATURE OF IL-1

Beginning in the 1940's, researchers had described a substance from acute inflammatory tissue which when injected into rabbits or humans would produce fever (reviewed in 1). This material was a small protein (mol. wt.: 10-20,000) and was called endogenous pyrogen. It was later shown that endogenous pyrogen was not pre-formed in cells but rather synthesized *de novo* when phagocytic cells, primarily macrophages, were stimulated with small (pg/ml) concentrations of endotoxins or by phagocytosis of a few (3-5) bacteria. Attempts to purify the endogenous pyrogen activity in inflammatory fluids or stimulated macrophage culture supernates proved difficult. Subnanogram amounts of protein seemingly possessed potent biological activity and during multiple purification steps, loss of activity was common. Nevertheless, purification procedures were published in the 1970's (2,3) and it was established that 30-50 ng/kg of homogeneous endogenous pyrogen produced monophasic fever in rabbits. No amino acid sequence for endogenous pyrogen activity was known at that time.

An important consequence of the development of purification methods was the demonstration that endogenous pyrogen did more than cause fever. Kampschmidt and his co-workers showed that endogenous pyrogen activity copurified with a substance called leukocyte endogenous mediator and that this material induced hepatic acute phase protein synthesis, caused a decrease in plasma iron and zinc levels and produced neutrophilia (4). Immunologists had also become interested in macrophage products and in 1972, lymphocyte activating factor was described as a 10-20,000 dalton protein which augmented T-cell responses to mitogens and antigens (5). The early reports on the chemical characteristics of lymphocyte activating factor suggested several similarities to endogenous pyrogen and in 1979 and 1980, the first reports were published on the ability of purified endogenous pyrogens to act as lymphocyte activating factor (reviewed in 1). The name lymphocyte activating factor was changed to interleukin-1 (IL-1) and included the originally described endogenous pyrogen and leukocytic endogenous mediator activities.

This concept that a molecule which acted on the brain to produce fever or the liver to induce acute phase protein synthesis was the same molecule which stimulated T-cells and B-cells was not easily accepted by immunologists. However, preparations of endogenous pyrogens which were homogeneous by SDS-PAGE induced T-cell proliferation (6,7). The dilemma was resolved by the molecular cloning of IL-1. Two IL-1 forms have been cloned, IL-1-beta from human monocytes (8) and IL-1-alpha from a murine macrophage line (9). The two forms had originally been identified as pI 7 (beta) and pI 5 (alpha) by isoelectric focusing. Homologues of IL-1-beta and -alpha have been found in other species. The cDNA's for IL-1 resulted in two important advances: 1) the entire amino acid sequence was deduced from the nucleotide sequence; and 2) expression of the cDNA's resulted in ample quantities of recombinant IL-1's for biological experiments.

The two forms of IL-1 are initially synthesized as 31,000 precursor polypeptides and share only small stretches of amino acid homology (26 % in the case of human IL-1). Neither form contains a signal peptide sequence which would indicate a cleavage site for the N-terminus. But amino acid sequence analysis of IL-1 which was purified as a 17,500 dalton molecule from human blood mononuclear cells revealed that the mature peptide N-terminus was at position 117 (alanine) of the 269 amino acid precursor (10). The N-terminus of the IL-1-alpha was in a similar position. There is now evidence that elastase is one of the enzymes which cleaves the IL-1 precursor at the alanine position yielding the mature peptide.

3. BIOLOGICAL PROPERTIES OF RECOMBINANT IL-1

The expression of IL-1 (alpha and beta) has been accomplished by several groups and, in general, there does not seem to be any difference in the spectrum of biological activities of either form. However, IL-1-beta is vulnerable to oxidation and thus, biological specific activities of IL-1-beta can be lower than IL-1-alpha which is highly stable. This is supported by experiments on naturally occurring IL-1 pI 7 (beta) and pI 5 (alpha). If cara is taken to express and purify recombinant IL-1-beta without oxidative conditions, the IL-1-beta form is equipotent with the alpha form in a variety of *in vivo* and *in vitro* assays.

One of the first and important experiments to be performed employing recombinant IL-1's was to confirm the ability of these molecules to induce the same changes or possess the same biological properties as had been reported for naturally occurring IL-1's. For the most part, the recombinant IL-1's have confirmed the previous studies (reviewed in 11). Both recombinant human IL-1-beta and -alpha augment T-cell responses to antigens or mitogens and cause fever in rabbits and other species (12). Both forms induce hepatic acute phase protein synthesis and cause sleep. However, some biological properties claimed to be due to natural IL-1 have not been demonstrated with either recombinant forms. These are the ability of IL-1 to cause neutrophil enzyme degranulation and muscle proteolysis *in vitro*. It appears that these apparent biological properties of IL-1 may have been due to contaminating protein(s) or due to the ability of IL-1 to augment the biological property of another protein. One likely candidate for contaminating protein is cachectin, also known as tumor necrosis factor (discussed below).

Although recombinant IL-1's do not induce neutrophil degranulation, they do induce eosinophil degranulation and histamine release from human blood basophils. In addition to confirming the biological activities of natural IL-1's, the recombinant molecules have been used to demonstrate new biological properties. These include increases in ACTH and corticosteroid release, increased sodium excretion, inhibiting of lipoprotein lipase, increased neutrophil and monocyte thromboxane A_2 synthesis, and the ability of IL-1 to cause hemodynamic shock (13).

4. SYSTEMIC VS. LOCAL EFFECTS OF IL-1

As shown in Table 1, IL-1 induces a variety of changes when injected into the host. These types of studies are carried out by injecting either intravenously or intraperitoneally recombinant or purified IL-1's into experimental animals, and in many cases, these changes mimic the animal's response to an injection of endotoxins or other microbial toxin or product. To date, these include fever, increased ACTH, corticosterone, slow wave sleep, insulin, copper levels, hepatic acute phase protein synthesis and blood neutrophils (late). *In vivo*, IL-1 decreases circulating neutrophils (early), lymphocytes and monocytes (early), systemic arterial pressure, systemic vascular resistance, serum albumin, iron, zinc and cytochrome P-450 enzyme activity. When injected intradermally, IL-1 induces an intense neutrophilic infiltration and can substitute for endotoxin in the local Swartzman response. These systemic effects ar probably the "tip of the iceberg" in that no systematic study has yet to be made on the metabolic, endocrinologic, hematologic and nervous system changes induced by a systemic injection of IL-1. However, those changes that have been reported are not due to "industrial" doses of IL-1. In fact, the half-life of IL-1 in rabbits after an intravenous injection is 5 minutes. The plasma level of IL-1 45 minutes after an intravenous injection which results in a 1°C fever is approximately 100 pg/ml and this concentration is similar to the *in vitro* concentration for one half-maximal lymphocyte activating unit/ml (12).

When compared to whole animal studies, there have been considerably more studies on the *in vitro* effects of IL-1. Some of the *in vitro* effects are probably relevant to systemic responses. For example, the effects of IL-1 on endothelial cell arachidonic acid metabolism and procoagulant activity *in vitro* likely explain the shock-like state that IL-1 produces *in vivo*. In addition, the ability of IL-1 to induce various lymphokines *in vitro* likely occurs *in vivo*, but it is difficult to demonstrate circulating levels of interferons (beta or gamma) and interleukins (2 and 3) un-

TABLE 1. Systemic effects of recombinant human IL-1.

Central nervous system	Metabolic
Fever	Hypozincemia, Hypoferremia
Brain PGE-2 synthesis	Decreased cytochrome P450 enzyme
Increased ACTH	Increased acute phase proteins
Increased corticosteroid	Decreased albumin synthesis
Increased slow wave sleep	Increased survival rate in mice
(Decreased appetite)	Increased bacterial clearance
	Increased (at high dose decreased) insulin production
	Lipoprotein lipase inhibition
	Increased sodium excretion
Hematologic	**Vascular wall**
Neutrophilia	Increased leukocyte adherence
Lymphopenia	Increased PGI and PGE synthesis
Neutrophil TxA	Increased platelet activating factor
Tumor necrosis	Increased procoagulant activity
Bone marrow release	Increased plasminogen activator inhibitor
Hemodynamic effects	
Hypotension	
Decreased systemic vascular resistance	
Decreased central venous pressure	
Increased cardiac output	
Increased heart rate	
Decreased blood pH	

less large doses are given. Some human studies, however, have employed large amounts of recombinant IL-2 and interferon-gamma and some of the systemic effects of these therapies may be due to IL-1 production.

An approach to interpreting the large body of evidence for the multiple biological effects of IL-1 is to view local production and action as the "autocrine" action of IL-1 and the systemic effect as the "hormonal" property. At present, it seems that autocrine effects of IL-1 predominate in some diseases, whereas systemic effects are characteristic of IL-1 produced as a result of toxemia, septicemia, widespread tissue damage or intravenous antigenic challenge. What seems increasingly clear is that tissues that produce IL-1 are either themselves the targets of local IL-1 effects or are capable of acting on adjoining tissue. For example, IL-1 produced by macrophages in the lymph nodes likely induces IL-2, IL-2 receptors, IL-3, hybridoma growth factor and IFN-gamma which exert their effects on nearby T-, B- and NK cells. In addition, in a dense cellular response, even low doses of IL-1 probably synergize with IL-2 and IL-4 for enhancement of lymphocyte responses. Table 2 summarizes some immunologic properties of IL-1.

IL-1 produced by microglia and astrocytes in the brain has its effect on local gliosis but may induce no systemic responses. A similar case can be made for IL-1's ability to stimulate fibroblast granulocyte-macrophage colony stimulating factor, interferon-beta (10) and hybridoma growth factor (14). IL-1 produced by keratinocytes in the skin is another example.

TABLE 2. Autocrine effects of interleukin-1.

Attraction of neutrophils, lymphocytes, monocytes (in vivo)
Basophil histamine release
Eosinophil degranulation
Proliferation of dermal fibroblasts
Increased collagen synthesis
Increased collagenase production
Chrondrocyte protease release
Induction of fibroblast and endothelial GM-CSF activity
Production of PGE-2 in dermal and synovial fibroblasts
Increased neutrophil and monocyte thromboxane synthesis
Cytotoxic for human melanoma cells
Cytotoxic for human beta islet cells (insulin-producing)
Cytotoxic for thymocytes
Increased bone resorption (osteoclast activating factor)
Stimulation of fibroblast interferon-beta synthesis
Stimulation of fibroblast hybridoma growth factor synthesis
Keratinocyte proliferation
Mesangial cell proliferation
Gliosis

However, local production and biological activity of IL-1 in the joint space has attracted considerable attention because of a role for IL-1 in the pathogenesis of various joint diseases, such as rheumatoid and osteoarthritis.

The joint macrophage has been shwon to produce IL-1, and IL-1 has been measured in a variety of human arthritides. Special dendritic fibroblasts cultued from human arthritis synovia produce large amounts of collagenase and PGE_2 in response to IL-1 (15) and this local property of IL-1 likely contributes to pain and destructive disease. In addition, IL-1 is a direct stimulator of _de novo_ collagen synthesis and induces fibroblast proliferation. Thus, there is also a role for IL-1 in the process of thickening synovial membrane. IL-1 has been shown to stimulate chondrocyte metalloproteinases and PGE_2 production, and IL-1 induces the resorption of bone in its role as an osteoclast activating factor. Thus, the autocrine effects of IL-1 may play a major role in the pathogenesis of several disease processes in addition to being a prominent response to local infection and injury. Table 3 lists some possible autocrine effects of IL-1.

5. HOW IL-1 MEDIATES HOST DEFENSE MECHANISMS

From the preceding discussion, it seems clear that IL-1 affects many tissues and organ systems. Are these effects beneficial to the host in the struggle to overcome microbial invasion ? IL-1-mediated responses appear to fall into two types : (_1_) acute phase changes such as fever, sleep, ACTH release, effects on vascular tissue, hepatic acute phase protein synthesis, decreased plasma iron and zinc, increased neutrophils and increased interferon and colony stimulating factor production, and (_2_) general immunologic stimulation leading to the development of specific antibodies or specific cytotoxic T-cells which are in charge of the ultimate eradication of the invader.

TABLE 3. Immunological effects of IL-1.

T-cell activation
(IL-2 production, increased IL-2 receptor number or binding)

B-cell activation
(Synergism with B-cell growth factor (BSF-1/IL-4))
(Induction of interferon-beta-2/hybridoma growth factor/BSF-2)

Natural killer cells
(Synergism with IL-2 and IFN for tumor killing)
(Increased natural killer cell-tumor binding)
(Production of IL-1 from NK cells)

Increased lymphokine production
(IL-2, IL-3, GM-CSF, interferon-beta-1, interferon-beta-2, interferon-gamma, leukocyte inhibitory migration factor)

Macrophage cytotoxicity
(Increased IL-1 production)

The acute phase changes are produced within hours after the onset of infection. Fever may play various roles. Elevated body temperature itself reduces the replication rates of some bacteria and several viruses. Moreover, several studies have demonstrated that immunologic responses such as helper T-cell activation, the generation of cytotoxic T-cells, B-cell activation and antibody synthesis are augmented by elevated temperature. Considering the fact that most lymph nodes are located close to the surface of the skin and are at 34-36°C, the rise in temperature due to the pyrogenic effect of IL-1 may play a direct role in host defense. There is also evidence that neutrophil phagocytosis and some oxidative processes are enhanced by elevated temperature. *In vivo* studies have also provided substantial evidence that elevated temperature in fish, reptiles and mammals augments various host defense mechanisms resulting in increased survival (16).

IL-1 increases the synthesis of several proteins in isolated hepatocytes and, at the same time, decreases the synthesis of albumin. Some of the hepatic proteins are synthesized normally and include various antiproteases, several complement components, fibrinogen, haptoglobin, ceruloplasmin and others. In these cases, the action of IL-1 on the hepatocyte is to increase the rate of synthesis of the proteins at the transcriptional level. However, IL-1 also initiates gene expression of new products, generally proteins not synthesized in healthy subjects but only in association with infection or injury or other pathological processes. The IL-1-induced increases in normal hepatic proteins is usually within the range of 2-3 fold, but the IL-1-initiated synthesis of pathological proteins can be 100-1000 fold. Two such proteins, serum amyloid A (SAA) protein and C-reactive protein are classical "acute phase reactants", and markers of disease. SAA contributes to the development of secondary amyloidosis. Other pathological acute phase proteins include, alpha macroglobulin and acid-1-glycoprotein.

Do these hepatic proteins play a role in host defense mechanisms ?

The strongest implication that they serve an important function comes from data which show the presence of acute phase reactants such as the pentaxins SAA and CRP in the invertebrate *Limulus* horseshoe crab. Not only is their presence for 400 million years of evolution an indication of their importance to the host, but, in addition, their structure has been amazingly conserved in that primary structure of the *Limulus* pentaxin is nearly identical to that of the human. Moreover, the starfish coelomocyte, which behaves like a macrophage, synthesizes large amounts of a protein called sea star factor (17), which shares certain similarities with IL-1.

The overwhelming physical property of CRP and SAA is their ability to bind to lipids. In addition, others such as alpha-1-macroglobulin and ceruloplasmin act as oxygen scavengers. Because many of the acute phase proteins are large glycoproteins, they bind to bacterial surfaces and this physical property allows them to act as all-purpose, non-specific opsonins. Acute phase proteins also include a series of anti-proteases and these may play a role in off-setting the action of some bacterial proteases. Finally, the liver is the source of other proteins which bind divalent cations such as iron and zinc. These metalloproteins are usually not secreted but bind plasma iron and zinc and localize these in the liver, spleen and bone marrow. Of major benefit to the host is that the lowered tissue levels of iron and zinc result in reduced microbial replication, particularly at febrile temperatures. It appears that the acute phase proteins served evolution and host defense well enough that many species survive without T-cells, B-cells or the ability to make specific antibody.

The clinical association of increased sleepiness with infectious disease has received clarification and new information from studies on IL-1. The so-called "sleep factor", isolated from the urine of sleep-deprived humans and the cerebrospinal fluid of animals, has been chemically identified as an N-acetylated muramic acid linked to a tetra-peptide structure, similar to the peptidoglycan units of muramic acid-tetrapeptide found in all bacterial cell walls (18). Sleep factor purified from either human urine or peritoneal dialysis fluid is, like other muramyl peptides, an inducer of IL-1 (19). Like sleep factor, both natural and recombinant IL-1 induce increased slow wave sleep when injected either intravenously or intracerebroventricularly. The onset of IL-1-mediated increases in slow wave sleep is considerably faster than that of sleep factor, and there is evidence that sleep factor induces slow wave sleep *via* the production of IL-1 either peripherally or from astrocytes and microglia within the central nervous system. The benefit to the host of increased sleep may be one of conserved energy and metabolic resources at a time when these are needed to fight infection. Therefore, sleep and decreased appetite, also an IL-1-induced change associated with acute phase responses, may be considered part of the behavioral alterations serving a host defense mechanisms. The decrease in appetite would reduce the desire to seek food with its attendant dangers for much of the animal world. It is unclear how effective reduced food intake would be to specific host defense mechanisms. Nevertheless, we are reminded of Pasteur's comment that during infection, the host attempts to "starve out" the microbe as a mechanism of resistance.

The second phase of IL-1's role in augmenting host defense mechanisms is tied to its immunostimulatory properties. IL-1 stimulates both T-cells and B-cells. T-cells are involved in various host defense functions by one of the following mechanisms : (*1*) helper T-cells which activate B-cells and lead to antibody production; (*2*) cytotoxic T-cells which identify and kill virus-infected cells and tumor cells; and (*3*) the production of T-cell products, or "lymphokines". IL-1 plays a role in all three. There is ample evidence that IL-1, in the presence of mitogen or antigen, will increase T-cell production of IL-2 as well as IL-2 receptors. In addition, IL-1 stimulates T-cell production of IFN-gamma, B-cell growth factor (probably IL-4) and colony stimulating factors. IL-1 is also a potent augmenting factor for IL-4 (formally B-cell stimulating factor-1). In each of these roles of T-cell activation or lymphokine production, IL-1 seems to "augment" the response. There is an unresolved question whether IL-1 is absolutely needed for T-cell activation. Experiments which show T-cell independence for IL-1 employ T-cell hybridomas which may be unique (20). In addition, since membrane bound IL-1 functions in T-cell activation (21), one need not to show a role for soluble IL-1 in T-cell activation. Some investigators have shown that soluble IL-1 functions in a system of macrophage-depleted human peripheral blood T-cell activation by anti-T3 monoclonal antibody attached to Sepharose 4B beads (22). In this experimetnal design, the Sepharose-bound anti-T3 may act as a surrogate macrophage, and soluble IL-1 is provided exogenously. The result is increased production of IL-2 and IL-2 receptors. However, such a system is open to question since the presence of some macrophages and B-cell sources of antigen presentation or membrane bound IL-1 still cannot be ruled out. The final answer of whether there is an absolute requirement for IL-1 in T-cell activation remains unresolved. However, there is no question that IL-1 augments T-cell functions. In this role, whether an absolute requirement or not, it functions as an adjuvant.

IL-1 activates B-cells and contributes to formation of antibody. The first experiment to show the critical role for IL-1 in B-cell activation was performed with anti-IL-1 (23), then called anti-leukocytic pyrogen, since it was prepared and studied using the pyrogenic activity of IL-1. Adding this antibody early to human peripheral blood mononuclear cells stimulated with pokeweed mitogen completely prevented B-cell activation and subsequent antibody formation. Later, it was shown that IL-1 synergized with various B-cell growth and differentiation factors (24,25). Recently, recombinant IL-1 has been shown to synergize with recombinant B-cell stimulating factor-1 (also known as IL-4). Therefore, we can rule out the fact that the effects of natural IL-1 or B-cell growth factor were due to other molecules. However, it is unclear whether these B-cell augmenting activities of IL-1 are the result of a direct action of the IL-1 or the result of IL-1's ability to induce other B-cell stimulating factors, including IL-2, since IL-2 also activates B-cells.

Recent evidence demonstrates that IL-1 induces the production of another B-cell stimulating factor. This factor was originally described as a 26 kilodalton protein from IL-1-stimulated fibroblasts which had hybridoma growth factor activity. The 26 kD protein also had the ability to stimulate plasmacytoma cell growth. The N-terminal sequence of the homogeneous 26 kD factor matches that of an IFN-beta-2 (14). Others also reported on a B-cell stimulating factor (BSF-2) which is the same molecule as the 26 K IFN-beta-2 (96). Thus, it seems that IL-1's role in augmenting B-cell function and ultimately leading to the production of protective or antimicrobial antibodies may be achieved through its ability to induce B-cell stimulating factors. However, this does not rule out a role for IL-1 in augmenting BSFs activity or upregulating their receptors.

IL-1 also plays a role in host defense against tumor cells. Several reports have demonstrated that natural killer cells, also called large granular lymphocytes, produce IL-1 (27). There is experimental evidence that IL-1 increases the binding of natural killer cells to tumor targets and that tumor cells induce IL-1 from natural killer cells (28). In addition, natural killer cells from patients with large tumor burdens produce significantly less IL-1 and have decreased killing ability than cells from healthy individuals. However when incubated with exogenous IL-1, the impaired natural killer cells are restored. Exogenous IL-1 also augments the binding of natural killer cells from healthy donors to tumor targets. Although IL-1 added exogenously does not enhance natural killer cell activity as does IL-2 or IFN-alpha, it does synergize with these cytokines and result in enhanced cytotoxicity by large granular lymphocytes. Since IL-1 induces IFN-beta (29) and IFN synergizes with the effects of IL-1 on natural killer cells, one could view both mechanisms as an efficient aspect of host defense against tumors. However, unlike T and B cell responses which are enhanced by IL-1 at febrile temperatures, the effect of IL-1 on natural killer cells is reduced at febrile temperature (30).

A new and potentially important role for IL-1 in host defense against tumors is the observation that during IL-2-induced generation of lymphokine-activated killer T-cells (LAK cells), IL-1 is produced by both monocytes and natural killer cells (31). To this end, high concentrations (1000 units/ml) of IL-2 are required, and it is unlikely that IL-2 induction of IL-1 occurs under normal circumstances of host challenges from disease. The observation is consistent with IL-2 receptors on monocytes. However, the IL-1 which is produced as a result of high dose IL-2 may help explain the toxicity observed with high dose IL-2 and LAK cell therapy. This also includes the presence of acute phase proteins and the hemodynamic shock.

The ability of IL-1 to activate the vascular endothelium may explain the toxicity of IL-2 or LAK cell therapy. We have recently shown that rabbits injected with recombinant IL-1-alpha or -beta manifest the typical hemodynamic changes characteristic of septic shock (13). IL-1 (5 μg/kg) induces a sudden and dramatic fall in arterial blood pressure, systemic vascular resistance and central venous pressure, whereas a compensatory increase in cardiac output and heart rate occurs. During this time there is also a massive leukopenia (70 % reduction) and thrombocytopenia (40 % reduction). The IL-1 induced shock-like state is averted by a previous intravenous injection of a cyclooxygenase inhibitor and is reversed by this inhibitor when injected during the hypotensive state. These observations are consistent with a fundamental property of IL-1 : its ability to induce rapid increases in cyclooxygenase products, particularly prostaglandin E_2. In fact, the onset and maximum effect of intravenous bolus injection of IL-1 on the hemodynamic parameters is nearly identical with that observed for fever (12) and neutropenia (29).

6. COMPARISON BETWEEN IL-1 AND CACHECTIN/TUMOR NECROSIS FACTOR

Another monocyte/macrophage product, synthesized and released as a consequence of microbial stimulation, is tumor necrosis factor (TNF). TNF is identical to a substance described as cachectin which is a key mediator of the "cachexin" observed in animals with parasitic disease. The subject of TNF/cachectin has recently been reviewed (32). The biological properties of TNF/cachectin share remarkable similarities with those of IL-1, with the notable exception that TNF/cachectin, at concentrations which induce acute phase changes in many tissues, has no immunostimulatory effects. However, nearly every biological property of IL-1 has been observed with TNF/cachectin. These include fever, the induction of PGE_2 and collagenase from a variety of tissues, bone resorption, inhibition of lipoprotein lipase, increases in hepatic acute phase reactants and a decrease in albumin synthesis. Slow wave sleep is also observed following the injection of TNF or IL-1. Both molecules also induce fibroblast proliferation and new collagen synthesis. The cytotoxic activity of TNF/cachectin differs from that of IL-1. For example, IL-1 is cytotoxic for the beta cells of the pancreatic islets of Langerhans and also for human melanoma cells; but, in the same assays, TNF/cachectin has no effect. Likewise, IL-1 is inactive on a variety of tumor cell targets for which TNF is a potent cytotoxin. In fact, the cell line most often used in cytotoxic assay for TNF, L929 fibroblasts, is unaffected by IL-1.

In vivo, IL-1 and TNF/cachectin induce fever by their direct ability to induce hypothalamic PGE_2. In addition to fever, these cytokines will result in hypotension, leukopenia and local tissue necrosis. On a weight basis, rabbits are equally sensitive to the shock-inducing properties of IL-1 and TNF/cachectin. These responses likely reflect the effects of these two cytokines on the vascular endothelium. Both IL-1 and TNF/cachectin stimulate PGI_2, PGE_2 and platelet activating factor production from cultured endothelium. In addition, both cytokines stimulate procoagulant activity, leukocyte adherence and plasminogen activator inhibitor action on these cells. Despite these similarities, receptors for TNF/cachectin have been shown to be specific and distinct, and receptor binding is unaffected by IL-1. Thus, the most likely explanation is that TNF/cachectin and IL-1 stimulate similar intracellular messages and alter cellular metabolism in a similar way. Table 4 lists the biological similarities between TNF/cachectin and IL-1. As noted, both TNF/cachectin and IL-1 stimulate the production of more IL-1. This can be shown *in vivo* and *in vitro* from human monocytes and human endothelium. Substances that activate macrophage cytotoxicity are thought to accomplish this through TNF/cachectin. Furthermore, animal models of *in vivo* tumor necrosis are clearly observed following injection of TNF/cachectin and less so following comparable doses of IL-1.

TABLE 4. Comparison of biological properties between IL-1 and TNF/cachectin.

Biological property	IL-1	TNF/Cachectin
Endogenous pyrogen fever	+	+
Slow wave sleep	+	+
Hemodynamic shock	+	+
Increased hepatic acute phase protein synthesis	+	+
Decreased albumin synthesis	+	+
Activation of endothelium	+	+
Decreased lipoprotein lipase	+	+
Decreased cytochrome p450	+	+
Decreased plasma Fe/Zn	+	+
Increased fibroblast proliferation	+	+
Increased synovial cell collagenase and PGE	+	+
Induction of IL-1	+	+
T/B cell activation	+	-

At present, there are no data to suggest that IL-1 (alpha or beta) and TNF are related at the amino acid level or even secondary or tertiary structural level. However, it does seem clear that the cytokines are biologically related molecules. One can only speculate that acute phase responses such as fever, hepatic protein synthesis, decreases in plasma Fe/Zn, *et alia*, were of such importance to host survival that nature imparted these responses to more than a single molecule.

7. PRODUCTION OF IL-1

It has been proposed that IL-1 is important to the host's ability to fight an invading microbe. This concept is supported by the fact that a considerable amount of the polyadenylated RNA found in macrophages following stimulation codes for IL-1. As much as 2-5 % of the total poly A is IL-1-beta. The predominance of the IL-1 suggests that the host requires a significant amount of IL-1 at a critical time following the onset of infection.

Recently, we have developed a radioimmunoassay specific for human IL-1-beta. We have determined the levels of IL-1 produced by human blood monocytes stimulated with heat-killed *Staphylococcus albus* or endotoxin using this specific assay. The IL-1 activity as measured by this method is free of interferring substances (both enhancing as well as inhibiting) and hence reflects the amount of IL-1 as antigen. In general, we have found that human blood monocytes produce approximately 5-10 ng/million monocytes over an 8 hour period. The intracellular IL-1 levels are high after 4 hours but these are reduced as the extracellular levels increase with time. The amount of IL-1-alpha has not been determined. Thus these results are consistent with the prominence of IL-1-beta mRNA in a stimulated monocyte and suggests that IL-1 is a major product of these cells. The conclusion is that IL-1 production and function are linked to host defense mechanisms not only because of its multiple biological activities but also because the amount made is hardly an epi-phenomenon of cell activation but rather a major product of the cell.

7.1. Fixed macrophagic cells

These cells are often located in strategic blood-filtering organs and are involved in primary phagocytic defense mechanisms for invading microorganisms. For example, there are fixed macrophages lining the alveolar space, in the lamina propria and in the dermis. Fixed macrophagic cells also make up the lining of blood-filtering organs, for example, the Kuppfer cells of the liver, and the splenic sinusoidal cells and lymph node macrophages. Peritoneal macrophages are also part of the primary phagocytic defense mechanism since these cells would be involved with microorganisms derived from intestinal perforations. Several studies have shown that peritoneal, splenic, and alveolar cells produce IL-1 when stimulated *in vitro*. The different methods used to remove and purify these cells from other tissue cells may affect IL-1 production. Most studies involving IL-1 production from peritoneal cells employ agents such as oil or thioglycolate and these agents may not only raise the total number of peritoneal macrophages in the peritoneal cavity, but may also "prime" the macorphage for subsequent stimulation *in vitro*. Several biologically active molecules such as tuftsin, while not an IL-1 inducer, lower the threshold of peritoneal macrophages to exogenous stimuli like endotoxin. Such "priming" may also be part of the mechanism by which pretreatment of blood monocytes with interferons increases endotoxin-induced IL-1 production.

Langerhans' cells isolated from the skin can be separated from keratinocytes by adherence techniques and when stimulated, these cells produce

IL-1 (33). Using a similar method, dentritic cells from human synovium can also be separated from other synovial cells and produce IL-1 upon stimulation.

7.2. Keratinocytes

Several keratinocyte lines of human or mouse origin produce IL-1 without any apparent stimulant but increase production when incubated with endotoxin or toxic shock syndrome toxin-1. Keratinocyte-derived IL-1 exists in multiple weight species (50-15,000) and 3 different isoelectric points of 7.4, 6.1, and 5.1, characteristic of human blood monocytes. Keratinocyte IL-1 also shares with monocyte-derived IL-1 such multiple biological activities as fever induction. Keratinocyte-derived IL-1 is the source of IL-1 found in the cornified epidermis. Rabbits irradiated with ultraviolet light have circulating levels of IL-1 and it seems probable that this may be keratinocyte-derived. It is also speculated that keratinocyte IL-1 production may be involved in a variety of skin lesions, including acute sun-burn, or as a result of ultraviolet B wave phototherapy for certain skin diseases.

7.3. B-cell IL-1

Scala and co-workers were the first to demonstrate that Epstein-Barr virus-infected human B-cell lines produce IL-1 (34). Circulating B-cells isolated from the blood of apparently healthy humans can be stimulated by a variety of agents to produce IL-1. Many of these agents, such as *Staphylococcus aureus* Cowan's strain, anti-mu and endotoxin are used as stimulators of B-cell activity and part of their stimulation may be mediated *via* IL-1 induction. A considerable amount of the B-cell-derived IL-1 may be membrane-bound (21) and this finding is consistent with several studies implicating B-cells as antigen-presenting cells in which IL-1 could not be detected. An explanation for this may be that the IL-1 is bound and stimulates T-cell in its bound form.

7.4. Large granular lymphocytes

These cells are also known as natural killer cells and have been purified from human blood and shown to release IL-1 upon stimulation with endotoxin or tumor cell target (27,31). Patients with large tumor burdens have circulating NK cells which produce little IL-1 and have markedly reduced ability to bind and lyse tumor targets.

7.5. Mesangial cells

Rat mesangial cell lines produce IL-1 and this IL-1 is active in both the thymocyte and fever assay. The rat IL-1 derived from mesangial cells has been subjected to several purification procedures and this mesangial cell IL-1 has multiple molecular weight and charged forms. In addition to stimulating thymocytes and the thermoregulatory center, mesangial cell IL-1 has also been shown to act as a growth factor for itself.

7.6. Astroglia and microglia

Brain astrocytes present antigen and express MHC class II antigens on their cell surface. In addition, these neural cells produce IL-1. Human glioma cell lines produce an IL-1 which is antigenically related to human monocyte IL-1. Mice injected with endotoxin intraperitoneally develop fever for several hours and astrocytes separated from other neural tissue contain large amounts of intracellular IL-1 (35). In fact, astrocyte IL-1 may play an important role in fever induction. IL-1 also stimulates glial cell proliferation and locally produced IL-1 may be involved with central nervous system tissue scarring, also known as gliosis (36). Recent studies have also focused on another central nervous system cell, the microglia, as a potent source of IL-1. The relative production of IL-1 from microglia seems 10-50 greater than that from astrocytes. This is not surprising since microglia are considered brain tissue macrophages.

7.7. IL-1 production from blood vessel cells

The two major blood vessel cells, endothelium and smooth muscle both produce IL-1. Endothelial cell IL-1 has been reported by several groups to produce IL-1 and recently shown at the level of gene expression (37). In general, IL-1 production from endothelial cells is an inducible event and endotoxin is particularly effective when compared to other well-known inducers of monocyte IL-1. Freshly isolated adult human saphenous vein endothelium or newborn umbilical artery endothelium and cell lines derived from these cells have been used.

Smooth muscle cell lines of human origin produce IL-1 which is similar to monocyte IL-1 by immunoprecipitation and mRNA hybridization (38). The findings that both blood vessel-derived cells produce IL-1 is important to the understanding of pathological processes in vasculitis since endothelial cells are activated by IL-1 to increase their adhesiveness for leukocytes and pro-coagulant activity. Therefore, in disease processes such as antigen-antibody complex-mediated diseases or tissue injury and vessel disruption, endothelial and/or smooth muscle IL-1 likely contributes to the progression of the lesions.

7.8. Epithelial cells

Epithelial cell IL-1 production has been reported in two studies : 1) IL-1 from the gingival epithelial cell and 2) IL-1 produced by corneal epithelium. In the gingiva, local IL-1 production clearly plays an important role since IL-1 stimulates bone resorption and osteoclast activating factor has been shown to be the same polypeptide sequence as IL-1-beta. IL-1 produciton by corneal epithelium is also of considerable clinical significance since IL-1 is chemotactic for neutrophils, monocytes and lymphocytes.

ACKNOWLEDGEMENTS

These studies are supported by NIH grant AI 15614. The author wishes to thank Dr. Sheldon M. Wolff for his support in these studies.

REFERENCES

1. Dinarello CA. Rev. Infect. Dis. 6:51-95, 1984.
2. Murphy PA, Chesney J, Wood WB Jr. J. Lab. Clin. Med. 83:310-322, 1974.
3. Dinarello CA, Renfer L, Wolff SM. Proc. Natl. Acad. Sci. USA 74:4624-4627, 1977.
4. Kampschmidt RF. in Physiol & Metab Responses of the Host (Powanda MC, Canonico PG, Eds). Elsevier/North-Holland, Amsterdam, pp 55-74, 1981.
5. Gery I, Waksman BH. J. Exp. Med. 136:143-155, 1972.
6. Dinarello CA, Bernheim HA, Cannon JG, LoPreste G, Warner SJC, Webb AC, Auron PE. Brit. J. Rheumatol. 24 (suppl):59-64, 1985.
7. Hanson DF, Murphy PA. Infect. Immun. 45:483-490, 1984.
8. Auron PE, Webb AC, Rosenwasser LJ, Mucci SF, Rich A, Wolff SM, Dinarello CA. Proc. Natl. Acad. Sci. USA 81:7907-7911, 1984.
9. Lomedico PT, Gubler U, Hellmann CP, Dukovich M, Giri JG, Pan Y-CE, Collier K, Semionow R, Chua AO, Mizel SB. Nature 312:458-462, 1984.
10. Van Damme J, De Ley M, Opdenakker G, Billiau A, De Somer P, Van Beeumen J. Nature 314:266-268, 1985.
11. Dinarello CA. in Year in Immunology 1985-1986. Karger, Basel, vol 2, pp 68-89, 1986.

12. Dinarello CA, Cannon JG, Mier JW, Bernheim HA, LoPreste G, Lynn DL, Love RN, Webb AC, Auron PE, Putney SD. J. Clin. Invest. 77:1734-1739, 1986.
13. Okusawa S, Gelfand JA, Connolly RA, Dinarello CA. Fed. Proc., in press, 1987.
14. Van Damme J, Opdenakker G, Simpson RJ, Rubira MR, Cayphas S, Vink A, Billiau A, Van Snick J. J. Exp. Med., in press, 1987.
15. Dayer J-M, de Rochemonteix B, Burrus B, Demczuk S, Dinarello CA. J. Clin. Invest. 77:1734-1739, 1986.
16. Bernheim HA, Bodel PT, Askenase PW, Atkins E. Brit. J. Exp. Pathol. 59:76-84, 1978.
17. Marcum JA, Levein J, Prendergast RA. Thromb. Haemost. 52:1-3, 1984.
18. Krueger JM, Karaszewski JW, Davenne D, Shoham S. Fed. Proc. 45:2552-2555, 1986.
19. Dinarello CA, Krueger JM. Fed. Proc. 45:2545-2548, 1986.
20. Watts TH, Brian AA, Kappler JW, Marrack P, McConnell HM. Proc. Natl. Acad. Sci. USA 81:7564-7568, 1984.
21. Kurt-Jones EA, Kiely JM, Unanue ER. J. Immunol. 135:1548-1550, 1985.
22. Williams JM, DeLoria D, Hansen JA, Dinarello CA, Loertscher R, Shapiro HM, Strom TB. J. Immunol. 135:2249-2255, 1985.
23. Lipsky PE, Thompson PA, Rosenwasser LJ, Dinarello CA. J. Immunol. 130:2708-2714, 1983.
24. Falkoff RJM, Muraguchi A, Hong J-X, Butler JL, Dinarello CA, Fauci AS. J. Immunol. 131:801-805, 1983.
25. Falkoff RJM, Butler JL, Dinarello CA, Fauci AS. J. Immunol. 133:692-696, 1984.
26. Hirano T, Yasukawa K, Harada H, Taga T, Watanabe Y, Matsuda T, Kashiwamura S, Nakajima K, Koyama K, Iwamatsu A, Tsunasawa S, Sakiyama F, Matsui H, Takahara Y, Taniguchi T, Kishimoto T. Nature 324:73-76, 1986.
27. Scala G, Allavena P, Djeu JY, Kasahara T, Ortaldo JR, Herberman RB, Oppenheim JJ. Nature 309:56-59, 1984.
28. Herman J, Dinarello CA, Kew MC, Rabson AR. J. Immunol. 135:2882-2886, 1985.
29. Van Damme J, Opdenakker G, De Ley M, Heremans H, Billiau A. Clin. Exp. Immunol. 66:303-311, 1986.
30. Dinarello CA, Dempsey RA, Allegretta M, LoPreste G, Dainiak N, Parkinson DR, Mier JW. Cancer Res. 46:6236-6241, 1986.
31. Numerof R, Dinarello CA, Mier JW. Lymphokine Res., in press ,1987.
32. Beutler B, Cerami A. Nature 320:584-588, 1986.
33. Sauder DN. Lymphokine Res. 3:145-151, 1985.
34. Scala G, Kuang YD, Hall RE, Muchmore AV, Oppenheim JJ. J. Exp. Med. 159:1637-1652, 1984.
35. Fontana A, Grob PJ. Lymphokine Res. 3:11-16, 1984.
36. Giulian D, Lachman LB. Science 228:497-499, 1985.
37. Libby P, Ordovas JM, Auger KR, Robbins AH, Birinyi LK, Dinarello CA. Am. J. Pathol. 124:179-186, 1986.
38. Libby P, Ordovas JM, Birinyi LK, Auger KR, Dinarello CA. J. Clin. Invest. 78:1432-1438, 1986.

SECTION IV : TUMOR IMMUNITY

SECTION CHAIRMAN : PROF. M. VANDEPUTTE

M. Vandeputte

INTRODUCTION TO SECTION IV : TUMOR, VIRUS AND TRANSPLANT IMMUNITY

M. VANDEPUTTE
Rega Institute, Katholieke Universiteit Leuven, B-3000 Leuven, Belgium

I have the pleasure to introduce two very good friends of the Rega Institute, Professor A.C. Allison and Professor S. Strober. They have both made major contributions to the field of tumor immunity, a field which is characterized by periods of great expectation as well as periods of profound frustration.

ANTHONY C. ALLISON'S career as a person and scientist is marked by a great degree of versatility. He was born in Kenya and after a couple of years sent to school in Britain where he obtained his M.D. degree at Oxford University. He then went to work with Sir Christopher Andrews, the discoverer of the influenza virus, at the National Institute for Medical Research in Mill Hill. Later he moved to the Clinical Research Center where he became one of the leading scientists attracting many British and foreign investigators including some of our own Institute. During that period we first met and sympathized. He became a regular guest of the Rega Institute and since he shared so many scientific interests with our group, his suggestions were of great help to all of us. Dr. Allison also played an important role as a member of several scientific committees such as the ASLK-CGER Cancer Fund Committee and other Belgian committees for the attribution of scientific degrees or awards.

Driven by his love for the tropics, Dr. A.C. Allison left behind his laboratory in Britain to start a new life in the country where he was born, Kenya, where he was appointed Director of the International Laboratory for Research on Animal Diseases in Nairobi. It was for him a period of great moral challenge and, to a certain extent, if I may say so, a period of desillusion. After a few years in Nairobi he made the difficult decision to move again, this time, to a country at the top of modern science and technology, California, where he was appointed Director of the Division of Biological Sciences at Syntex Corporation where he is still working.

During his many visits to Leuven Dr. Allison was a regular guest at the home of Professor P. De Somer. I have always been struck by the similarity in character and interest of these two men. They shared their joy in a nice meal served with excellent wines, their interest in art and music but also their allergy for self-sufficient and dull people. This, together with a common scientific interest made them good friends.

Dr. Allison's scientific career has covered many different fields and disciplines. He made major contributions to areas as diverse as interferon, complement, transplantation, immunology, virology, lysosomes, tolerance, virus immunity and tumor immunity. The latter two topics will be the subject of Dr. Allison's review lecture.

SAMUEL STROBER is Professor at Stanford University, and presently Head of the Division of Immunology. Born in New York he graduated as medical doctor at the Harvard Medical School in Boston. Interested in immunology, he first went to work in Oxford with Prof. Gowans, one of the most distinguished experts in the biology of lymphocytes. Back in the United States, he spent then several years at N.I.H. in the laboratory of Dr. Lloyd Law, a pioneer in the field of tumor immunity. Having acquired by then a solid reputation as immunologist he was asked to join the staff of the Department of Medicine at Stanford University.

It was at Stanford University that Prof. Strober investigated together with his colleagues S. Slavin and Z. Fuks the immunologic aspects of total lymphoid irradiation, a therapy originally developed in the Department by Professor H.S. Kaplan for the treatment of Hodgkin's disease. His studies led to new applications of the total lymphoid irradiation in medicine, i.e. in transplantation and the treatment of rheumatoid arthritis and autoimmune disease.

These new applications led to a close collaboration between Prof. Strober's group and our own group in Leuven. This collaboration has been further strengthened by regular contacts between the two groups at several international meetings and by the opportunity given to Dr. M. Waer to work in Prof. Strober's Department. We look forward to learning about Prof. Strober's views on the immunological effects of total lymphoid irradiation and its possible clinical applications.

MECHANISMS OF RESISTANCE TO VIRUSES AND TUMORS, AND THE DEVELOPMENT OF VACCINES

A.C. ALLISON
Department of Immunology, Institute of Biological Sciences, Syntex Research, 3401 Hillview Avenue, Palo Alto, California 94304, U.S.A.

1. INTRODUCTION

My research in the area of virus and tumor immunology began in the 1960s, when I was working in the Division of Virology of the National Institute for Medical Research, Mill Hill. The Institute was at that time a major world centre of immunology research, and since most other virologists were becoming molecular biologists I chose (as usual) to do something different : using contemporary immunological methods to define major mechanisms of resistance to infectious agents, especially viruses and virus-induced tumors. At the same time we began research on adjuvants, which had long been used empirically to increase humoral and cell-mediated immune responses to a variety of antigens. The whole program was motivated partly by scientific curiosity about mechanisms of immunity and immunoregulation, and partly by the practical objective of using viral subunits and defined bacterial antigens in vaccines. In this way I hoped to contribute to the development of a new generation of vaccines which would avoid the problems associated with live virus vaccines (severe or persistent infections in some persons) and with crude bacterial vaccines such as that used for pertussis. Now, twenty years later, that goal is at last being realized and I am becoming involved in molecular biology research after all. It is the only way to produce many antigens in quantity and to dissect mechanisms of immunoregulation. One cannot escape destiny.

Our interests soon began to overlap with those of Piet De Somer and his colleagues at the Rega Institute. In fact, my research programme has paralleled in a remarkable way that of Michel Vandeputte. We might have became rivals but instead all became, and remained, friends. As well as being a good microbiologist, Piet was an administrator who combined vision with practical ability and business sense. Building up the Rega Institute, and staffing it with able young investigators, was an example. Piet was also one of the two persons who persuaded the Belgian National Savings Bank to establish a fund to support cancer research. I have been a member of the committee reviewing the grant applications from its inception, and on periodic visits to Belgium observed how the funds allowed several first-rate cancer research teams to develop in that country. I watched with admiration the skill with which Piet, as Rector of the Katholieke Universiteit Leuven (K.U.L.), managed the separation from the Université Catholique de Louvain (U.C.L.). I also recall with pleasure staying in Piet's house, drinking excellent wine from Bordeaux and discussing African and Khmer sculptures as well as those clever Belgian surrealists, Delvaux and Magritte. Piet was above all a civilized gentleman and a loyal friend.

A.C. Allison

2. ANALYSING PRINCIPAL MECHANISMS OF IMMUNITY AGAINST INFECTIOUS AGENTS AND TUMORS

Our general strategy was to compare the effects of viruses and other infectious agents in laboratory animals with intact immune systems and those with more or less selective immunodeficiencies, humoral and cell-mediated, congenital or induced. The findings could, in turn, be correlated with those in humans with immunodeficiencies, recently reviewed by Hayward (1). A remarkably consistent pattern emerged.

The role of antibodies in resistance to bacteria and viruses was reviewed as the original findings accumulated (2,3) and again recently (4,5). Antibodies, provided that they bind antigens with sufficient affinity, can protect by themselves. Examples are the neutralization of bacterial toxins and of infectivity of enteroviruses. However, if antibodies activate complement following antigen binding, their efficacy is often increased. Such complement binding increases opsonization of bacteria for uptake by phagocytic cells and neutralization of herpes simplex virus, for instance. The importance of all three interacting components is illustrated by the increased severity of many bacterial infections in children with defects of any one of these functions (deficiency of IgG or C3, or defects in the capacity of leukocytes to kill bacteria). We were especially interested in the latter, defining the enzyme, cytochrome b_{245}, that generates superoxide in neutrophils and its absence or abnormal function in chronic granulomatous disease (4,6).

We provided evidence that antibodies act synergistically with mononuclear phagocytes in resistance to viruses (7). Other types of antibody-dependent effector cell activity probably also contribute to immunity against infectious agents and tumors (8). In contrast, non-neutralizing antibodies can actually enhance infections of monocytes, as we showed for dengue viruses (9). The obvious conclusion is that to produce protective immunity by vaccines it is, in general, desirable to elicit the formation of antibodies of isotypes that efficiently activate complement and function synergistically with phagocytic and other antibody-dependent effector cells. Examples are the IgG2a isotype in mice and IgG1 and IgG3 isotypes in humans. Selective induction of the formation of antibodies of isotypes giving maximal protection is discussed below.

3. CELL-MEDIATED IMMUNITY (CMI)

T-lymphocytes have many roles, including that of functioning as helpers for antibody formation, as we first showed for viruses (10). We also found that a product of activated T-lymphocytes, interferon-γ (IFN-γ), activates macrophages, thereby increasing resistance to virus multiplication in those cells (11). The classical studies of Zinkernagel and Doherty (12) showed that the cytotoxic subset of T-lymphocytes lyses virus-infected cells in a genetically restricted fashion. When clones of cytotoxic T-cells meet syngeneic virus-infected cells they also release IFN-γ (13), which, perhaps indirectly, has antiviral activity. Activated T-lymphocytes produce TNF-β (lymphotoxin) which can lyse tumor cells.

One of our interests has been the recognition by T-lymphocytes of core antigens of viruses not present in their envelopes. We found that the nucleoprotein antigen of influenza virus is expressed on the surface of infected cells (14). It is a major target of cytotoxic T-lymphocytes (15). Vaccination with cloned core antigens of hepatitis B virus can protect chimpanzees from the infection (16) and core antigens of retroviruses may become important components of vaccines.

These are some of the mechanisms by which CMI is effected against viruses and virus-induced tumors. When we began our analysis very little was known about the role of CMI. Children with severe hypogammaglobulinemia but without obvious defects of CMI were reported to recover from smallpox and BCG vaccinations, whereas children with defective CMI had severe, often fatal, vaccinial and BCG infections. Our laboratory models behaved in the same way. We found (17) that if guinea pigs were injected intradermally with vaccinia virus coated with antibody, CMI could be elicited in the absence of demonstrable circulating antibody. In such immune animals, replication of the virus in the skin was suppressed. Even in the more susceptible rabbits, circulating antibody had little effect on replication of vaccinia virus in the skin. From these observations, and studies of ectromelia virus infections in mice (18), it is clear that CMI plays an important role in terminating poxvirus infections. We also studied the effect of immunosuppression on fibroma virus infections in rabbits (19). In normal rabbits this poxvirus elicits at the site of injections fibromas which eventually regress, whereas in methotrexate-treated adult rabbits we observed lethal, generalized infections. The same was true of fibroma infections of newborn rabbits (20).

Soon after Jacques Miller made his pioneering observation that neonatal thymectomy of mice suppresses CMI, as shown by later skin allograft retention, he came over from the Institute for Cancer Research to tell us about it. We realized that this would be a valuable tool for testing the hypothesis that CMI limits virus-induced oncogenesis. An interesting model is polyoma virus (21). This is a common infection of wild mice, but the natural infection is not associated with tumor development. However, when the virus is injected into newborn mice with an uninfected mother, they subsequently develop many tumors. The question we posed was whether adults are resistant because of effective immune surveillance or because adult tissues are less susceptible to malignant transformation.

We found that neonatally thymectomized mice, infected as adults with polyoma virus, even by contact with infected animals, developed tumors (22). Similar observations were reported by Law (23). Later we followed neonatal thymectomy with repeated administration of anti-lymphocytic serum (ALS), which further suppresses T-cell-mediated immunity. This regimen markedly increased polyoma tumors, even in genetically resistant C57BL mice. All thymectomized ALS-treated mice infected as adults developed tumors (3). Serum from immune donors administered 24 hours after infection markedly reduced the incidence of tumors, presumably because it limited spread of virus. When given 7 days after infection, immune serum had little effect. In contrast, T-lymphocytes from immune donors given as long as 7 weeks after infection reduced the incidence of tumors. Later we (24) and Vandeputte _et al_. (25) found that nude mice infected as adults with polyoma virus developed multiple tumors of hair follicles and other sites. Vandeputte (26,27) found that neonatal thymectomy and ALS increased polyoma oncogenesis in rats. Vandeputte _et al_. (28) also found that inoculation of polyoma virus into the uteri of pregnant rats after removal of the foetuses elicited haemangiosarcomas.

All these observations, as well as analogous observations that we made on induction of leukemias in mice (29), established that T-lymphocyte-mediated immunity does indeed provide a powerful surveillance mechanism against malignancies induced by different types of viruses.

4. SURVEILLANCE AGAINST MALIGNANCIES IN HUMANS

Three human malignancies that have a high incidence in certain populations are known to be associated with virus infections. Particularly in South-Eastern China and Taiwan, primary carcinoma of the liver is associated with chronic hepatitis B infection, which usually follows early exposure to the virus (30). High hepatitis B virus carrier rates may also explain the prevalence of primary carcinoma of the liver in many African populations. Nasopharyngeal carcinoma in Africa and China, and Burkitt's lymphoma in Africa, are associated with Epstein-Barr (EB) virus infection. The same virus can induce lymphomas in immunosuppressed persons (31). HTLV-I infections are also associated with some adult T-cell leukemias in the southern islands of Japan and some other countries.

The question of whether T-lymphocyte-mediated surveillance also operates against non-virus-induced malignancies is too broad to discuss here. Reference can be made to the studies of Kripke (32) showing that T-lymphocytes can prevent the growth of skin tumors induced by ultraviolet radiation in mice. A similar mechanism may contribute to ultraviolet radiation-induced skin tumors in humans.

5. THE DEVELOPMENT OF AN ADJUVANT FORMULATION THAT ELICITS FORMATION OF ANTIBODIES OF PROTECTIVE AND CELL-MEDIATED IMMUNITY

To apply the abovementioned findings for prevention of virus infections and virus-induced tumors, it is necessary to have virus antigens in sufficient quantity as well as a means of administering them that elicit the formation of protective antibodies and CMI. Methods for producing sufficient quantities of inactivated viruses and their subunits have become available in recent years. Using recombinant DNA technology, many antigens can be produced relatively cheaply. We have concentrated our efforts on developing an adjuvant formulation that will allow these antigens to be used successfully (33). This is the culmination of a long research programme on adjuvants that showed their first effects to be exerted on macrophages and mediated through helper T-lymphocytes (34).

As a counterpart of the mycobacterium in Freund's complete adjuvant (FCA) we use a synthetic analogue of muramyl dipeptide (MDP). This analogue, N-acetylmuramyl-L-threonyl-D-isoglutamine (Fig. 1), is a potent adjuvant but is much less active than CFA in inducing undesirable effects, including fever, vasculitis, anterior uveitis and arthritis. The latter are components of Reiter's syndrome, to which persons of the HLA-B27 haplotype are prone. For an adjuvant formulation to be used in humans, adequate separation of adjuvant effects from such side effects is esssential. The MDP analogues that are good adjuvants are all inducers of IL-1 production, and capacity to induce the release of growth factors for lymphocytes may be one of their mechanisms of action.

A substitute for the mineral oil component of Freund's adjuvant was also needed. We first introduced liposomes as vehicles for antigen (35). However, a more versatile vehicle was subsequently identified. The second component of the adjuvant that we now use is an emulsion of squalane and a non-ionic detergent, Pluronic polymer L-121 (Fig. 2). This is a single, unbranched chain consisting of a central polymer of polyoxypropylene, which is hydrophobic, flanked by two polymers of polyoxyethylene, which is hydrophilic because of its hydrogen-bonding capacity. Such Pluronic polymers concentrate at the interface between emulsions of oil or squalane and aqueous media. They retain antigens at this interface through their surface activity and hydrogen-bonding capacity. Pluronic polymers also activate complement by the alternative pathway, and C3b on the surface of the

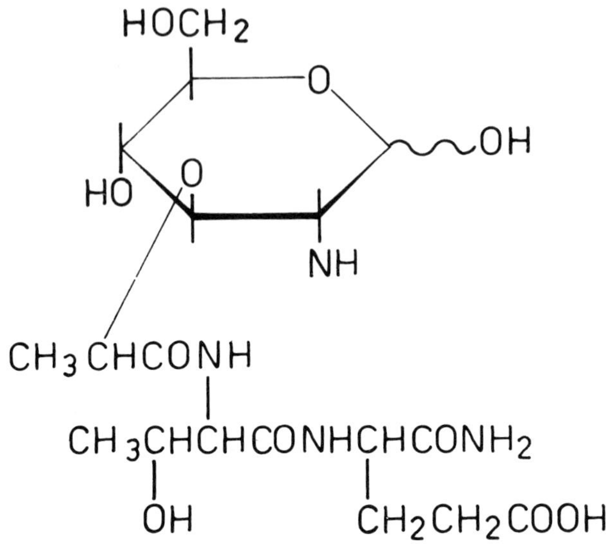

FIGURE 1. Structure of N-acetylmuramyl-L-threonyl-D-isoglutamine.

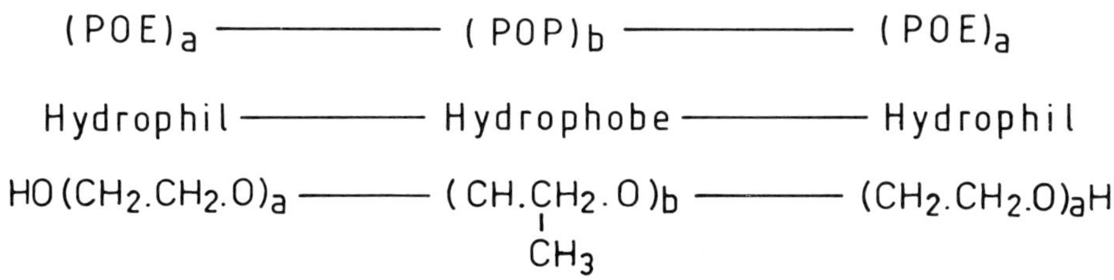

FIGURE 2. Structure of Pluronic polyol.

antigen-coated microspheres facilitates their attachment to C3b receptors on antigen-presenting cells (interdigitating cells and follicular dendric cells in lymphoid tissues). Thus the function of the vehicle (squalane/ Pluronic formulation) is to transfer antigens efficiently to the surface of antigen-presenting cells. The combination of the threonyl analogue of MDP and squalane Pluronic polymer formulation is termed "Syntex adjuvant formulation-1" or SAF-1.

A variety of viral and non-viral protein antigens, administered in this adjuvant formulation, elicit high titers of antibodies in primary and secondary responses as well as CMI. The latter is shown not only by delayed hypersensitivity but also by T-lymphocyte proliferative responses, and interleukin-2 production, in the presence of antigen. In the mouse antibodies are mainly of the IgG2a isotype, which activates complement and acts synergistically with antibody-dependent effector cells.

6. EXAMPLES OF THE USE OF SAF-1

SAF-1 allowed us to develop the first highly efficacious vaccine against feline leukemia virus, using inactivated virus as antigen (36). The importance of SAF-1 was shown by the fact that the same antigen, tested simultaneously with other adjuvants, did not elicit good protection. A similar protocol elicited in monkeys excellent protection against a simian AIDS retrovirus (37). It should be possible to extend the protection to human retroviruses, such as HTLV-I, without changing protocol. Alternatively, the adjuvant formulation could be used with cloned surface and core antigens.

Influenza virus vaccines likewise often elicit poor antibody responses in young children and old people, in whom the virus produces more severe disease than in other age groups. It would be desirable if an adjuvant could be used to improve responses to the vaccine, especially in susceptible age groups. Fig. 3 shows the effect of SAF-1 on immunizing guinea pigs with influenza virus vaccine (N. Byars, A.C. Allison and A. Kendall: unpublished data). The primary and secondary antibody responses to the virus were markedly increased, even when using a low dose of antigen. Antibody responses of young and aged mice to influenza virus vaccine were also greatly augmented by SAF-1.

The first generation vaccine against hepatitis B virus was surface antigen prepared from plasma. This vaccine is safe (38) and after three or four doses elicits antibody levels believed to be protective in 60 to 95 % of normal persons. In susceptible persons, e.g. newborns, renal dialysis patients and drug addicts, responses are less satisfactory; so there is room for improvement. Low responses might be due to inherited unresponsiveness (as seen with HBsAg in mice), to the use of a suboptimal adjuvant formulation, or both. The second generation vaccine is the HBsAg produced by recombinant technology in yeast (16,39), which is the first recombinant vaccine to be licensed by the Food and Drug Administration. This vaccine is administered in alum, and we have compared responses of several experimental animals to HBsAg and HBcAg in different adjuvants (K. Murray, N.E. Byars and A.C. Allison: unpublished data). Primary and secondary responses of guinea pigs to HBsAg in SAF-1 were much greater than in alum, and of rabbits and calves to HBsAg and HBcAg in SAF-1 were comparable to those with FCA. The antibody levels obtained were greatly in excess of the WHO standards which are believed to be protective. Thus, it is possible to envisage a third-generation vaccine based on administration of a wider antigen repertoire (which should reduce the likelihood of inherited unresponsiveness), such as S+preS antigens or S+C antigens, in a highly effective adjuvant formulation such as SAF-1. This might not only prevent hepatitis in susceptible subjects in developed countries but also transmission of the infection from mothers to newborn children in Asian and African countries where the carrier rate of the virus is high. Successful vaccination could eventually lead to a reduction in virus carriage and incidence of primary carcinoma of the liver.

Berman et al. (40) found that cloned herpes simplex virus type 1 glycoprotein D in Freund's complete adjuvant efficiently protected guinea pigs from genital herpes infection. A glycoprotein surface antigen of EB virus in SAF-1 elicited in mice high antibody responses (A.J. Morgan, M.A. Epstein and A.C. Allison: unpublished data). It will soon be tried in cottontop tamarins to ascertain whether it can protect them from EB virus-induced lymphomas. Protection of Chinese and Africans against EB virus-induced nasopharyngeal carcinomas and Burkitt lymphomas may, in fact, be feasible.

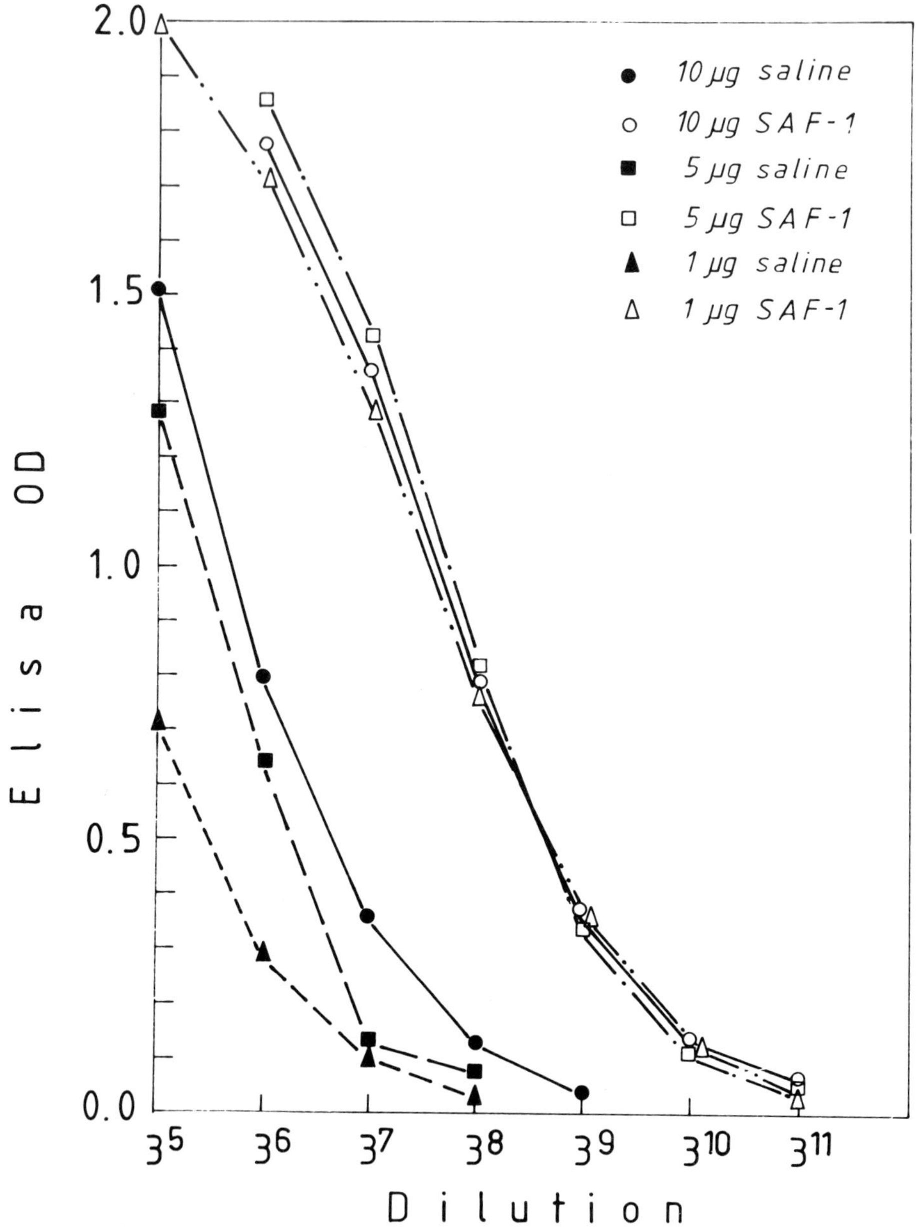

FIGURE 3. Antibody responses of guinea pigs to standard influenza virus vaccine compared with the same antigen in SAF-1. Antibody titers measured 4 weeks after vaccination.

The success of the adjuvant formulation in eliciting protection against leukemias and sarcomas in cats led us to examine the possibility that it might be useful also in comparable malignancies bearing non-viral antigens. Ronald Levy and his collaborators at Stanford University have studied B-lymphomas in mice and humans, using their monoclonal immunoglobulin (McIg) products as tumor-specific antigens. Immunization of mice with McIg in SAF-1 elicited high titers of antibodies against the antigen and protection against challenge with the cells (M. Campbell, R. Levy, N. Byars and A.C. Allison: unpublished data). Attempts to immunize mice with established tumors, and produce cures by a combination of active immunization and chemotherapy, are in progress.

Using small doses of human monoclonal antibody with SAF-1 in cynomolgus monkeys, we have elicited high titers of specific (anti-idiotypic) antibodies and delayed hypersensitivity. Hence it seems likely that anti-idiotypic immunization will be feasible in humans, and that could be useful not only in lymphomas but also using anti-idiotypes or antibodies against T-cell receptors as internal images of poorly immunogenic molecules, e.g. bacterial capsular polysaccharides in children under two years of age.

These examples suffice to illustrate the potentialities of a good adjuvant formulation. Many other applications can be envisaged, including better vaccines against bacterial infections of children, herpes viruses and parasites. The possibility of active immunization against solid tumors, using surface membrane components or growth factors as antigens, can also be considered. Already it seems likely that a new generation of vaccines based on antigens produced by recombinant DNA technology administered in SAF-1 will have a major impact. If immunization against human immunodeficiency virus and solid tumors proves feasible, our hopes and dreams when initiating this research program will have been exceeded. Robert Louis Stevenson wrote "To travel hopefully is better than to arrive". We are certainly travelling hopefully, although we envisage a journey with many points of arrival and departure rather than a single destination.

ACKNOWLEDGMENTS

I am glad to acknowledge the contributions of many collaborators, who are listed as co-authors. In particular, Noelene Byars has been a major collaborator in the development and exploitation of the adjuvant formulation.

REFERENCES

1. Hayward A. in Clinical Aspects of Immunology (Lachmann PJ, Peters DK, Eds). Blackwell Scientific Publications, Oxford, vol. 2, pp 1658-1712, 1982.
2. Allison AC, Burns WH. in Immunogenicity (Borek F, Ed). North Holland Press, Amsterdam, pp 155-179, 1972.
3. Allison AC. Transplant. Rev. 19:3-55, 1974.
4. Allison AC. in Clinical Aspects of Immunology (Lachmann PJ, Peters DK, Eds). Blackwell Scientific Publications, Oxford, vol. 1, pp 101-129, 1982.
5. Allison AC. in Immunological Adjuvants and Modulators of Non-Specific Resistance to Microbial Infections (Majde J, Ed). Alan R. Liss Inc., New York, in press, 1986.
6. Segal AW, Allison AC. in Oxygen Free Radicals and Tissue Damage. Ciba Foundation Symp. 65 (new series). Excerpta Medica, Amsterdam, pp 205-219, 1979.

7. Ragar-Zisman B, Allison AC. J. Gen. Virol. 19:339-351, 1973.
8. Lopez C. in Immunobiology of Herpes Simplex Virus Infection (Rouse BT, Lopez E, Eds). CRC Press, Boca Raton, pp 45-57, 1984.
9. Halstead SB, O'Rourke EJ, Allison AC. J. Exp. Med. 146:21229, 1977.
10. Virelizier JL, Allison AC, Schild GC. J. Exp. Med. 140:1571-1578, 1974.
11. Virelizier JL, Allison AC, de Maeyer E. Infect. Immun. 17:282-285, 1977.
12. Zinkernagel RM, Doherty PC. Adv. Immunol. 25:52-76, 1979.
13. Morris AG, Lin Y-L, Askonas BA. Nature 295:150-152, 1982.
14. Virelizier JL, Allison AC, Oxford JS, Schildt GC. Nature 266:52-54, 1977.
15. Townsend ARM, Rothbard J, Gotch EM, Bahader G, Wraith D, McMichael AJ. Cell 48:459-468, 1966.
16. Murray K, Bruce SA, Hinnen A, Wingfield P, van Erd PM, de Reus A, Schellekens H. EMBO J. 3:645-650, 1984.
17. Turk JL, Allison AC, Oxman MN. Lancet i:405-407, 1962.
18. Blanden RV, Pang TE, Dunlop MBC. in Cell Surface Reviews (Poste A, Nicolson GL, Eds). North Holland, Amsterdam, vol; 2, pp 249-268, 1977.
19. Allison AC, Friedman RM. J. Natl. Cancer Inst. 36:859-868, 1966.
20. Allison AC. J. Natl. Cancer Inst. 36:869-876, 1966.
21. Allison AC. in Viral Oncology (Klein G, Ed). Raven Press, New York, pp 481-487, 1980.
22. Allison AC, Taylor RB. Cancer Res. 207:703-707, 1967.
23. Law LW. Cancer Res. 29:1-21, 1969.
24. Allison AC, Monga JN, Hammond V. Nature 252:746-747, 1974.
25. Vandeputte M, Eyssen H, Sobis H, De Somer P. Int. J. Cancer 14:445-450, 1974.
26. Vandeputte M, Denys P, Leyten R, De Somer P. Life Sci. 7:475-478, 1963.
27. Vandeputte M. Transplant. Proc. 1:100-105, 1969.
28. Vandeputte M, Meyer C, Sobis H. J. Natl. Cancer Inst. 56:517-521, 1976.
29. Allison AC, Law LW. Proc. Soc. Exp. Biol. Med. 127:207-212, 1968.
30. Beasley RP, Hwang LY. Semin. Liver Dis. 4:113-121, 1984.
31. Crawford DH, Edwards JM, Sweny P, Hoffbrand AV, Janossy A. Int. J. Cancer 28:705-709, 1981.
32. Kripke ML. Immunol. Rev. 80:87-102, 1984.
33. Allison AC, Byars N.E. J. Immunol. Methods, in press, 1986.
34. Allison AC. Ciba Found. Symp. on Immunopotentiation 18:72-94, 1973.
35. Allison AC, Gregoriadis G. Nature 252:252-254, 1974.
36. Braemer AC, Peterson M, Renneke G, Bass E, Allison A, Byars N, Fraser D. in Proceedings of the 65th Conference on Researcher Workers in Animal Disease, Chicago, 1984.
37. Marx PA, Pedersen NC, Lerche NW, Osborn KG, Lowenstine LJ, Lackner AA, Maul DB, Kwang H-S, Kluge JD, Zaiss CP, Sharpe V, Spinner AP, Allison AC, Gardner MB. J. Virol. 60:431-435, 1986.
38. Francis DP, Feorino PM, McDougal S, Warfield D, Getchell J, Cabradilla C, Tong M, Miller WJ, Schultz LD, Bailey FJ, McAleer WJ, Scolnick EM, Ellis RW. JAMA 256:869-872, 1986.
39. Scolnick EM, McLean AA, West DJ, McAleer WJ, Miller WJ, Bunyak EB. JAMA 251:2812-2815, 1984.
40. Berman PW, Gregory T, Crase D, Lasky LA. Science 227:1490-1492, 1985.

TOTAL LYMPHOID IRRADIATION

S. STROBER
Division of Immunology, Stanford University School of Medicine, Stanford, California 94305-5111, U.S.A.

1. INTRODUCTION

Total lymphoid irradiation (TLI) has become a useful tool for developing strategies to induce tissue specific transplantation tolerance and to treat autoimmune diseases in laboratory animals and humans. TLI was originally developed as a treatment for Hodgkin's disease, and has been widely used during the past 20 years. Approximately 90 % of patients with early stages of the disease are cured after TLI (1). Extensive follow-up observations of more than 1,000 patients in remission for up to 10 years have shown that there is no increased risk of hematologic malignancy or non-Hodgkin's lymphoma after treatment with TLI alone (2-4). The incidence of severe complications with TLI, including bacterial infection, is approximately 1 % (1).

2. IMMUNE DEFICIENTS AFTER TLI

Before radiotherapy, total lymphocyte counts and percentage of T and B cells in patients with Hodgkin's disease were not significantly different from those observed in normal controls (5). However, at the completion of treatment, the mean total peripheral blood lymphocyte (PBL) count ($503/mm^3$) was 4 standard deviations below the mean of normal controls. Recovery of the lymphocyte count began shortly after completion of radiotherapy and reached pretreatment levels about two years later (5). However, after full recovery of the total count, the percentage of T cells as measured by an *in vitro* cytotoxicity assay with anti-T cell antiserum was approximately half the pretreatment value (5). In addition, there was a doubling of the percentage of B (Ig-bearing) cells. The T lymphocytopenia and B lymphocytosis persisted for at least 10 years after treatment in the absence of recurrent disease. Although suppressor/cytotoxic T cells recovered fully, helper T cells did not (6).

The peripheral blood lymphocytes of untreated patients with Hodgkin's disease showed a decreased response to PHA *in vitro*, measured by ^{3}H-thymidine or ^{3}H-leucine incorporation, as compared to normal controls (5). The response returned to normal levels by incubation of the cells for 24 hours in tissue culture medium with fetal calf serum before stimulation with PHA (5). The reversible deficit was thought to be due to the presence of an inhibitory serum factor (5).

After radiotherapy, the PHA response of patients with Hodgkin's disease fell significantly below that of the pretreatment level. The reduction in the PHA response persisted for at least 10 years after radiotherapy without disease recurrence (5). The reduced PHA response after treatment could not be reversed by incubation of the PBL in cell culture medium, and appeared to be an intrinsic functional deficit (5).

The mixed lymphocyte reaction (MLR) of untreated patients with Hodg-

S. Strober

kin's disease was similar to that of normal controls as measured by incorporation of ^{3}H-thymidine (5). However, after radiotherapy, the MLR was virtually eliminated, and fell to background levels for about 20 months. Thereafter, a slow recovery was observed such that normal responses were uniformly observed 5 or more years after treatment (5).

The minority ($\sim$ 27 %) of untreated patients with Hodgkin's disease showed a delayed hypersensitivity skin reaction to CNCB. Of the responding group, almost all lost their skin reactivity immediately after the completion of TLI. More than half of these treated patients showed a delayed hypersensitivity response to DNCB one year after treatment. However, 29 % of these patients remained anergic for at least eight years after radiotherapy (5).

The main advantages of TLI as compared to other modalities of irradiation which can suppress the immune system are : 1) the targeting of high-energy x-rays to the lymphoid tissues while shielding the vital nonlymphoid organs so as to minimize toxicity; 2) fractionation of the total dose of radiation such that multiple small doses are used to minimize side effects of each treatment and reduce blood element depressions; and 3) careful control of the dose of radiation absorbed by the targeted lymphoid tissues such that radiation treatments can be discontinued with the appearance of side effects and subsequently resumed when the side effects have subsided. The main disadvantage of total body irradiation is the toxicity to nonlymphoid tissues, including the bone marrow. Although injection of radioactive compounds that are selectively taken up by the immune system may eliminate this disadvantage, it is difficult to control the dose of irradiation delivered or to discontinue irradiation.

3. EASE OF TOLERANCE INDUCTION IN ADULT MICE GIVEN TOTAL LYMPHOID IRRADIATION (TLI)

Fractionation and shielding techniques similar to humans can be used to deliver 3,400 rads (17 fractions of 200 rads each) to the lymph nodes, spleen, and thymus of adult mice without substantial morbilidy and mortality (7). The immune system of the treated mice has many similarities to those of the neonatal mice, including the ease of tolerization and the high proportion of "null" cells (8). Thus, allogeneic bone marrow injected intravenously into mice within one week after TLI results in permanent chimerism without the development of graft-versus-host disease (GVHD) (7,9-11). The resultant chimeras are specifically tolerant of the tissues of the marrow-donor strain and will permanently accept donor-type skin grafts and promptly reject third-party grafts (7,9-11).

As in the case of the neonate, the ease of tolerance induction for alloantigens is short-lived. Although about 90 % of irradiated mice will accept allogeneic marrow grafts within one to two days after the completion of irradiation, only about 50 % will accept grafts transplanted seven days after irradiation. By day 21, none of the recipients will accept marrow grafts (12). The tolerogenic effects of the radiotherapy are related to the total dose of radiation, the dose per fraction, and the amount of lymphoid tissue enclosed within the radiation ports (12,13). A cumulative dose of at least 3,000 rads is required to achieve tolerance in a high proportion of marrow recipients (12,14). Although the major lymph nodes above and below the diaphragm as well as the spleen must be irradiated in order to prepare mice for successful marrow transplantation, irradiation of the thymus is not essential (12). Rats given TLI have also been successfully tolerized to alloantigens, and permanently accept skin and heart transplants following marrow transplantation (15).

4. TOLERANCE IN OUTBRED DOGS AND BABOONS GIVEN TLI

The use of TLI alone (1,800 rad total dose) or in combination with a brief course of other immunosuppressive agents was studied in adult mongrel dogs given heterotopic heart allografts from unmatched donors (8). Maximum allograft survival with TLI alone was 28 days, and with six intramuscular injections of rabbit anti-dog thymocyte globulin (ATG) alone was 33 days. However, marked synergy was observed with the combination of the two regimens such that four of eight allografts survived 200 days or more. Three to eight dogs given TLI and ATG maintained their allografts with minimal or no cellular infiltrate during the entire observation period (360 to 495 days). All three animals rejected a third party heart within two weeks. Thus, the latter recipients were specifically unresponsive to the initial allograft despite the withdrawal of all immunosuppressive reagents after day 10.

Long term acceptance of liver and kidney allografts in outbred baboons treated with TLI in the absence of marrow transplantation has been reported by Myburgh and his colleagues (16). The most successful regimen involved the use of enlarged radiation fields which included the whole abdomen as in previous rodent studies, but with a cumulative radiation dose of 800 rad given in 100 rad fractions. Using this procedure, unmatched kidney allografts survived more than 200 days without post-transplant immunosuppressive drugs in 80 % of recipients (16). Some animals in these studies were followed for as long as four years with intact kidney allografts (16). Specific unresponsiveness in the latter graft recipients was demonstrated by the rapid rejection of a third party graft transplanted one year after the initial graft (16).

5. STUDIES IN MAN

The first clinical study of TLI was performed at the University of Minnesota in high-risk cadaver renal transplant recipients who had already rejected at least one transplant (17). Patients were given up to 4,500 rads before transplantation, and prednisone and azathioprine as the maintenance immunosuppressive therapy after transplantation. One-year graft survival in this group was approximately 30 % higher than that in a historical control group given prednisone and azathioprine without TLI. A subsequent study of the use of cyclosporine and prednisone in a similar group of high-risk patients showed similar patient and graft survival in the TLI and cyclosporine groups (18).

The design of the University of Minnesota study may have limited the effectiveness of TLI, since difficulty in obtaining crossmatch negative donor kidneys for the highly sensitized recipients frequently resulted in long gaps between the completion of TLI and renal transplantation. In addition, the use of post-transplant azathioprine may have been deleterious, since the canine studies showed increased morbidity and mortality and shorter graft survival when azathioprine was added to the TLI-ATG regimen.

In a subsequent study from the University of Leuven, diabetic recipients were prepared with TLI (2,000-3,000 rads) and received primary cadaveric renal transplants within two weeks (19). Post-transplant maintenance immunosuppression was limited to low-dose prednisone only. One-year graft survival was more than 80 %. In view of the synergy between TLI and ATG in canine studies, and the capacity to maintain patients on low-dose prednisone after TLI, a protocol combining all three modalities was developed by investigators at Stanford University (S. Strober, R.T. Hoppe and H.S. Kaplan) and at the Pacific Medical Center, San Francisco (D. Sampson, B. Levin and G. Collins) (20). Patients were prepared with TLI (2,000 rads) and given an unmatched primary cadaveric renal transplant within two weeks in almost all cases. After transplantation, six doses of rabbit anti-human thymocyte globulin was administered on alternate days, and maintenance immunosuppression was restricted to low-dose prednisone (0.1-0.2 mg/day).

Our previous studies in laboratory animals showed that TLI can induce specific tissue transplantation tolerance in rodents and dogs when grafting occurs shortly after the completion of radiotherapy (7-9,15). To reduce the interval between the completion of TLI and renal transplantation in human studies, only those patients with low levels of cytotoxic serum antibodies were recruited for the study and no attempt was made to match the HLA types of the donor and recipient. Only 1 of the 28 patients had a previous transplant (HLA-matched sibling).

Graft survival in the Stanford-Pacific Medical Center study was 77 % at 12 months as judged by Kaplan-Meier analysis. In 89 primary (first-graft) cadaveric renal transplant recipients given cyclosporine and prednisone during the same time interval at the same institution, the 12-month graft-survival was 67 % (Kaplan-Meier plot) ($p > 0.5$, χ^2 analysis) (21). The percentage of patients with several risk factors for poor graft survival (diabetes, age > 50 years, and lack of blood transfusions) was higher in the TLI than in the cyclosporine group. The results with cyclosporine were similar to that (∿ 73 %, 1 year graft survival) reported in two multicentre studies (22,23). The percentage of patients with functioning grafts who had at least one rejection episode was 47 % in the TLI group and 58 % in the cyclosporine group in our institution. The mean concentration of serum creatinine at the last observation point was significantly lower in the TLI group (1.5 mg/dl ± 0.20), compared with the cyclosporine group (1.98 mg/ml ± 0.12) ($p < 0.05$; Student's t test). Actuarial patients survivals at 1 year in both groups were similar (91 % TLI; 96 % cyclosporine).

Our results show that cadaveric kidney graft function can be maintained with low-dose prednisone as the sole chronic immunosuppressive drug regimen in recipients prepared with TLI. Graft and patient survivals seem to be at least as good as those obtained with prednisone plus cyclosporine given to patients in the same or other institutions (22,23). The advantages of the TLI over cyclosporine include superior graft function as judged by the level of serum creatinine, and reduced cost of maintenance drug therapy. In addition, prednisone alone should produce less long-term toxicity than would cyclosporine plus prednisone. The combination of cyclosporine and prednisone has been associated with progressive nephrotoxicity in heart transplant recipients (24). Long-term observations of the patients enrolled in the TLI study are required to adequately assess the potential complications of the TLI regimen, and controlled trials are required to compare the efficacy, side-effects, and costs of TLI and cyclosporine. However, the TLI study demonstrates the feasibility of markedly reducing post-transplant immunosuppression as compared to conventional regimens, and may provide the basis for transplantation tolerance in humans.

6. UNCONTROLLED AND CONTROLLED STUDIES OF TLI IN THE TREATMENT OF INTRACTABLE RHEUMATOID ARTHRITIS

Two uncontrolled studies suggested that TLI is effective in the treatment of patients with intractable rheumatoid arthritis (25,26). The efficacy of TLI was subsequently determined in a controlled trial. Twenty-six patients with intractable rheumatoid arthritis were randomized to receive either a high (immunosuppressive) dose (2,000 rad total; 200 rad per fraction) or a low dose (200 rad total, 20 rad per fraction) (27). Patients and observers were blinded with regard to the treatment group. Of the patients entered, 13 completed high dose and 11 completed low dose therapy. The patients in the two groups were well matched prior to treatment in regard to age, duration of disease, disease activity (i.e., morning stiffness, joint scores, and global composite score), prednisone dose, functional class (American Rheumatism Association), and rheumatoid factor titre.

Patients were evaluated twice before and at three and six months after TLI by blinded observers (27). Significant improvements in morning stiffness, joint tenderness, joint swelling, and global composite scores were observed in the high dose group at three and six months after TLI as compared to pretreatment values. No significant change was observed in any parameter in the low dose group. When the percentage improvements in the high and low dose groups were compared, the former showed significantly greater improvement in three of the four clinical parameters. While ten of 13 patients in the high dose group had more than a 25 per cent improvement in three of the four parameters, only three of 11 patients in the low dose group had such improvement ($P < 0.05$). These results indicate that TLI (2,000 rad) is an effective treatment for patients with intractable rheumatoid arthritis.

Both patients given high dose and low dose irradiation experienced constitutional symptoms during TLI, which resolved within two months after TLI. Furthermore, three patients in each group developed herpes zoster. Other complications observed in the high dose group included transient severe leukopenia with white cell counts less than 2,000/mm^3 (three patients), transient thrombocytopenia (one patient), pleuritis (one patient), and gastric ulcer (one patient). Two deaths in the high dose group 9 and 18 months after TLI (secondary to myeocardial infarction and pulmonary embolism) appeared to be unrelated to radiotherapy.

As expected, marked changes of _in vitro_ lymphocyte function were noted in the high dose group. Proliferative responses of peripheral blood lymphocytes to mitogens, allogeneic cells, and soluble antigens were reduced by 60 to 90 per cent. _In vitro_ immunoglobulin synthesis [pokeweed mitogen (PWM)-stimulated] was also reduced by 90 per cent in the high dose group. In contrast, no significant changes of _in vitro_ T cell function were observed in the low dose group. The functional deficits in the high dose group were associated with a marked T lymphopenia, primarily due to the reduction of the number of helper/inducer (Leu-3 cells). While the Leu-2/Leu-3 ratio rose from 0.72 to 1.95 after high dose TLI, no decrease in Leu-3 cells or increase in this ratio occurred after low dose TLI.

7. TREATMENT OF LUPUS NEPHRITIS WITH TLI

TLI (total dose 2,000 rad) was used to treat ten patients with systemic lupus erythematosus (SLE) who had severe, intractable glomerulonephritis (28). All patients had biopsy-proven diffuse proliferative disease and the nephrotic syndrome. Before TLI, all patients could not be tapered below 0.5 mg/kg prednisone per day for six months without an exacerbation of nephritis. Azathioprine was discontinued at the initiation of TLI in three patients who were receiving this drug in combination with prednisone. Patients were followed for up to 48 months after the initiation of TLI.

All ten patients had a decrease in 24-hour protein excretion and/or an increase in serum albumin concentration. The serum creatinine values improved in five patients and remained unchanged in three patients. The daily prednisone dosage was tapered by a mean of 60 per cent in six patients followed for more than one year. Side effects of radiotherapy were confined to herpes zoster in two patients.

In contrast to the changes in autoantibody levels in rheumatoid arthritis after TLI, elevated anti-DNA antibody levels in these SLE patients returned to the normal range immediately following treatment with TLI (28). This suggests that anti-DNA antibody production may be T helper cell dependent. Similar to the above studies of rheumatoid arthritis patients treated with TLI, a number of alterations of cellular immune function were noted after TLI, including (1) T cell lymphocytopenia, (2) a decrease in the number of helper (Leu-3) cells and a reversal of the Leu-2/Leu-3 ratio, (3) a decrease of *in vitro* lymphocyte proliferative responses to mitogens and allogeneic cells, (4) a decrease of spontaneous and PWM-stimulated immunoglobulin production *in vitro*, and (5) the appearance of circulating suppressor cells of PWM-stimulated immunoglobulin production.

8. CONCLUSION

TLI has been used as an immunosuppressive and tolerogenic regimen in autoimmune disease and organ transplantation. The rationale for its use originated from studies of Hodgkin's disease patients, where this radiotherapy regimen was noted to induce profound and long-lasting immune suppression and yet was well tolerated with few long-term side effects. TLI is a unique immunosuppressive regimen which produces a selective (and long-lasting) reduction in the number and function of helper T cells and certain subsets of B cells. It is paradoxical that remissions in Hodgkin's disease were associated with prolonged deficits in T cell immunity. This suggests that T cells may not play a major role in immune surveillance for this type of tumor. Conventional immunosuppressive drugs show little selectivity, and their effects are short-lived. The most important aspect of TLI is the potential for achieving transplantation tolerance and permanent remissions in autoimmune disease in laboratory animals. Attempts are being made to achieve similar goals in humans given TLI such that immunosuppressive drugs can be ultimately withdrawn from transplant recipients and patients with lupus nephritis.

REFERENCES

1. Kaplan HS (Ed). Hodgkin's Disease. Harvard University Press, Cambridge, pp 366-441, 1980.
2. Pederson-Bjergaard T, Larsen SO. New Engl. J. Med. 307:965-971, 1982.
3. Valagussa P, Santoro A, Kenda R, Fossanti Ballanti F, Frachi F, Banfi A, Rilke F, Bonadonna G. Brit. Med. J. 280:216-219, 1980.
4. Coleman CN, Burke JS, Varghese A, Rosenberg SA, Kaplan HS. *in* Malignant Lymphomas : Etiology, Immunology, Pathology, Treatment (Rosenberg SA, Kaplan HS, Eds). Academic Press, New York, pp 259-276, 1982.
5. Fuks S, Strober S, Bobrove AM, Sasazuki T, McMichael A, Kaplan HS. J. Clin. Invest. 58:803-814, 1976.
6. Haas GS, Halperin E, Doseratz D, Lingood R, Russell PS, Colvin R, Barrett L, Cosimi AB. J. Immunol. 132:1026-1030, 1984.
7. Strober S, Slavin S, Gottlieb M, Zan-Bar I, King DP, Hoppe RT, Fuks Z, Grumet FC, Kaplan HS. Immunol. Rev. 46:87-112, 1979.
8. Strober S, Modry DL, Hoppe RT, Pennock JL, Bieber CP, Holm BI, Jamieson SW, Stinson EB, Schroder J, Soumalainen H, Kaplan HS. J. Immunol. 132:1013-1018, 1984.
9. Slavin S, Strober S, Fuks Z, Kaplan HS. J. Exp. Med. 146:43-48, 1977.

10. Slavin S, Fuks Z, Kaplan HS, Strober S. J. Exp. Med. 147:963-972, 1978.
11. Strober S, Gottlieb M, Slavin S, King DP, Hoppe RT, Fuks Z, Bieber CP, Kaplan HS. Transplant. Proc. 12:477-482, 1980.
12. Gottlieb M, Strober S, Kaplan HS. J. Immunol. 123:379-383, 1979.
13. Waer M, Ang KK, van der Schueren E, Vandeputte M. J. Immunol. 132: 985-990, 1984.
14. Waer M, Ang KK, van der Schueren E, Vandeputte M. J. Immunol. 132: 991-996, 1984.
15. Slavin S, Reitz B, Bieber CP, Kaplan HS, Strober S. J. Exp. Med. 147: 700-707, 1978.
16. Myburgh JA, Smith JA, Stark JH, Browde S. J. Immunol. 132:1019-1025, 1984.
17. Najarian JS, Ferguson RM, Sutherland DER, Slavin S, Kim T, Kersey T, Simmons RL. Ann. Surg. 196:442-452, 1982.
18. Sutherland DER, Ferguson RM, Aeder MI, Lewis WI, Bentley FR, Ascher NL, Simmons RL, Najarian JS. Transplant. Proc. 15:2881-2888, 1983.
19. Vanrenterghem Y, Waer M, van der Schueren E, Ang K, Lerut T, Vandeputte M, Gruwez J, Bouillon R, Michielsen P. Dialysis Transplant. 12: 104-105, 1983.
20. Sampson D, Levin BS, Hoppe RT, Bieber CP, Miller E, Waer M, Kaplan HS, Collins G, Strober S. Transplant. Proc. 17:1299-1303, 1985.
21. Levin B, Hoppe RT, Collins G, Miller E, Waer M, Bieber C, Girinsky T, Strober S. Lancet ii:1321-1325, 1985.
22. European Multicentre Trial Group. Lancet ii:986-989, 1983.
23. Canadian Multicener Transplant Study Group. New Engl. J. Med. 309: 809-815, 1983.
24. Myers BD, Ross J, Newton L, Luetscher J, Perlroth M. New Engl. J. Med. 311:699-705, 1984.
25. Kotzin BL, Strober S, Engleman EG, Calin A, Hoppe RT, Kansas GS, Terrell CP, Kaplan HS. New Engl. J. Med. 305:969-976, 1981.
26. Trentham DE, Belli JA, Anderson RJ, Buckley JA, Goetzl EJ, David JR, Austen KF. New Engl. J. Med. 305:976-982, 1981.
27. Strober S, Tanay A, Field E, Hoppe RT, Calin A, Engleman EG, Kotzin B, Brown BW, Kaplan HS. Ann. Intern. Med. 102:441-449, 1985.
28. Strober S, Field E, Hoppe RT, Kotzin BL, Shemesh O, Engleman E, Ross JC, Myers BD. Ann. Intern. Med. 102:450-458, 1985.

SECTION V : NEW HORIZONS IN VIRUS RESEARCH

SECTION CHAIRMAN : PROF. E. DE CLERCQ

E. De Clercq

INTRODUCTION TO SECTION V : NEW HORIZONS IN VIRUS RESEARCH

E. DE CLERCQ
Rega Institute, Katholieke Universiteit Leuven, B-3000 Leuven, Belgium

In this last section of our Symposium on "Frontiers in Microbiology", three eminent scientists will lead you through the mysterious world of viruses, thereby touching upon three major aspects of present-day virology : (i) prevention of virus infections by the use of vaccines, (ii) chronically persistent virus infections, and (iii) AIDS ("Acquired immune deficiency syndrome") and related retrovirus infections.

The vaccination procedure was introduced almost two centuries ago by Jenner who succeeded to confer protection against smallpox by inoculation with cowpox. Thanks to the worldwide vaccination with cowpox, one of the major plagues of all times, smallpox, has been eradicated, an achievement which is still unequalled in medicine.

Our Institute has played a pioneering role in the development of vaccines, as Prof. P. De Somer was the first to produce the live (oral) polio vaccine at a large scale, thereby allowing our country to be the first where poliomyelitis has been completely eradicated. Also, the basic work that led to the development of the rubella vaccine (Cendehill strain) was started at our Institute. As a matter of fact, I remember Prof. De Somer's enthusiasm when I told him that the rubella strain we had managed to grow in the laboratory (it must have been in 1965) induced a neutralizing antibody titer in rabbits of up to 10,000. The meaning of this high antibody titer and the reason for De Somer's excitement escaped me at that time, but, in retrospect, it must have been the start of the rubella vaccine.

The world authority on vaccines is undoubtedly MAURICE R. HILLEMAN, who, as Director of the Merck Institute for Therapeutic Research at Merck Sharp and Dohme Research Laboratories, developed the majority of virus vaccines which are currently used, i.e. for the prevention of measles, mumps, rubella and hepatitis B. Dr. M.R. Hilleman is also Adjunct Professor of Pediatrics at the School of Medicine of the University of Pennsylvania in Philadelphia. He has been a member of the Expert Advisory Panel of the World Health Organization since 1952. He received the Lasker Medical Research Award in 1983. He is a member of the American Academy of Arts and Sciences, and also a member of the National Academy of Sciences of the United States. He holds several honorary doctorate degrees, including a doctorate degree from the K.U.Leuven. We are very glad and grateful that Dr. Hilleman has come back to his old Alma Mater to join us at this unique occasion to review the progress that has been made in the vaccine field, to assess the present state of the art and to speculate in how it should further evolve.

Inherent to many virus infections is their propensity to remain latent or chronic after the acute phase has subsided. The latent virus may be reactivated periodically as demonstrated by the recurrences of cold sores (also referred to as fever blisters or "herpes labialis"), or it may be exacerbated by an underlying immune disorder, as demonstrated by the devastating course an otherwise asymptomatic cytomegalovirus infection may assume in severely immunocompromised patients. Several viruses, including hepatitis B virus and human immune deficiency virus (AIDS virus) are notorious for their tendency to enter into a chronic state, and, one of the sanctuaries for such chronic persistent virus infections is quite often the central nervous system (CNS). Chronic virus infections have, in turn, been etiologically linked to both cancer (i.e. hepatoma, cervix carcinoma) and CNS disorders (i.e. multiple sclerosis).

Professor FRED RAPP is one of the world's leading experts in both virus and cancer research, and the authority on persistent virus infections. He is American Cancer Society Professor of Virology, Evan Pugh Professor of Microbiology and, almost within living memory, Chairman of the Department of Microbiology at Pennsylvania State University in Hershey. Prof. F. Rapp has served as member or chairman on numerous advisory boards and committees, and is presently the Editor of the Journal "Virology" and Editor-in-chief of the Journal "Intervirology".

I have had the pleasure to collaborate with Dr. Rapp for many years, and, based on the use of our potent and selective anti-herpes drug, bromovinyldeoxyuridine (BVDU), we worked out a system to keep the replication of herpesviruses, i.e. herpes simplex and varicella-zoster virus, under control. What has so far proven to be too formidable a task is the eradication of the latent virus from its reservoir(s), a problem we have to face not only with herpes, but also with several other viruses, including the AIDS virus. We are anxious to know whether the wizard Dr. Rapp has any clues on how to grasp and eliminate the virus from its sanctuary.

The third lecturer in our Section and last speaker at this Symposium needs little, if any, introduction. His name has almost become a myth, Bob Gallo. I should point out, however, that his fame dates back to many years before the discovery of the AIDS virus, namely to 1970 when Dr. ROBERT C. GALLO, almost simultaneously with H. Temin and D. Baltimore, described the existence of the "reverse transcriptase" (RNA-dependent DNA polymerase) in retroviruses (oncogenic RNA viruses) and leukemic cells harboring these viruses. This discovery generated in our own laboratory an intensive search for inhibitors that may specifically inhibit the reverse transcriptase and thereby block retrovirus replication. From this work originated several potent reverse transcriptase inhibitors; and one of these compounds, namely suramin, was later found to inhibit the replication of AIDS virus. It became the first drug to be used in the treatment of AIDS, although it may soon be replaced by more effective and more selective anti-AIDS agents.

Tribute should be paid to Dr. Gallo for his pioneering research into the origin of human leukemia, for which he was awarded together with Dr. Rapp, the first Ciba-Geigy Award in 1977, for the discovery of human T-cell growth factor, that, in turn, formed the basis for the discovery of the first human retrovirus [HTLV-I (human T-cell leukemia virus)], for which he was awarded the Albert Lasker Medical Research Award in 1982 and the General Motors Charles S. Mott Prize in 1984. Today, 22 November 1986, only one day after he has been awarded, for the second time, the Albert Lasker Award, now in recognition of the role he has played in the discovery of the AIDS virus (HTLV-III), we are particularly proud to have Dr. R.C. Gallo as our distinguished guest.

Like Dr. Hilleman and Dr. Rapp, Dr. R.C. Gallo is an old friend of the Rega Institute, where he gave in 1979 a highly appreciated lecture on the "Origin of naturally occurring leukemia of animals and man". This must have been very shortly before HTLV-I was discovered. Since then, mankind has witnessed the emergence of a new plague, AIDS, at about the same time another one, smallpox, was eradicated. We look forward to hearing from Dr. Gallo how he is envisaging the evolution of this new and frightening disease.

M.R. Hilleman

PERSPECTIVES IN DISEASE PREVENTION BY VACCINES

M.R. HILLEMAN
Merck Institute for Therapeutic Research, Merck Sharp & Dohme Research Laboratories, West Point, Pennsylvania 19486, U.S.A.

1. INTRODUCTION

It is an honor and a pleasure for me to join with you today to honor the memory of Professor Piet De Somer. We remember him with reverence and with affection for he was both a great scientist and a delightful friend, whom we all knew simply as Piet. Some might consider him a modern day Renaissance Man because his knowledge, his activities and his accomplishments were so diverse. He was a scientific researcher of distinction; the founder of two great institutions - the Rega Institute and the RIT Laboratories; a beloved teacher; and a great and distinguished rector of the Katholieke Universiteit Leuven. There are few among us who would excel or even match the achievements made by Professor De Somer in his single lifetime.

The paths of those of us who spent our careers in biomedical research crossed in many fields with those of Professor De Somer - in antibiotic and antiviral research, in interferon and interferon induction, in cancer etiology, and especially in disease prevention by vaccines. Vaccines were near and dear to Prof. De Somer and his development of poliomyelitis and rubella vaccines give cause to remember him as an innovator in preventive medicine. I think it fitting, therefore, to say a few words today about the past, present, and future of viral vaccines.

2. PROPAGATION SYSTEMS AS THE KEY TO VACCINE DEVELOPMENT

The science of vaccines (Table 1) took its roots from ancient medical arts that consisted either of contrived exposure of the subject to natural infection at an optimal time (usually age-related), or of the inoculation of fully virulent virus by an unnatural route such as was practiced by the ancient Chinese for preventing smallpox. The science of immunology and the vaccine era began almost two centuries ago with Jenner's development of smallpox vaccine using an attenuated pox virus grown in calf skin. It soon became evident that antigens were the element that made vaccines work. From that time on, the object has been to develop systems for producing antigens that can be delivered either as non-living vaccines or as attenuated live virus vaccines. The use of infected rabbit brains made Pasteur's rabies vaccine possible in the late 1800's and chick embryo tissue culture opened the door to attenuated yellow fever vaccine in the late 1930's. The use of embryonated hen's eggs made killed influenza virus vaccine possible in the early 1940's. The new and sophisticated cell culture technology of the late 1940's made possible vaccines such as those against poliomyelitis, measles, mumps and rubella prepared from virus that was grown in cell cultures. The modern era of subunit viral vaccines began with hepatitis B viral surface antigen extracted from the plasmas of human carriers of the infection. This was followed quickly by another vaccine

TABLE 1. Dependence of vaccine development on the evolution of antigen production systems

	Example
Pre-vaccine era	
Contrived exposure	Rubella
Variolation	Smallpox
Viral vaccine era	
Cow skin	Jenner's vaccinia
Animal organ	Rabies (brain)
Tissue culture (chick embryo)	Yellow fever
Cell culture	Poliomyelitis
Plasma from human carriers (subunit)	Hepatitis B
Recombinant (yeast)	Hepatitis B
Future vaccine era	
Synthetic oligopeptides	Experimental
Recombinant benign vector viruses	"
Anti-idiotype antigens	"
Genetically altered live viruses	"

using antigen made by recombinant technology employing a vector carrying surface antigen gene that is expressed in ordinary yeast cells. What the future holds is uncertain; but, very short peptide antigens produced synthetically, benign vector viruses such as vaccinia virus carrying genes that express desired antigens, anti-idiotypes that are anti-antibodies having the same three-dimensional structure and antigenic specificity as the original antigen, and live viruses genetically altered to bring about attenuation, might all find a place in future vaccines. In the near term, however, subunit vaccines produced by recombinant technology in vector expression systems will likely be the dominant methodology. Human hepatitis B vaccine has been the model for the development and establishment of the precedent for new era vaccines and is worthy of special review.

3. PLASMA-DERIVED HEPATITIS B VACCINE AS A MODEL FOR VIRAL SUBUNIT VACCINES

Hepatitis B (Table 2) is an extremely important and worldwide health problem of modern man. The importance of the disease is not because of the acute illness. It lies, instead, in late sequelae among persons who recover from the initial stages of the infection but become persistent carriers of the virus. The carrier state follows about 10 % of all hepatitis B virus infections and typically occurs in 70 % to 90 % of infants who are born to mothers who are heavily infected carriers. About 1 in 5 carriers die ultimately from cirrhosis of the liver and 1 in 20 die of liver cancer. The enormity of the problem is emphasized by the fact that there are about 200 million hepatitis B carriers in the world population. Of these, 40 million may die of cirrhosis and 10 million of hepatocarcinoma - a staggering but predictable event !

The development of any vaccine to prevent a disease necessarily depends upon the availability of an antigen of the virus against which a

TABLE 2. Plasma-derived hepatitis B vaccine

Hepatitis B virus infection commonly leads to a carrier state.

Twenty percent of carriers may die eventually of liver cirrhosis and 5 % of liver cancer.

There are about 200 million hepatitis B carriers in the world.

Plasma-derived vaccine against hepatitis B is prepared using subunit viral surface antigen purified from the blood of healthy carriers.

The vaccine induces antibody and is highly protective against infection, disease, and the carrier state. It is safe.

More than 2 million persons have received the vaccine. It is considered to be the first vaccine against human cancer.

---

protective immune response can be mounted. Most present-day viral vaccines are made using viruses grown in cultivated cells, but this is not possible for hepatitis B since the virus does not propagate in cell culture. It was found, however, that the surface antigen of hepatitis B virus which contains the protective immunologic determinants may circulate in large amount in the blood of otherwise healthy viral carriers. Starting in 1968, our laboratories developed a system for extracting and purifying this viral subunit antigen that was the source of antigen for present plasma-derived vaccine (1). The vaccine, given in alum adjuvant, induced antibodies in animals and man and afforded high-level protection against infection and disease on exposure to the virus. This vaccine, for which clinical tests were started in 1975, was licensed for general use in 1981. It has proved highly safe and effective in the more than 2 million persons, worldwide, who have received it to date. Since the vaccine prevents infection with the virus, it must certainly prevent its sequelae too. Because of this, the hepatitis B vaccine is considered to be the first vaccine against human cancer.

4. YEAST-DERIVED HEPATITIS B VACCINE AS A MODEL FOR VIRAL SUBUNIT VACCINES

Though entirely satisfactory in its performance, the production of hepatitis B vaccine from human plasma is limited by (a) the available supply of suitable plasma from chronically infected carriers of hepatitis B virus and by (b) the need to apply highly technical procedures to purify the antigen as well as to insure inactivation of all infectious agents that might be present in human plasma. For this reason, a new vaccine has been developed in our laboratories (2) that is prepared using hepatitis B surface antigen produced by recombinant technology in yeast.

This new vaccine was made possible by important new advances in molecular genetics whereby the genes that specify or encode the desired antigens are inserted into plasmid vectors which, in turn, cause the synthesis of the antigens in host cells such as those of yeast or mammalian tissues.

The deoxyribonucleic acid or DNA of hepatitis B virus (Figure 1) is in a partially double-stranded circular form and carries the code for a number of antigens of the virus such as the core antigen that lies within

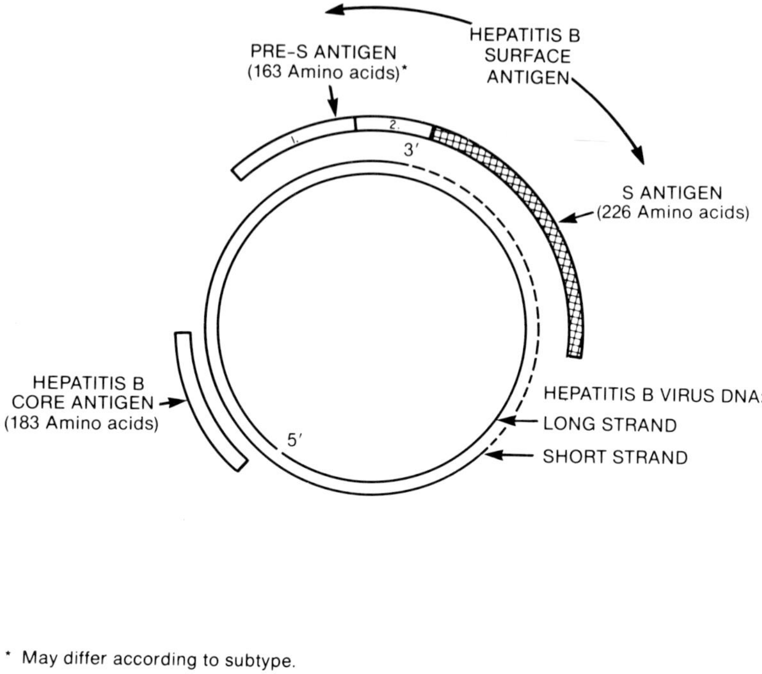

FIGURE 1. Hepatitis B virus genome and defined antigens that are produced.

the virus and the surface antigen that is embedded in the lipids of the viral surface membrane. The surface antigen is composed of two regions - the S region and the pre-S region. The gene that encodes the S region alone is used in the plasmid construct for making recombinant hepatitis B antigen. Pre-S antigen is under study at present to determine the advantage, if any, and safety for adding it to the vaccine.

The hepatitis B virus S antigen gene was inserted (Figure 2) by Valenzuela et al. (3) into a circular ring of DNA that is capable of independent replication in both bacterial and yeast cells. The S gene was inserted between a GAPDH promoter and an ADH-1 terminator for gene expression to form what is called the S antigen "cassette". Yeast cells that are transfected with the vector produce very large amounts of surface antigen when grown in fermentation tanks.

Purification of the expressed antigen (Table 3) was achieved by disruption of the yeast cells followed by a series of chromatographic separation steps. The purified antigen was formulated into an alum adjuvant, just as is the plasma-derived vaccine. The recombinant vaccine was found in very extensive laboratory and clinical studies to perform equally as satisfactorily as the plasma-derived vaccine. Most importantly, its safe-

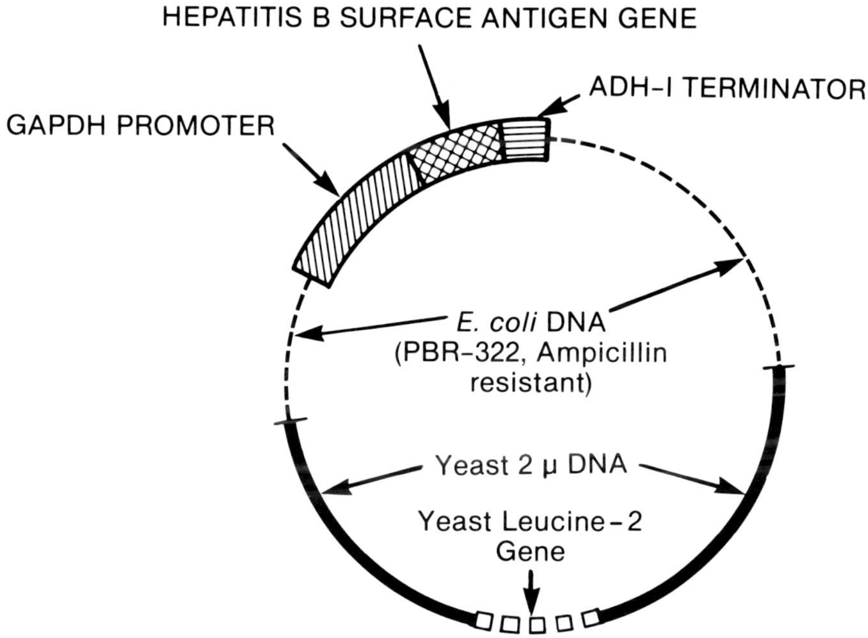

FIGURE 2. Construct of plasmid (pHBS56-GAP347/33) used to produce hepatitis B surface antigen in yeast.

TABLE 3. Yeast recombinant hepatitis B vaccine

Yeast cells transfected with the plasmid vector carrying the S antigen gene produce large amounts of surface antigen.
The surface antigen is purified by cell disruption followed by chromatographic separation steps.
Recombinant hepatitis B vaccine in alum adjuvant performs the same as plasma-derived vaccine.
Safety, immunogenicity and protective efficacy were proved in extensive field trials in man.
The vaccine was licensed in West Germany in May 1986 and in the U.S.A. in July 1986.

ty, immunogenicity and protective efficacy were proved in very extensive field trials in man and the vaccine was licensed for general use in West Germany in May 1986 and in the United States in July 1986. It is important to note that the RIT Laboratories which were founded by Professor De Somer have also developed a recombinant yeast vaccine against hepatitis B that has performed very well in the human trials that have been reported to date.

The yeast recombinant hepatitis B vaccine represents the world's first recombinant vaccine, and it will establish precedents for all such vaccines that are to follow. Quality Control has been an important aspect of the vaccine development since tests of the final product alone may not be adequate to establish the fidelity of the polypeptides and proteins as presented to the immune system. For this reason, both the host-vector system and the antigen product of that system are analyzed. Tests are performed to assure preservation of the integrity of the DNA construct, faithful translation of the surface antigen message, assumption by the polypeptide of its appropriate secondary and tertiary structures, presentation by the antigen of its appropriate immunologic determinants (epitopes), and an adequate level of the purity of the product. Finally, the host responses to the antigen are measured and these must be of adequate quality and quantity.

5. POLYTOPIC VACCINES

In its simplest sense (Table 4), one might consider the new science of vaccines as a new era in which the ultimate target will be to define the shortest length of an antigen that carries the wanted specificity and that can impart a useful immune response for an adequate period of time. In effect, this is the search for epitopes and will be far simpler to achieve for continuous epitopes, that are contained within a single amino acid sequence, than for discontinuous epitopes that may be made of distal parts of the protein that are brought together by secondary folding of the molecule. The important matter is that elimination of the great bulk of immunologically useless substance opens the way to including highly diverse collections of epitopes in simple polytopic vaccines (4) that protect against a large number of infectious agents. The immune system is ca-

TABLE 4. Polytopic vaccines

The ultimate target is to define the shortest sequences of amino acids that carry desired specificity and immunogenicity of a number of important antigens.
Continuous epitopes will be far easier to produce than discontinuous epitopes.
Eliminating immunologically useuless bulk opens the way to simple polytopic vaccines.
Antigen presentation and potentiation are critical to the success of polytopic vaccines.
Practical limitations to the number of doses of vaccine a person may receive makes complex polytopic vaccines especially important.

pable of almost unlimited response to a diversity of epitopes but it can also be quite capricious in what it accepts or rejects as an antigen. For this reason, much attention is being focused on the form of the presentation of antigens to the cells of the immune system. Potential of the immune response by presentation in micelles (5), in liposomes (6), in immunostimulatory complexes (6,7) and in ordinary adjuvants such as alum may serve satisfactorily for some antigens, but it stands to be improved and augmented, perhaps, by such devices as coadministration of immunoregulatory substances that are still to be discovered and developed.

It is important, in the practical sense, that there is a limit in the number of injections of vaccine that will be given to any one person. The merit of polytopic vaccines lies in the feasibility for protecting against many diseases after the fewest number of injections. These complexes of epitopic mixtures might be created by recombinant technology or even synthetically and this, I believe, will be an important direction for future vaccine research.

6. RECENT ACCOMPLISHMENTS OF VACCINE USE

Perhaps the most exciting accomplishment in the area of viral diseases during the past 3 decades has been the accelerated development of new vaccines whereby diseases can be prevented or largely eliminated from heavily immunized population groups.

The effect of application of 7 principal vaccines in the U.S.A. on reducing disease and mortality is illustrated quite dramatically in a table prepared recently by the U.S. Centers for Disease Control in Atlanta (8). In Table 5, the maximum number of cases of the 7 corresponding diseases in the U.S.A. in the peak year prior to starting vaccination is compared with the number of cases that occurred in 1985. The reduction in morbidity for each disease was in excess of 98 % except for tetanus that was 89 %. Neonatal tetanus, however, has been all but eliminated.

The individual reductions in measles, mumps and rubella are shown in Figures 3, 4 and 5. Measles vaccine, first licensed in 1963, reduced the number of reported cases from almost 900,000 to less than 3,000 in 1985.

Mumps vaccine, licensed in 1967, brought the number of reported cases from more than 150,000 to about 3,000 per year. Rubella vaccine was licensed in 1969 and brought the reported cases from almost 60,000 to less than a thousand. Simultaneously, the congenital rubella syndrome has all but disappeared.

TABLE 5. Reduction in morbidity of vaccine-preventable diseases

Disease	Maximum cases		Reported cases 1985	% Reduction
	Number	Year		
Diphtheria	206,939	1921	2	99.99
Measles	894,134	1941	2,704	99.70
Mumps	152,209	1968	2,886	98.10
Pertussis	265,269	1934	3,275	98.77
Poliomyelitis	21,269	1952	5	99.98
Rubella	57,686	1969	604	98.95
Congenital	20,000	1064-65	0	100.00
Tetanus	601	1948	71	89.19

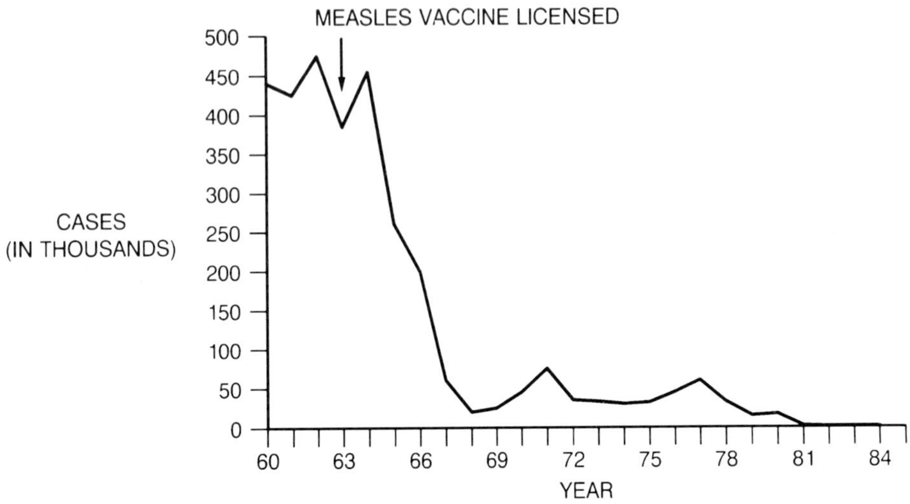

FIGURE 3. Reported measles cases, United States, 1960-1985 (provisional data). (Source : Dr. A. Hinman, CDC.)

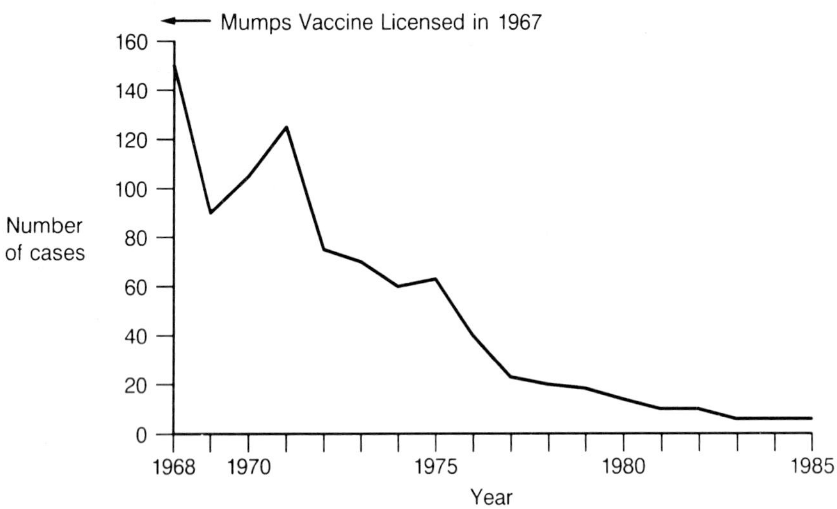

FIGURE 4. Reported mumps cases, United States, 1968-1985 (provisional data). (Source : Dr. A. Hinman, CDC.)

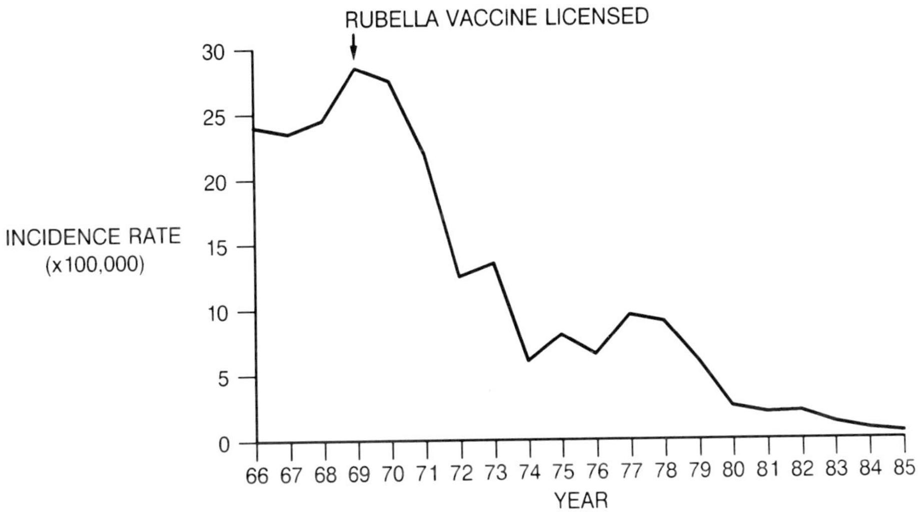

FIGURE 5. Rubella incidence rate per 100,000 population, United States, 1966-1985 (provisional data). (Source : Dr. A. Hinman, CDC.)

The meaning of vaccination can sometimes be told more forcibly in relation to reduction in the burden of illness and its resulting costs (Table 6) than in numbers of cases. In a recent study by White and his colleagues (9) at the Centers for Disease Control, it was calculated that the combined costs for measles, mumps and rubella in the U.S.A. in the absence of vaccine would be almost 1.4 billion dollars per year. The net estimated savings for the single year, 1983, were almost 1.3 billion dollars with a benefit-to-cost ratio of 14:1. The social benefits from extended productive life need also to be considered as exemplified in the use of pneumococcal vaccine in the elderly (10).

TABLE 6. Benefits and costs of the measles, mumps and rubella vaccination program in the U.S.A.

Vaccine	Disease savings	Program cost Vaccine & reactions (combined vaccine)	Benefit/ cost ratio
Measles	$ 669,312,000		
Mumps	$ 339,455,000		
Rubella	$ 379,398,000		
Total	$ 1,388,165,000	$ 96,192,000	14:1
	Savings in 1983 : $ 1,291,973,000		

7. THE APPLICATION OF VACCINES

The economic and social benefits that can be gained by the development and application of vaccines is enormous compared with that for other health procedures. Logic alone would suggest that vaccines be universally applied. Unfortunately, it is at least as difficult to accomplish the use of vaccines as it is to develop them. The application of vaccines involves a disparate complex of factors that involve social, political, economic and individual considerations, with contrasts that are conspicuously evident between the developed and the developing nations of the world. All nations have much to gain through immunization, but those in greatest need are least likely to be immunized. In contrast, nations that are most capable economically are most likely to neglect to immunize.

7.1. Developing nations (Table 7)

One of the most outstanding conditions in developing nations is the limitation of economic resources and health delivery capabilities that restrain the application of health procedures that may be useful or even indispensable. The total expenditure for all health care in many Third World countries may be no more than two or three dollars (U.S.) per person per year. Much of the vaccination initiative in developing nations is given by the World Health Organization's EPI, or Expanded Program for Immunization (11) that seeks to immunize all children against diphtheria, pertussis, tetanus, poliomyelitis, measles, and tuberculosis by 1990. Though in effect for several years, only 20 % to 30 % of the total EPI goal has been achieved. Vaccines are generally available without cost from UNICEF, WHO, governments, and charitable organizations, but they are poorly utilized because of lack of local resources, and the basic health delivery infrastructure needed to vaccinate the population.

The matter of costs for vaccines has been complicated by the recent incorporation of hepatitis B vaccine in the EPI program. The initial objective will be to vaccinate all newborns in high prevalence areas in order to halt the transmission of hepatitis B virus infection from carrier mothers to their offspring. Hepatitis B vaccine is very expensive to produce compared with other EPI immunogens and it is unlikely that it can be furnished at low cost or without charge as are the other EPI vaccines. For

TABLE 7. Application of vaccines in developing nations

Economic and social benefits of vaccines can be enormous.
Application of vaccines depends on a disparate complex of factors involving social, political, economic, and individual considerations.
Developing nations are limited by economic resources and health delivery capabilities.
The EPI program seeks to immunize all children against 6 diseases by 1990.
Incorporation of hepatitis B vaccine imposes large cost burdens that might be overcome by local vaccine production.
Hepatitis B vaccination is supported by economic incentives since the disease usually occurs in adulthood in the most economically productive years.

this reason, some nations of the Third World are seeking to manufacture their own hepatitis vaccine, taking advantage of their local labor costs and using manufacturing technology licensed from the outside. The development of the new yeast recombinant technology should go far toward making vaccine production feasible in the practical sense. Immunization against hepatitis B is strongly supported by economic incentives in developing nations since the late and important manifestations of the disease, i.e., cancer and cirrhosis, occur usually in adulthood during the most economically productive years.

7.2. Developed nations (Table 8)

The contrast is seen in developed nations where resources are large but overall motivation for vaccination is disproportionately low by comparison. One notable exception to this is the accomplishment by the public and private sectors in immunizing children in the U.S.A. This success was assisted by the traditional interests of pediatricians to prevent disease, by immunization programs carried out under governmental initiative, and by the legal requirement for immunization in order to attend school.

There is no similar record of success in immunizing adults in the U.S.A. Only a small proportion of adults at high risk to influenza, pneumonia, and hepatitis B have been vaccinated and there are deficiencies in immunity against tetanus, mumps, and rubella. The reason for underutilization of vaccines lies principally with apathy, cost, lack of an immunization tradition among adult health care providers, and lack of any legal compulsion. Many people do not wish to be "treated" to prevent diseases they might never experience anyway. The outlook for immunization of adults in the U.S.A. may be improving, however, as a result of public education in a new health initiative being carried out by government, academia, private and public health care providers, institutions, and foundations. The wheels move slowly, but deliberately !

TABLE 8. Application of vaccines in developed nations

Unlike Third World nations, Developed nations have large resources but low motivation for vaccine use, especially in adults.

Pediatric immunization in the U.S.A. has been exemplary because of interest by pediatricians, governmental initiatives, and legal requirements.

Adult immunization for influenza, pneumonia and hepatitis B in high risk groups is poor.

Apathy, cost, and lack of an immunization tradition among adult health care providers are contributory.

The situation is improving because of new initiatives by government, academia, private and public health care providers, institutions and foundations.

---

8. CLOSING REMARKS

It seems fair to conclude that the new era of genetic biotechnology has opened the door to disease control that will dwarf the achievements of the past. Disease prevention by vaccination may become imperative based on a need to eliminate the costs for illnesses that might have been averted. Together with expanded immunization, we hear of increasing consideration for national and international immunization programs to eliminate disease and perhaps eventually to eradicate (see 12,13) infections such as poliomyelitis, measles, rubella, and hepatitis, following the successful world eradication of smallpox that was achieved by 1979 (14). It is expected that production of new subunit vaccines will follow in the footsteps of recombinant hepatitis B vaccine as the first and most applicable approach to new vaccines for the immediate future. The development of the recombinant vaccine may be especially timely in providing the most likely source of antigens for use in vaccines against the acquired immune deficiency syndrome (AIDS) once the meaningful antigens have been identified. An effective vaccine against AIDS, I believe, might serve as a powerful stimulant of interest for immunization of adults against the presently underutilized vaccines, as well as against AIDS itself.

I think that Professor De Somer would have enjoyed the succeeding decades that are to come, watching the development of new vaccines and new initiatives for their use. We all shall miss his wise counsel, but we shall be benefited by the heritage he left behind in his scientific work, in his direction of the University, and in the Rega Institute and the RIT Laboratories that he founded.

REFERENCES

1. Hilleman MR, Buynak EB, McAleer WJ, McLean AA, Provost PJ, Tytell AA. in Viral Hepatitis (Szmuness W, Alter H, Maynard J, Eds). The Franklin Institute Press, Philadelphia, pp 385-397, 1982.
2. Hilleman MR, Weibel RE, Scolnick EM. J. Hong Kong Med. Assoc. 37:75-85, 1985.
3. Valenzuela P, Medina A, Rutter WJ, Ammerer G, Hall BD. Nature 298: 347-350, 1982.
4. Hilleman MR. Dev. Biol. Stand. 63:5-13, 1986.
5. Zuckerman AJ. in The Liver Annual 1 (Arias I, Frenkel M, Wilson JHP, Eds). Elsevier, Amsterdam, pp 81-107, 1981.
6. Allison AC. in New Approaches to Vaccine Development (Bell R, Torrigiani G, Eds). Schwabe & Co. AG, Basel, pp 133-156, 1984.
7. Morein B, Simons K. Vaccine 3:83-93, 1985.
8. Hinman AR, Bart KJ, Orenstein WA. Int. J. Epidemiol. 14:502-503, 1985.
9. White CC, Koplan JP, Orenstein WA. Am. J. Pub. Hlth. 75:739-744, 1985.
10. Sisk JE, Riegelman RK. Ann. Int. Med. 104:79-86, 1986.
11. Henderson RH. Rev. Infect. Dis. 6 (suppl. 2):S475-S479, 1984.
12. Stuart-Harris Sir Charles. Rev. Infect. Dis. 6:405-411, 1984.
13. Evans A. Am. J. Epidemiol. 122:199-207, 1985.
14. Behbehani AM. Microbiol. Rev. 47:455-509, 1983.

THE CHALLENGE OF CHRONIC VIRUS INFECTIONS

F. RAPP
Department of Microbiology, The Pennsylvania State University, College of Medicine, Hershey, Pennsylvania 17033, U.S.A.

1. INTRODUCTION

The career and contributions of Prof. Piet De Somer have bridged the era between the solution of major problems concerning acute virus infections and the newly emerging problems concerning the control of chronic infections due to microorganisms and especially those caused by viruses. His work on the ribonucleic acid of poliovirus hallmarked the beginning of molecular virology, and helped related investigations to stay the fear of poliomyelitis epidemics. He then pursued investigations on antiviral drugs and interferon, work still of great interest and importance. His strong interest in cancer viruses at a time when it was widely believed that viruses played no role in human neoplastic disease, has been amply justified by the discovery of a class of retroviruses associated with human T-cell lymphomas (HTLV) and acquired immunodeficiency syndrome (AIDS) (1,2) which predispose individuals to a variety of malignancies. I have no doubt that were he still alive today, his research interest would focus on chronic virus infections and ways to control or prevent disease by these viruses.

Acute diseases due to viruses have been known for millenia. The early work of Jenner, which resulted in the development of a vaccine against smallpox, was followed by that of Pasteur and the production of an anti-rabies vaccine. These pioneering studies stimulated a variety of investigations demonstrating that various diseases of plants, animals, and man were due to "filterable agents", subsequently referred to as viruses. The famous observations of Ellermann and Bang (3) and of Rous (4) suggesting that chicken leukemia and chicken sarcomas, respectively, could be induced by viruses undoubtedly focussed attention on the possibility that malignancy might be caused by this class of microorganisms.

These observations, coupled with developments in cell culture and media to support these cultures, subsequently led to the large-scale introduction of a vaccine against yellow fever (5) and to vaccines against a number of acute virus infections (i.e., poliomyelitis, rubeola, rubella, and mumps). While some of these vaccines, including the more controversial influenza virus vaccine, have had variable success, those against poliomyelitis, measles (rubeola), and rubella have been especially successful (Table 1). The most successful has been the smallpox vaccine, which has effected the total eradication of this disease from the world population. The elimination of smallpox was aided by the absence of an animal reservoir for the virus and by judicious use of the vaccine by public health officers on a global scale. Nobel prizes were awarded to many of the key investigators for these studies concerning the properties of viruses and their control (Table 2).

The successful control of acute virus infections, reductions in bacterial diseases because of improvements in sanitary engineering, and the

F. Rapp

TABLE 1. Development of successful virus vaccines for human use.

Virus	Year	Type of vaccine
Smallpox	1798	Attenuated
Rabies	1885	Inactivated
Yellow fever	1940	Attenuated
Polio	1953	Inactivated
	1955	Attenuated
Rubeola	1963	Attenuated
Rubella	1969	Attenuated
Mumps	1968	Attenuated
Hepatitis B	1981	Inactivated

widespread use of antibiotics suggested to many that chronic virus infections could be controlled in similar fashion. Far-reaching discoveries have prompted such hopes in every decade (Table 3). Regrettably, the poor record of medicine in controlling chronic infections due to any microbial organism, including malaria, tuberculosis, syphilis, gonorrhea, as well as many parasitic infections in Third World countries, was not taken into account in these calculations.

In the face of the first increase in the incidence of gonorrhea in the United States in ten years, with resistance of the disease to tetracycline and sulfonamide on the rise and because the organism poses a serious fertility problem in the Third World, the development of vaccines against *N.gonorrhea* have received a new impetus. Regrettably, protection by antipilus and anti-P-1 protein vaccines against *N.gonorrhea* has not been realized despite the induction of neutralizing antibody. Syphilis remains a public health problem and astonishingly, the role of the immune system in moderating or exacerbating the disease remains a mystery. Equally disappointing is the knowledge that systemic vaccination against chlamydial infection may cause tissue damage. Vaccines against these chronic infections were marginal at best and for many virus diseases, useful vaccines await the future (Tables 4 and 5). Chronic infections have posed far greater problems than envisioned and will continue to resist efforts for control despite the impact of molecular biology, biotechnology, and advances in immunology.

Chronic virus infections can be subdivided into those involving the central or peripheral nervous systems, the vascular system, and mucosal cells lining various body tissues. Of special interest are viruses affecting the vascular and nervous systems because they are generally more severe and have more far-reaching consequences to the patient. In addition, they flare up most readily in patients on immunosuppressive therapy to prevent transplant rejection or as part of chemotherapeutic regimens for the treatment of cancer. These patients, already at high risk, are further endangered by viruses activated by such treatment.

Diseases now known or thought to be associated with chronic virus infections include subacute sclerosing panencephalitis (SSPE, due to rubeola virus); neurologic symptoms due to human immunodeficiency virus (HIV); multiple sclerosis (perhaps due to an HTLV-like virus); Alzheimer's and Creutzfeldt-Jakob disease (unproven but suspected virus etiology). Viruses found in the vascular system include, among others, those associated with

TABLE 2. Nobel Prize research that impacted on virus discoveries.

Year	Recipient(s)	Contribution
1946	Wendell Stanley	Crystallization of tobacco mosaic virus
1951	Max Theiler	Yellow fever and attenuated vaccine
1954	Thomas Weller, John Enders, Frederick Robbins	Cultivation of polioviruses in cell culture; advent of modern virology
1958	Joshua Lederberg, Edward Tatum, George Beadle	Discovery that genes act by regulating specific chemical processes; genetic recombination; bacterial genetics
1959	Arthur Kornberg	Mechanisms in biological synthesis of nucleic acids
1965	Jacques Monod, Francois Jacob, André Lwoff	Genetic regulation of enzyme synthesis and lysogeny
1966	Charles Huggins, Peyton Rous	Research on chemical and viral etiology and cancer
1969	Max Delbrück, Alfred Hershey, Salvador Luria	The role of genetic material in bacterial and viral replication
1975	David Baltimore, Howard Temin, Renato Dulbecco	Characterization of the interaction between tumor viruses and the genetic material of the cell
1976	Baruch Blumberg, D. Carleton Gajdusek	Delineation of long incubation diseases caused by viruses
1978	Daniel Nathans, Hamilton Smith, Werner Arber	Development of restriction endonuclease technology; basis of bacterial resistance to bacteriophages
1980	Paul Berg, Walter Gilbert, Frederick Sanger	Pioneering work in recombinant DNA technology; methodologies for DNA sequencing
1982	Aaron Klug	Development of crystallographic electron microscopy elucidating the structure of viruses
1983	Barbara McClintock	Discovery of transposable genetic elements
1984	Cesar Milstein, Georges Köhler, Niels Jerne	Development of technique of monoclonal antibody formation and immuno-plaque technique
1986	Rita Levi-Montalcini, Stanley Cohen	Discovery of factors controlling cell growth and development

TABLE 3. Some important milestones in the development of virology.

Date	Contributor(s)	Contribution(s)
1798	Jenner	Developed vaccine against smallpox
1885	Pasteur	Produced rabies vaccine
1892	Ivanowski	Recognized a "filterable agent" as responsible for tobacco mosaic disease
1898	Löeffler and Frosch	Reported that the causative agent of foot-and-mouth disease was a virus
1899	Beijerinck	Demonstrated multiplication of tobacco mosaic virus in infected plant tissues
1902	Reed and Carrol	Discovered yellow fever virus
1908	Ellermann and Bang	Demonstrated transfer of chicken leukemia with cell-free extracts
1911	Rous	First evidence of a virus producing a solid tumor in an animal (chicken)
1915, 1917	Twort and d'Herelle	Separately discovered bacteriophage: d'Herelle developed the plaque assay and introduced the term bacteriophage
1928	Elford	Developed ultrafiltration methods providing comparative data on virus size
1931	Woodruff and Goodpasture	Introduced the use of embryonated eggs to propagate viruses
1933	Shope	Produced papillomas in rabbits with a filterable agent
1933	Ruska	Built first electron microscope
1935	Stanley	Crystallized tobacco mosaic virus
1936	Bittner	Described mouse mammary tumor virus and its transmission by milk
1939	Delbrück	Began systematic studies of bacteriophage
1940	Theiler	Developed effective yellow fever vaccine
1941	Hirst	Demonstrated that influenza virus would agglutinate red blood cells
1946	Lederberg and Tatum	Reported genetic recombination
1949	Enders, Weller and Robbins	Developed human cell cultures for quantitating replication of poliovirus by cytopathology
1950	Coons	Developed immunofluorescence technique
1951	Gross	Induced lymphoid leukemia in mice with cell-free extracts of tumors from leukemic mice

TABLE 3. (continued).

1952	Dulbecco	Developed the plaque technique for assaying animal viruses
1952	Hershey and Chase	Showed that only the DNA of bacteriophage is required for replication
1952	Zinder and Lederberg	Discovered transduction
1953	Lwoff	Discovered lysogeny
1955	Salk; Sabin; Koprowski	Developed effective poliomyelitis vaccines
1957	Isaacs and Lindenmann	Described interferon as antiviral
1960	Sweet and Hilleman	Discovered simian virus 40 (SV40)
1964	Epstein, Achong and Barr	Described the presence of a herpes-type virus (EBV) associated with Burkitt lymphoma
1964	Temin	Postulated that viral RNA can direct the synthesis of DNA provirus
1965	Blumberg	Identified hepatitis B virus
1966	Benjamin	Identified virus nucleic acid in transformed cells
1970	Baltimore; Temin and Mizutani	Independently found evidence of RNA-directed DNA polymerase (reverse transcriptase) in virions of RNA tumor viruses
1971	Gajdusek	Established Kuru as a viral disease
1965-1972	Arber; Smith and Wilcox; Danna and Nathans	Developed restriction endonuclease technology
1975	Berg; Maxam and Gilbert; Sanger	Developed recombinant DNA technology and methodologies for DNA sequencing
1975	Milstein and Köhler	Monoclonal antibody production
1977	Brugge and Erikson	Identified _scr_ gene of Rous sarcoma
1980	Gallo	Isolated first human retroviruses (HTLV-I and II)
1982	Weinberg; Cooper; Wigler	Oncogenes described and partially characterized
1983	Montagnier; Gallo	Separately isolated human immunodeficiency virus

TABLE 4. Marginally effective human virus vaccines.

In use
Adenoviruses (military recruits)
Influenza virus (older population)
Experimental
Cytomegalovirus (transplant recipients)
Varicella-zoster virus (selected children)

TABLE 5. No vaccines available.

Virus	No. of types
Herpes simplex virus	2
Epstein-Barr virus	1
Papillomaviruses	> 45
Polyomaviruses (BK, JC)	2
Creutzfeldt-Jakob virus	1
Respiratory syncytial virus	1
Rhinoviruses	> 100
Coxsackie virus (A, B)	> 30
Hepatitis A virus	1
Coronaviruses	2
Human T-cell lymphoma virus	2
Human immunodeficiency virus	> 4

AIDS (now known to be caused by HIV); cytomegalovirus, which can multiply in many tissues but quite often is implicated in interstitial pneumonia; and hepatitis B virus. Of the many viruses affecting other tissues, their association with human cancers has still to be ruled out especially in light of the presence of virus genetic material in many cancer tissues. Most prominent are the papillomaviruses, which cause a variety of warts and are associated with genital neoplasia and neoplastic diseases of other organs.

A number of viruses are spread by sexual contact (Table 6) : hepatitis B virus, herpes simplex virus type 2, cytomegalovirus, HIV, and certain papillomaviruses. Diseases caused by these viruses are governed by social and demographic factors (Table 7) and often affect the relatively young.

TABLE 6. Sexually transmitted viruses.

Herpes simplex virus types 1 and 2
Cytomegalovirus
Papillomavirus types 6, 11, 16, 18, 31, 33, and 35
Hepatitis A and B
Human immunodeficiency virus

TABLE 7. Social and demographic factors influencing trends in STDs.

Age distribution of population
Use of contraceptives
Sexual behavior

2. PROBLEMS ASSOCIATED WITH HERPES SIMPLEX VIRUSES (HSVs) AND CYTOMEGALOVIRUS

The herpesvirus group comprises numerous structurally related and morphologically indistinguishable viruses that infect humans and probably all animal species. The members that infect humans include HSV types 1 and 2 (HSV-1 and HSV-2), varicella-zoster virus (VZV), human cytomegalovirus (HCMV), Epstein-Barr virus (EBV). All can be life-threatening, reside in the host for extended periods of time, and can cause continued recurrent disease long after initial infection of the human host.

The DNA of herpesviruses has a molecular weight ranging from 80 to 150 million, with HSV having a DNA of 95 million, and the largest, HCMV, having a DNA of 150 million. All have a capsid surrounded by a lipid envelope containing a large number of proteins, of which several are glycosylated. Although morphologically indistinguishable, the DNA of these viruses varies both in size as well in structure (Fig. 1). The DNA of HSV-1, HSV-2, and HCMV can exist in four isomeric forms (6). The long and short unique regions are covalently linked and the junction region contains reiterated inverted repeats that are reproduced at both ends of the genome. The long and short components can invert independently producing four isomeric structures. The unique DNA structure of these three herpesviruses is by no means a requirement, as many other herpesviruses have only one isomeric form and cannot invert either of the two components. It is not known why evolutionary forces have enabled HSV-1, HSV-2, and HCMV to produce these four forms when only one is needed for infectivity and the rest appears to be of little consequence for infection or for virus survival.

Herpesviruses have a similar method of replication involving coordinated temporal synthesis (7). The genes involved in replication are designated immediate early (α), early (β), and late (γ) genes (Fig. 2). As a rule, one class of genes needs to be transcribed and translated for the second class to begin synthesis; synthesis by the next class generally shuts off expression of the earlier genes. There are five immediate early genes, at least one of which is required to activate all subsequent transcription. The early and late genes are a far larger and heterogeneous

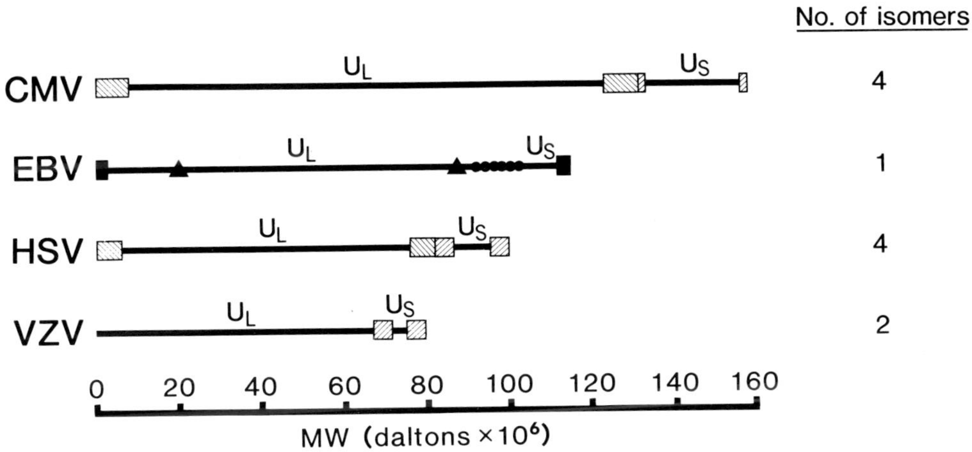

FIGURE 1. Comparative DNA structure of human herpesviruses.

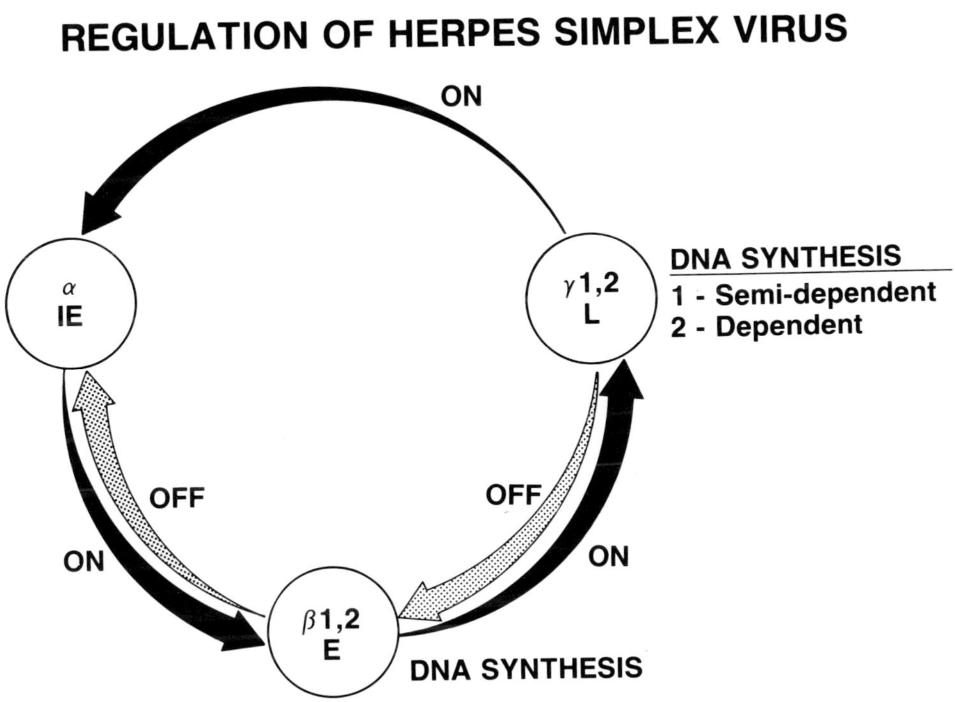

FIGURE 2. Regulation of replicative cycle of herpesviruses.

group; the early genes are primarily involved in the synthesis of virus DNA and include such well characterized virus genes as DNA polymerase, thymidine kinase, ribonucleotide reductase, as well as a number of major DNA binding proteins. The late genes generally specify capsid proteins including the glycoproteins found in the virion.

Of great importance is the fact that differences in the endonuclease digestion patterns of HSV-1, HSV-2, and HCMV DNA allow one to trace the virus during infection (8). This is mirrored by differences in polypeptide patterns. Whether such variations will ultimately confuse vaccine protocols and lower the efficacy of vaccines remains to be further investigated.

Genital herpes now has a reservoir of 40 to 60 million infected individuals in the United States. The disease has been spreading dramatically for ten years. There also is a major asymptomatic and undiagnosed reservoir, with an unknown percentage of these individuals shedding virus and therefore representing putative sources of infection. This has amplified the risk to newborns, resulting in an increase in neonatal infection from 4.8 to 14 cases per 100,000 live births during the last twenty years in the Seattle, Washington area (K. Holmes, personal communication).

A major problem associated with HSVs is the fact that they are introduced into the population at a fairly young age: HSV-1 generally infects children before they reach the age of 6 and HSV-2 often is introduced shortly after the onset of sexual activity. Following such introduction, both viruses can remain latent in the peripheral neurons, thus escaping elimination by both antibody and specific T-cell responses. An important fact is that there is 50 % base sequence homology between the nucleic acids of HSV-1 and HSV-2 (9), which is reflected in antigenic similarities in a number of HSV proteins, including some of the glycoproteins. This leaves some doubt as to the efficacy of vaccination. An individual infected with HSV-1 earlier in life and carrying antibodies capable of neutralizing both type 1 and 2 viruses is still susceptible to infection of the genital tract by HSV-2. Thus, natural immunity does not prevent reinfection by a virus that can be neutralized by circulating antibody and against which a T-cell-mediated response is present. This suggests that an innovative approach to vaccination against HSV is needed if the related diseases are to be prevented. Of further concern is the fact that the latent virus can travel down the anoxal pathway and cause recurrent disease in the face of an active immune response. Thus, any vaccine that fails to prevent the establishment of latency will not fulfill the purpose of preventing either virus infection or virus activation. Therefore, it becomes important to study how viruses remain latent and what causes activation.

In studying problems concerned with virus latency, we have chosen to establish _in vitro_ cell culture systems to investigate the molecular events connected with virus latency. The problems posed include a better definition of the state of the viral genome during latency, the molecular steps required to activate the virus, and hopefully some way to prevent activation or to eradicate the virus in latently infected cells.

With the help of Professor Erik De Clercq, systems have been established in human embryo fibroblasts and in human neurons obtained from aborted fetuses (10,11). These studies have revealed that it is possible to establish virus latency [i.e., no virus can be detected (12,13)], that the latent virus is present as DNA and does not appear to be integrated in the host cell chromosomes (10), and that it is possible to activate the virus by a variety of techniques including superinfection with genetically unrelated but morphologically related HCMV (14). These studies established that the superinfecting virus can induce HSV to replicate in latently infected cells and that the portion of the HCMV genome involved is a region of at least 22 kilobases (unpublished data). The task now will be to further narrow that portion of genome involved and to determine the functions associated with the HCMV genome that activate HSV.

In accompanying experiments, attempts to eradicate the virus in cell culture generally have proven unsuccessful. Nevertheless, there are imaginative ways to approach this problem using newer methods in biotechnology and we hope to carry out such experiments in the future.

Latency studies have been carried out with HCMV with similar results. However, it is noteworthy that it is more difficult to establish latent HCMV infection, perhaps because the replication cycle of this virus is so much longer than that of HSV. This is mirrored in the human body where HCMV tends to persist with a small amount of virus being released over long periods of time and where true virus latency has not been firmly established. That the immune system helps to check the spread of HCMV as well as HSV-1 and HSV-2 has been clearly established by the fact that immunosuppressive therapy causes massive infection of many organs by HCMV and may be life-threatening to those being treated with such drugs. This resulted in attempts to produce an attenuated live virus vaccine (15). However, the use of an HCMV vaccine in the general population is contraindicated because of the transforming properties of HCMV (16) and the failure to establish a good animal model enabling safety testing in non-human hosts. It is pertinent to note that very similar problems are involved with VZV vaccines; experimental VZV vaccines already undergoing extensive clinical trials worldwide (17) suffer from the same deficiencies that the HCMV vaccine has associated with it.

Such studies are especially important because HCMV infections are common worldwide, and result in numerous clinical syndromes. However, of special concern is the ability of HCMV to cross the human placenta; fully 1 % of all live births are congenitally infected by this virus, with 5 to 10 % of these infants suffering clinically manifest disease and an additional 10 to 30 % eventually demonstrating developmental or physical handicaps. The virus also has emerged as a major pathogen in immunosuppressed patients and in those suffering from AIDS. Thus, attempts to control this pathogen should receive high priority.

3. PROBLEMS ASSOCIATED WITH PAPILLOMAVIRUSES

Papillomaviruses have long resisted cultivation in cell culture systems, thus thwarting investigations into the properties of these viruses. Reports on papillomaviruses reach back as far as 1907. Ciuffo (18) observed that human warts could be induced by filterable agents (now known as papillomaviruses) and the pioneering work of Shope and his colleagues (19) on cottontail rabbit papillomavirus provided the first model of a naturally occurring virus-induced neoplasm. The number of papillomaviruses associated with human diseases has been open to question. However, recent advances in biotechnology have greatly catalyzed progress in this field.

There are approximately 45 human papillomaviruses (HPVs), based on characterization by DNA homology; if an HPV shows less than 50 % homology with existing types, the HPV is classified as a new type. Many HPV DNAs have been reproduced _in vitro_ and their genes have been translated. Thus, the genome of papillomaviruses follows a closed circle of double-stranded DNA and is well characterized (Fig. 3). Also well established are the tissue locations of some of the papillomaviruses. For example, it is now known that HPV types 6, 11, 16, 18, 31, 33 and 35 are involved in the

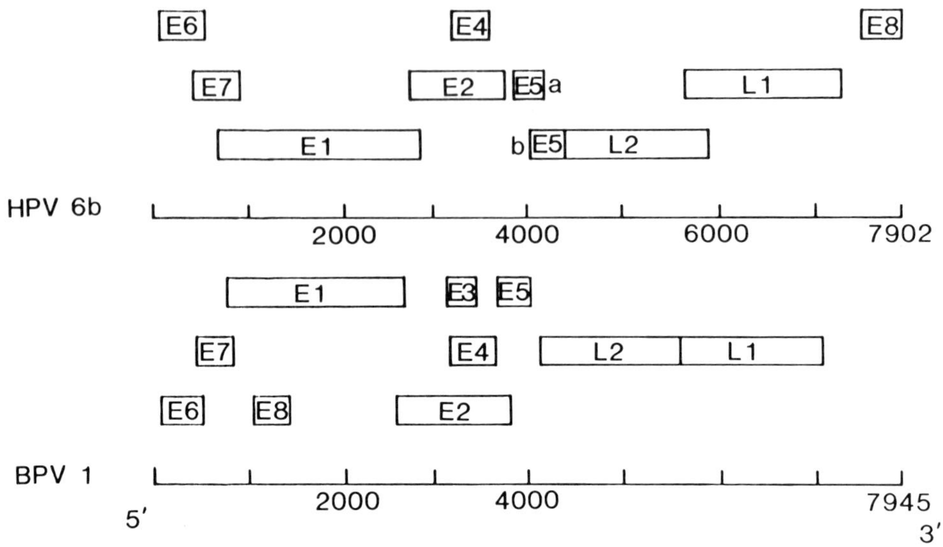

FIGURE 3. Genome organization of papillomavirus DNAs; human papillomavirus type 6b (HPV 6b) and bovine papillomavirus (BPV 1).

etiology of genital warts and that HPV types 6 and 11 are associated with respiratory papillomas which are often acquired at birth. Other HPV types are the causative agents of warts at other sites (Tables 8 and 9). The DNA of these viruses is known to persist and the virus is produced only in the keratinizing layers of the wart. However, the underlying squamous cells appear to contain the virus genome, which appears to be under the control of the cell as no virus particles are produced even though small amounts

TABLE 8. HPVs associated with benign warts.

HPV type	Lesions
1a, b, c	Plantar warts
2a-e	Common hand warts
3a, b	Flat warts/juvenile warts
4	Plantar warts
7	Butchers warts
10a, b	Flat warts
13	Oral focal hyperplasia (Heck lesions)
34	Bowenoid papulosis
36	Actinic keratosis
37	Keratoachanthoma
39	Cutaneous lesions
41	Flat warts

TABLE 9. HPVs associated with epidermodysplasia verruciformis.

HPV type	Lesions
5a, b	Macules
8	Macules
9	Warts and macules
12	Warts and macules
14a, b	Skin lesions
15	Skin lesions
17a, b	Skin lesions
19-29	Warts and hyperplastic lesions

of viral DNA are retained. On occasion, some benign warts progress to carcinomas (20). It is now known that some HPV DNA types are consistently associated with specific carcinomas (Table 10). Thus, there is strong evidence that the etiology of certain cancers involves particular HPV types, but more evidence is required to make such an association more than casual and raise it to the level of causation (Table 11). In this regard, the problems associating HPVs with cancer are not different from those associating the herpesviruses with cancer or hepatitis B virus with primary hepatocellular carcinoma.

TABLE 10. HPVs associated with cancer *in situ* or invasive carcinoma.

HPV type	Lesions
6a-f	Condylomata acuminata; cervical intraepithelial neoplasia I, II, III; vulvar intraepithelial neoplasia I, II, III; laryngeal papillomas
11a,b	Condylomata acuminata; cervical intraepithelial neoplasia I, II, III; laryngeal papillomas
16	Condylomata acuminata; cervical intraepithelial neoplasia I, II, III; vulvar intraepithelial neoplasia I, II, III; Bowenoid papulosis; malignant carcinoma, cervix and penis
18	Malignant carcinoma, cervix and penis
30	Laryngeal carcinoma (rare)
31	Cervical intraepithelial neoplasia; malignant carcinoma, cervix
33	Bowenoid papulosis; cervical intraepithelial neoplasia; malignant carcinoma, cervix
35	Cervical intraepithelial neoplasia; malignant carcinoma, cervix
38	Melanoma
40	Laryngeal carcinoma

TABLE 11. Association of human papillomaviruses with genital cancer.

Viral DNA in dysplasia, cancer *in situ*, carcinoma
Viral DNA in metastatic cells
Viral DNA; episomal in condyloma, integrated in carcinoma
Virus and DNA can transform cells
Transformed cells cause tumors

What would make such associations convincing would be the use of a vaccine to prevent virus infection and a subsequent reduction in the number of cancers associated with that particular virus. This may well be accomplished with hepatitis B virus and primary hepatocellular carcinoma since the hepatitis B vaccine is being administered in countries with a high incidence of liver cancer where hepatitis B is a major public health problem.

The certainty and the assurance that a vaccine against HPVs might be developed and prove to be efficacious are very low at the moment. Nevertheless, it is worth pursuing. However, better ways must be found to grow large quantities of papillomaviruses to better characterize the proteins and their functions. At the moment, only DNA homology studies separate these viruses and very little is known about their antigenic differences so that seroepidemiology is not feasible. This in turn means that the level of HPV infection in the general population remains unknown and the transmission of many HPVs is poorly understood.

4. PROBLEMS ASSOCIATED WITH HIV

The problems associated with this virus are immediate and serious and Dr. R.C. Gallo will confront these issues in great detail in the subsequent paper. HIV is generally introduced by sexual transmission or introduction of infected blood products, resists eradication, infects a number of cell types, which certainly include T4 helper cells or their precursors, and causes widespread damage to the immune system as well as to cells of the central nervous system.

Associated with AIDS are opportunistic infections by a great variety of organisms and neoplastic disease, which usually are controlled by the immune system but which flourish when that mechanism is impaired. Since HIV integrates its genome, after it has been transcribed to DNA by reverse transcriptase, into the host chromosomes, eradication will not be simple, and prevention of disease will rest on the ability to block the virus at the portal of entry or the reverse transcription step. In addition, the role of this and related viruses (many are yet to be isolated) in central nervous system disorders and other tumors or disease syndromes remains to be studied. Since it is now known that HIV undergoes antigenic changes, successful vaccine therapy will face the same handicaps that have prevented effective influenza virus vaccines. An animal model for testing safety using human viruses is not in sight, although there are animal models for related animal retroviruses. Whether these are relevant to human disease remains open to question.

5. THE CHALLENGE OF THE ETIOLOGY OF DISEASES NOW LINKED TO VIRUSES

The initial association of microorganisms, including viruses, with acute infections was fairly simple and direct. Following establishment of Koch's postulates linking microorganisms to disease, a large number of disease syndromes were attributed to bacteria, a few to viruses, and subsequent investigations have not changed many of these assignments. Unfortunately, Koch's postulates are difficult to apply directly to viruses, since viruses often do not cross species barriers and replication in another cell type (i.e. an animal host) is not always possible. Thus, Koch's postulates cannot be fulfilled for many of the diseases associated with viruses. The same problems that have haunted viral oncologists over the past 70 years in establishing a connection between viruses and cancer continue to plague those working on diseases with as yet unknown etiologies. Recent studies on multiple sclerosis by Prof. Hilary Koprowski and colleagues (21) suggest that an HTLV-like virus may be involved; however, this remains to be confirmed by other investigators. In addition, diseases of the central nervous system with suspected viral etiology, such as Alzheimer's, Creutzfeldt-Jakob, and other autoimmune disease complexes require further investigation. The mere presence of a virus or its nucleic acid in tissue is not enough to establish etiology, as such materials often represent "passenger" information that does not involve causation. This is akin to the problem with AIDS, which renders the patient susceptible to a variety of microorganisms that produce disease syndromes that are not caused directly by HIV. The challenge of virologists, molecular biologists, and investigators examining epidemiology and pathogenesis will be to try creative ideas to better understand the causation of these chronic infections. The ultimate hope is that such understanding will lead to therapy or preventive measures that will reduce the incidence of these diseases and the attendent physical, mental, and financial hardships.

6. CONCLUSIONS

Virologists and scientists in other disciplines have been instrumental in reducing the effects of acute infections on the health of the population. Especially impressive is the near eradication of infections that used to produce high morbidity and mortality in children. Many of these illnesses have now been relegated to rate events in advanced countries and incidences certainly will decline in the Third World. Better understanding chronic virus infections coupled with attempts to treat and prevent those infections will be the focus of investigations during the next decade. It is hoped that the massive effort now aimed at the AIDS virus (one of those establishing chronic infection) will spill over into research on other virus infections and so that at the end this century may be heralded by major health advances against chronic infections.

ACKNOWLEDGEMENTS

The authors thanks M.K. Howett for critical review of the manuscript and M. Clement for editorial assistance. New observations cited in this paper were supported in part by grants CA 27503, CA 34479 and CA 09124 awarded by the National Cancer Institute, and by BRSG grant S07 RR05680-18 awarded by the Biomedical Research Support Grant Program, Division of Research Resources, National Institutes of Health.

REFERENCES

1. Barré-Sinoussi F, Chermann JC, Ray F, Nugeyre MT, Chamaret S, Gruest J, Dauguet C, Axler-Blin C, Vézine-Brun F, Rouzioux C, Rozenbaum W, Montagnier L. Science 220:868-871, 1983.
2. Gallo RC, Sarin PS, Gelmann EP, Robert-Guroff M, Richardson E, Kalyanaraman VS, Mann D, Sidhu GD, Stahl RE, Zolla-Pazner S, Leibowitch J, Popovic M. Science 220:865-867, 1983.
3. Ellermann V, Bang O. Zentralbl. Bakteriol., Parasitkde. Abt. I (Orig.) 46:595-609, 1908.
4. Rous P. J. Am. Med. Assoc. 56:198, 1911.
5. Theiler M, Smith HH. J. Exp. Med. 65:787-800, 1937.
6. Sheldrick P, Berthelot N. Cold Spring Harbor Symp. Quant. Biol. 39: 667-678, 1975.
7. Honess RW, Roizman B. J. Virol. 14:8-19, 1974.
8. Buchman TG, Simpson T, Nosal C, Nahmias AJ. Ann. N.Y. Acad. Sci. 354:279-290, 1980.
9. Kieff E, Hoyer B, Bachenheimer S, Roizman B. J. Virol. 9:738-745, 1972.
10. Wigdahl B, Scheck AC, Ziegler RJ, De Clercq E, Rapp F. J. Virol. 49: 426-436, 1984.
11. Wigdahl B, Smith CA, Traglia HM, Rapp F. Proc. Natl. Acad. Sci. USA 81:6217-6221, 1984.
12. Wigdahl BL, Scheck AC, De Clercq E, Rapp H. Science 217:1145-1146, 1982.
13. Wigdahl BL, Ziegler RJ, Sneve M, Rapp F. Virology 127:159-167, 1983.
14. Colberg-Poley AM, Isom HC, Rapp F. Proc. Natl. Acad. Sci. USA 76: 5948-5951, 1979.
15. Plotkin SA, Furukawa T, Zygraich N, Huygelen C. Infect. Immun. 12: 521-527, 1975.
16. Albrecht T, Rapp F. Virology 55:53-61, 1973.
17. Takahashi M, Otsuka T, Okuno Y, Asano Y, Yazaki T, Isomura S. Lancet ii:1288-1290, 1974.
18. Ciuffo G. Giornale Italiano delle Malattie Veneree 42:12-17, 1907.
19. Shope RE, Hurst EW. J. Exp. Med. 58:607-624, 1933.
20. McCance DJ. Biochim. Biophys. Acta 823:195-205, 1986.
21. Koprowski H, DeFreitas EC, Harper ME, Sandberg-Wollheim M, Sheremata WA, Robert-Guroff M, Saxinger CW, Feinberg MB, Wong-Staal F, Gallo RC. Nature 318:154-160, 1985.

STRUCTURE AND FUNCTION OF THE GENOME OF HTLV-III AND RELATED VIRUSES: CURRENT CONCEPTS ON THE ROLE OF HTLV-III IN THE PATHOGENESIS OF AIDS*

C. GURGO[1] and R.C. GALLO[2]
[1]Centro di Endocrinologia ed Oncologia Sperimentale del C.N.R., Via S. Pansini 5, I-80131 Napoli, Italy, and [2]Laboratory of Tumor Cell Biology, National Cancer Institute, NIH, Bethesda, Maryland 20892, USA

1. INTRODUCTION

HTLV-III, the etiological agent of the acquired immunodeficiency syndrome (AIDS), is a member of the human T-lymphotropic retrovirus family which can be divided into two groups: A, which includes HTLV-I and HTLV-II, which are associated with human leukemias and lymphomas, and B, which, in addition to the widely distributed HTLV-III, includes a new group of human T-lymphotropic viruses recently isolated in West Africa, namely LAV-2, HTLV-IV and SBL 6669, which like HTLV-III are cytopathic _in vitro_ for T4+ T-cells and, in the cases of LAV-2 and SBL 6669, are associated in man with AIDS and ARC, respectively.

In this article we will present our current knowledge of the structure of HTLV-III and the functions of its genes; we will address the question of the genetic variability of the viral envelope and the impact of this problem on vaccine production. We will discuss the relatedness of the various recently isolated human T-lymphotropic viruses, comparing where possible the results of structural and serological studies, and extend the analysis to the related simian STLV-III which is not pathogenic in its natural host, but causes an AIDS-like disease when injected into macaques.

2. STRUCTURE OF THE HTLV-III GENOME

The genome of HTLV-III, in its integrated provirus form, is formed by two flanking long terminal repeats (LTRs), which contain regulatory elements and eight genes which code for structural and regulatory proteins (1). Like all animal retroviruses, the genome contains the _gag_, _pol_ and _env_ genes, which code for the structural core proteins, the viral polymerase and the envelope proteins, respectively. In addition, as was predicted by nucleotide analysis and subsequently confirmed by the isolation of the respective mRNAs and protein products, it contains other genes which utilize complex splicing mechanisms for their expression: _sor_, corresponding to a small open reading frame located between _pol_ and _env_, which codes for a 23 kd protein; and 3'_orf_, located downstream of the _env_ gene, which extends into the U3 element of the 3' LTR and codes for a protein of 27 kd. The functions of the _sor_ and 3'_orf_ proteins have not yet been identified.

Several other open reading frames have been localized in the central part of the genome, between the _pol_ and the _env_ genes. Three of these overlapping open reading frames (orfs) have been demonstrated to be viral genes: one of these, _tat_-III, has been identified by a functional assay

---

*Supported by the P.F. Oncologia of the C.N.R. of Italy and by the Italian-American Bilateral Agreement for Cancer Research.

R.C. Gallo

since its product, a protein of 14 kd, is able to posttranscriptionally transactivate the expression of viral genes and heterologous genes linked to the viral LTR. Overlapping the tat-III gene, but in a different frame, is the art/trs gene, which is involved in post transcriptional regulation of viral protein production. The eight viral gene, R, has only recently been identified (2). It overlaps the sor gene and terminates before the first coding exon of tat-III. A DNA insert corresponding to this open reading frame was subcloned into an expression vector and the protein produced was immunoreactive with most sera of individuals with AIDS, but not with sera of healthy controls. The region coding for this protein is highly conserved among HTLV-III isolates but its function remains to be elucidated.

3. THE VIRAL LTR

The long terminal repeat (LTR) of HTLV-III contains 634 base pairs (3). It is composed of the functional elements for the expression of viral gene products which are recognized by cellular and viral factors at the transcriptional and posttranscriptional levels. In the U3 region, which is 453 nucleotides long, transcriptional enhancer and promotor elements were initially located but not dissected in the region between nucleotides -137 and -17 from the CAP site (4). The enhancer was subsequently localized in a region spanning nucleotides -105 to -80, containing two direct repeats of 10 nucleotides (6).

A second region containing three closely spaced promoter elements for the binding of the cellular transcriptional factor SP1 was defined between nucleotides -77 and -45 in the proximity of the TATA box (5). However, only the second of these binding sites is critical for biological activity (6). Zones of limited homology with HTLV-I and -II, as well as sequences homologous with gamma interferon and IL-2, have been reported (3).

The R region, 93 nucleotides long, contains a sequence responsive to the viral transactivating factor (tat) (7). This responsive element was originally localized in a region from nucleotide -17 to +80 (4), and subsequently to nucleotide +56 (6).

The U5 region is 83 nucleotides long. The contiguous primer site shows complementarity to $tRNA^{lys}$.

The LTR of STLV-III (800bp) is considerably longer than that of HTLV-III (683bp) due to extra sequences not found in the human virus (8). The observed homologies between the two viral structures occur mainly in the known functional regions which include the two primer binding sites for the first and second DNA strand, the enhancer region and the TATA box of the RNA polymerase promotor. As in the case of HTLV-III, the primer binding site is complementary to $tRNA^{lys}$.

4. THE REGULATORY GENES

The human lymphotropic retroviruses HTLV-I, HTLV-II and HTLV-III share a unique feature, that is the ability to transactivate the expression of genes linked to their respective LTRs (9). This function is mediated by the translation product of a 2kb mRNA. However, the structural organization of the genes coding for the respective proteins and the mechanism of action of their products (tat-I, -II, -III) is different. The transactivation proteins of HTLV-I and -II (38-42kd) are encoded by a long open reading frame which is located in a highly conserved region of 1kb at the 3'end of a long stretch of nucleotides (1.6kb) between the env gene and the 3'LTR. Transactivation by these two proteins occurs at the level of transcription.

The effect of tat-III (14kd), the corresponding protein of HTLV-III, is mediated at the posttranscriptional level (7,10,11). Its mRNA is the result of the transcription of three different exons and involves double splicing. The major functional domain of this gene is the second exon which is located between the sor and env genes. The tat-III product is essential for virus production (12,13); thus, the deletion of the tat-III gene or the abrogation of normal splicing mechanisms in a biologically active clone of HTLV-III inhibits virus production in transfected cells. It is believed that the interaction of tat-III protein with a specific target in the leader sequence of the viral mRNA is required for the efficient synthesis of viral structural proteins.

A second regulatory gene has been identified which also acts at a posttranscriptional level. It partially overlaps the tat-III gene and the 3' env gene and is probably translated from the same double spliced mRNA as tat-III, but in a different frame. This gene product was detected by functional assays (14), since 1) production of gag and env proteins was inhibited by mutations near the two coding exons for tat-III and these mutations were complemented in trans by tat-III. They were, however, complemented by a plasmid, containing the two coding exons of this gene, that was designed to express the gene product in a system constitutively expressing tat-III. It was initially thought that this gene would function by derepressing negative regulatory sequences in the mRNA encoding the structural proteins, that is, by removing a block at the level of mRNA translation. It was designated antirepressor transactivator (art). Subsequently, an independent study proposed that the product of this gene, a protein of 25kd, functioned at the level of splicing; it would regulate the steady state levels of HTLV-III genomic and spliced subgenomic mRNA (15). More specifically, in the absence of its product, splicing would occur to generate only mRNA encoding regulatory proteins. In its presence, splicing of mRNA would be directed toward the production of structural proteins. It was alternatively designated trs (transacting regulator of splicing) in this study.

Structure studies of the simian STLV-III genome have localized nucleotide sequences highly homologous to the HTLV-III tat-III and art/trs genes (16). The tat-III gene has been identified by functional assays not only in the STLV-III but in all the members of the new group of West African isolates, that is HTLV-IV, LAV-2 and SBL 6669 (17).

5. THE VIRAL RECEPTOR

It is of general knowledge that viruses utilize for their entry cellular proteins which can be mediators of important physiological functions. HTLV-III binds to the T4 differentiation antigen of the helper/inducer subset of T-lymphocytes, and this binding is a necessary prerequisite for internalization and subsequent virus replication. For example, infection of T4+ T-lymhocytes by HTLV-III/LAV and penetration of pseudotypes of vesicular stomatitis virus and HTLV-III can be prevented by preincubation of the T-lymphocytes with several monoclonal antibodies specific for T4 (18,19). Moreover, the involvement of specific regions of T4 in the binding is demonstrated by the observation that binding of a particular T4 specific antibody (OKT4a) but not of another (OKT4) prevents the binding of HTLV-III and *vice versa* (20,21). The isolation of a complex of T4 and gp120 from infected cells further supports the idea that T4 is an essential part of the viral receptor (21).

The T4 molecule is also present on other types of cells. Neurological disorders in AIDS patients have been linked to the presence of HTLV-III in the brain (22,23). Recently, a protein immunoprecipitable by OKT4 has been located on human brain cells as well as on human T cells (24,25). In addition, the viral envelope protein gp120 could be covalently linked to the T4 molecule on brain cell membrane of three different mammalian species, namely man, monkey, and rat. The binding of the virus to the T4 molecule could be prevented by a polypeptide the structure of which is common to the viral envelope of another lymphotropic virus, EBV. Because HTLV-III is not known to cross species, it seems clear that binding to a T4 molecule is not sufficient for entry. Transfection experiments with mouse fibroblasts have shown that when membrane barriers are overcome, there are no other cellular restraints for HTLV-III/LAV replication (26). Thus, the possibility for HTLV-III to infect other T4+ cells or other types of cells seems to depend on its ability to adapt to cellular receptors.

T4 molecules have also been found on the membrane of dendritic cells, Langerhans cells, and macrophages (27,28). Various HTLV-III isolates can infect and replicate in human monocytes/macrophages of healthy donors and strains of HTLV-III isolated from macrophages of AIDS patients are able to replicate better in these cells *in vitro* than in T4+ T-lymphocytes, while HTLV-III prototypes selected for growth on T4+ T-lymphocytes can more easily infect and replicate in T4+ T-lymphocytes. It has been proposed that macrophages are the first site of infection (28). The observation that in the lymphadenopathy phase of AIDS the virus is confined to the follicular region of lymph nodes where it appears to replicate in dendritic cells, and not in typical T-cell areas, further strengthens this concept (29). As will be discussed later, replication of the virus in macrophages in the early phase of infection and continuous changes in the hypervariable regions of the envelope protein could result in the production of an heterogenous population of particles with various alterations in the external protein. This could provide an escape mechanism from immunosurveillance and the potential for viral adaption to various cell receptors (30).

6. GENETIC VARIABILITY OF THE *ENV* GENE OF HTLV-III

The *env* gene of HTLV-III codes for a large precursor protein, gp160, which is cleaved upon maturation to produce the extracellular envelope protein, gp120, and a smaller transmembrane protein, gp41. Nucleotide sequence analysis confirmed the evidence obtained by restriction nuclease mapping that a consistent genetic variability exists among the various isolates. When sequence comparison was made among five independent isolates, namely clones HTLV-III (BH10), HTLV-III(RF), HTLV-III(WMJ1), Lav-1a, and Arv-2 to determine the extent and the nature of the genetic variation, a broad spectrum of diversity was observed (31). The variation in the nucleotide sequences was not uniform over the whole genome; a remarkable pattern was observed in the distribution of sequence differences, with prevalent changes occurring in the *env* region. The *gag* and *pol* genes were the most conserved; nucleotide changes were mainly single point mutations, which occurred prevalently in the third codon position and were mostly silent, while the same type of mutation in the *env* region resulted in amino acid substitutions. In addition to point mutations, clustered nucleotide changes were common in the *env* gene, involving deletion/insertions and duplications. The variability was not evenly distributed throughout the *env* gene, and a pattern of constant and variable regions emerged.

The urge to understand the biological significance of the genomic heterogeneity of HTLV-III/LAV has stimulated studies aimed at better establishing the extent of polymorphism and geographical spread, the evolutionary relatedness of various isolates, and the rate of mutation. Thus virus has been isolated from different individuals from the same country or geographical area, from the same individual at different times, from individuals from different countries, and from monkeys affected with AIDS-like diseases.

When the sequences of two isolates from Zaire were compared, they showed the same genetic organization as the American isolates, indicating a common ancestor (32). However, they were more divergent between themselves than the American isolates, due perhaps to a wider diffusion of the virus in this endemic region of Africa.

Results with four new isolates from the U.S. confirmed the pattern of variability observed in the env region (33). When sequential isolates from the same individual were analyzed, the nucleotide changes were clustered in the same variable regions (34). They consisted of point mutations and insertions/deletions. A very high rate of mutation was observed. Comparison of nucleotide sequences revealed that viruses sequentially isolated from the same individual showed much less variation than viruses isolated from different individuals. The pattern did not support a model of progressive mutation, but rather parallel evolution from a common ancestor.

Elucidation of the primary amino acid sequence of the envelope proteins of several isolates has provided information useful for the prediction of secondary structure as well as for antigenic epitope localization (31,35). Potential antigenic epitopes have been identified with a computer assisted program, superimposing the predicted secondary structure on 1) hydrophilicity, 2) flexibility, and 3) surface probability profiles, using the criteria that hydrophilic domains of proteins containing a high number of beta turns, in addition to high values of flexibility and high probability of location on the surface of proteins, are considered candidates for antigenic sites, since they form accessible, mobile loops. With several HTLV-III isolates, the analysis predicted one epitope with high antigenic potential for each of the five variable regions. The epitopes predicted for the constant regions, due to the low number of calculated beta turns, or to the general low hydrophilicity of this region, have low antigenic potential, with the exception of one region, consisting of a stretch of hydrophilic amino acids, which fulfils all the prerequisites. In conclusion, considering that the pattern of cysteines which are involved in the folding of protein structures is highly conserved among various isolates and that the variable regions which are interspersed with constant regions are bounded by cysteines, it is possible to visualize the tertiary structure of the HTLV-III env protein, as a highly conserved folded core structure surrounded by flexible, hydrophilic, loop-like structures, i.e. the variable regions forming the antigenic epitopes.

7. VIRUS-SPECIFIC ANTIBODIES DURING THE COURSE OF THE DISEASE

Antibody production to HTLV-III antigens is an overt manifestation of viral infection and one of the first clinical signs of the development of the disease (36). Previous studies have consistently demonstrated that the first emerging antibodies are directed against the viral core proteins, namely the major core protein, p24, before full seroconversion occurs (37). Convincing evidence has been presented which demonstrates that the early selective detection of gag antigen is instead the result of technical limitations. Recently it was shown that, when the sera of an asymptomatic subject became positive with time for viral antibody, only anti-p24 antibody could be detected by the Western blot technique; the same serum, however, revealed a complete profile of reactivity against envelope and core proteins when assayed by radioimmunoprecipitation and polyacrylamide gel electrophoresis (RIP-PAGE) (38).

Attempts to correlate the appearance of specific antibody with disease progression have been unsuccessful, since no particular protein has been highlighted as diagnostic for the development of AIDS (39). However, some general features were derived in a recent study aimed at monitoring changes in antibody specificity with time in a group of HTLV-III infected individuals, including asymptomatic, ARC ,and AIDS patients (38). The results showed that the ratio of antibody to *gag* proteins to antibody to the *env* proteins increases steadily with time, reaches a plateau, then declines. Thus, while antibodies to the *env* proteins are always detectable, the antibodies to the *gag* proteins tend to disappear (the disappearance of antibodies to the core protein, which are the most conserved viral proteins, is consistent with the disappearance of T cells with specificity for the core protein epitopes, and consequent lack of amplification of B cells specific for these antigens). It has been clearly indicated that all viral encoded proteins are immunogenic *in vivo* (36,40,41). The possibility that some of the naturally occurring antibodies to HTLV-III/LAV in AIDS patients could have the ability of neutralizing viral infectivity was explored in the light of a possible therapeutic use of such antibodies (42). Indeed, low titers of antibody capable of blocking the infection of H9 cells *in vitro* by HTLV-III were found in the serum of patients with AIDS and ARC (43). Similar findings were reported by other laboratories using additional functional assays, such as the ability to inhibit syncytia formation and envelope neutralization of VSV(HTLV-III) pseudotypes (44).

Because the heterogeneity observed in the envelope region of HTLV-III isolates could be the result of combined effects of immune pressure and mutation, the possibility that the natural neutralizing antibodies could act as the selective force for the emergence of the variants was tested *in vitro*. Recent studies have shown that when variation occurs *in vitro* the presence of neutralizing antibodies is determinant for the emergence of variants (42). Therefore, while the finding of neutralizing antibodies opens the hope for the development of vaccines, the possibility that the virus is mutating under their pressure adds another order of magnitude to the difficulty of vaccine development. In fact, such preventive vaccines must be capable of neutralizing a wide spectrum of viral variants.

8. MECHANISMS OF CYTOTOXICITY

No viral gene has yet been directly implicated in T-cell killing. 3' *orf* and *sor*, the functions of which have not yet been understood, do not have direct influence on viral cytotoxicity, since deletions in these regions do not affect the ability of the virus to replicate and ultimately kill cells (45). Neither is *tat*-III directly responsible for cytotoxicity, since cells that constitutively express this product multiply in the presence of functional *tat*-III protein.

A mechanism used by enveloped viruses in cell cytotoxicity involves the formation of syncytia and is mediated by the *env* glycoproteins. It is also known that retroviruses can induce cell fusion using receptors present on cell surfaces.

Two independent studies have evaluated the possibility that a similar mechanism might determine the specific cytotoxicity of HTLV-III. With this objective (46), a plasmid vector constructed to express the *env* gene of HTLV-III in a compatible system was used to transfect a T4+ T-cell line. In the presence of the HTLV-III transactivating factor, high levels of *env* protein were produced and a large number of giant cells formed, resulting in cell death. When transactivation of the *env* gene was allowed to occur in an analogous B-cell system, production of envelope proteins occurred but did not result in cell fusion or cytotoxicity.

Using a different approach with a cell line infected and not infected with HTLV-III and a variant of this line negative for T4, Lifson *et al.* (47) evaluated the minimal requirements for fusion, using immunofluorescence techniques, antibodies to T4, and inhibitors of protein synthesis. It was demonstrated that in the fusion of the T4+ cells, only one cell need to be infected and, moreover, productive infection is not necessary, since the presence of the viral gp120 alone is sufficient for cell fusion.

Since syncytium formation can account for the death of non infected T4+ cells and only a low number of these cells appears to be infected *in vivo*, such mechanism has been proposed as a model for the depletion of helper cells observed in AIDS. However, there is no compelling evidence that this is the case *in vivo*, and other models have been proposed, based on the observation that long term cultivation of infected T4+ T-cells can be induced by immunostimulation. These cells release high amounts of virus and die without syncytium formation.

Experiments aimed to reproduce *in vitro* the conditions which lead to cell death *in vivo* (48) have demonstrated that upon stimulation with PHA, peripheral blood lymphocytes infected with HTLV-III are killed following the release of IL-2 and a burst of virus production. Within the limitations imposed on the interpretation of experiments carried out *in vitro* with regard to their relevance *in vivo*, the results essentially suggest that the death of infected helper T-cells occurs as a result of strong antigenic activation.

Thus, the events set into motion *in vitro* by the use of a strong, non-specific stimulation (PHA) would progressively occur *in vivo* via a series of specific antigenic stimulations which would cause a progressive depletion of infected helper T-cells engaged in the process of antigen recognition. In a normal situation, helper T-cells with a given specificity recognize antigen after its processing and presentation by macrophages or other antigen-presenting cells (APCs). This condition is fulfilled with T-cells that, through their specific receptor, bind simultaneously to the processed antigen and Ia, the class II protein coded for by the major histocompatibility complex (MHC), on the surface of macrophage membranes. When this occurs, the T-cells upgrade the expression of IL-2 receptors, and the macrophage increases its rate of secretion of IL-1. The latter in turn stimulates the production of IL-2 and the growth of those T-cells having IL-2 receptors on their surface.

In addition to the expansion of T4+ T-cells, specific cytotoxic and suppressor T-cells are also amplified. Cytotoxic T-cells are programmed to destroy cells carrying the antigen with which their T-cell receptor interacts, in the context of class I MHC proteins, which are present on all nucleated cells. Activated helper cells will also provide "help" for the differentiation and the proliferation of B-lymphocytes which produce immunoglobulin with specificity for the same antigen.

HTLV-III can enter the blood as free virus or inside allogeneic lymphocytes. In the latter case, it must elude the first immune defense, the natural killer cells, which does not require the fine mechanism of recognition used by cytotoxic T cells for killing. Several lines of evidence may be used in support of the idea that HTLV-III becomes established first in the cells of the monocyte/macrophage lineage, which paradoxically may constitute a reservoir for the virus from which it prepares the dismantling of the immune system.

Viruses enter cells by virtue of an interaction with a cellular receptor. The T4 differentiation protein present on helper/inducer T-cells is involved in the binding of HTLV-III/LAV. This molecule is also present on the membrane of macrophages and other cells, such as Langerhans cells, dendritic cells, and neurons. Other proteins surrounding the T4 molecule, which may differ from cell type to cell type and from one person to another (MHC proteins), may be involved in the formation of the receptor; they may affect the stringency of virus binding and subsequent internalization of the virus, and constitute a variable element to which the virus adapts in order to penetrate into the cells. It is believed that the highly conserved core structure of the major envelope protein contains a domain common to all isolates for the binding to T4. The predicted loops which cover this conserved structure may play a role in the binding to the viral receptor. The nucleotide changes that occur in the hypervariable regions of the env gene corresponding protein may allow the virus to better bind to the receptor present on the T-helper cells, as well as to infect other types of cells.

A situation in which the virus is replicating in monocytes/macrophages when the helper T-cells are not yet infected could correspond to the early stage of HTLV-III infection, when the immune system is not yet impaired or capable of responding with both its cellular and humoral components, as judged by the production of a large spectrum of antibodies against the viral antigenic determinants. Continuous replication in macrophages, with concomitant changes in the hypervariable regions of the env gene, could increase the opportunity for the production of a variant better able to bind to T4 and other proteins forming the viral receptor (30). Since the external loops on the env protein structure where variations occur are also the sites containing antigenic determinants, new epitopes will be generated with consequent amplification of new subsets of T4+ T-cells and increased antibody production against the variable viral epitopes. When conditions for efficient binding to the helper cells are attained, all the T4+ T-cells, among which are those that have specificity for the early viral envelope epitopes and for the mutated epitopes, may become susceptible to infection. In the light of the observation of Zagury et al. (48), these cells should die following specific interaction with the corresponding antigens. However, the helper T-cells which participated in the first immune response against the viral antigens and have specificity for viral epitopes that are no longer present, will survive in spite of infection for lack of specific antigenic stimulation, provide a proliferative advantage and a rationale for the selection of a virus that is highly specific for the helper T-cell virus receptor. This model is consistent with the persistence of a constant and relatively low number of infected T4+ T-cells in AIDS patients (68), the continuous production of antibodies against the viral envelope proteins, and the disappearance with time of antibodies made against the most conserved viral gene product, the gag proteins (38).

9. STLV-III AND WEST AFRICAN ISOLATES

The description of an immunodeficiency syndrome among captive macaques stimulated the search for an infectious agent which might be etiologically implicated (49-52). The idea that such virus might be related to HTLV-III was suggested by the analogy with another simian retrovirus, STLV-I (53), which is antigenically and genetically highly related to HTLV-I (9), the etiological agent of adult T-cell leukemia/lymphoma and also associated with neuropathies and immunological disorders. STLV-I was detected in Asian and African monkeys and in disease-free chimpanzees by virtue of an antigenic crossreactivity with HTLV-I (54,55). It has been associated with spontaneous lymphomas in captive macaques (56). Transmission of lymphoma from one macaque to another has occurred following injection of cell free filtrates from a tumor and a link between transmissible lymphoma and induction of an immunodeficiency syndrome in macaques was suggested (57). A simian virus, antigenically related to HTLV-III was subsequently identified in serological studies and isolated (58,59). The new virus, named STLV-III, was cytotoxic for T4+ T-cells _in vitro_ and morphologically similar to HTLV-III. It seemed reasonable to hypothesize that such virus might also be present in wild monkeys and that this infectious agent might have been transmitted to man, sometime in the past, setting up the condition for the emergence of AIDS. The implication that AIDS originated in Africa, where it has now reached epidemic proportions in some regions, is justified by retrospective serological studies which trace the presence of antibody to HTLV-III back to 1972 (60). In a survey of a variety of primates for an HTLV-III related virus, high levels of antibodies to the STLV-IIImac isolate were observed in wild African green monkeys, but not in chimpanzees or baboons, in Central Africa. A retrovirus, designated STLV-IIIagm, was then isolated and characterized (61). This virus was cytotoxic _in vitro_ and induced syncytium formation. Fifty percent of the wild African green monkeys appear to be infected with this virus, but seropositivity is not associated with immunodeficiency. The presence of antibody to STLV-III in captive macaque, however, is closely associated with the development of this syndrome.

Molecular hybridization studies have demonstrated a moderate homology between STLV-IIIagm and HTLV-III (62). Using the full length genome of HTLV-III or fragments corresponding to well characterized subgenomic regions, different degrees of homology were observed in different regions of STLV-III. The highest degree of homology was found in the region corresponding to the _gag_ and _pol_ genes, moderate homology in the 3' _orf_, and modest homology in the _env_ region. Analogously, crosshybridization with DNA fragments of STLV-III clones confirmed the homologies with the corresponding genetic regions of HTLV-III, while hybridization was not detected with the DNA of other retroviruses, namely HTLV-I, HTLV-II, BLV, and FeLV.

A virus highly related to STLV-III was found in serum samples of several people in a group at risk for AIDS who were undergoing routine screening for antibodies against HTLV-III (63). Serum samples of apparently healthy people which were positive for HTLV-III by an ELISA assay were successively screened for antibodies to the major antigens of STLV-III and HTLV-III by RIP-PAGE. A strong reaction was obtained with all the major antigens of STLV-III, the _env_ and _gag_ related gp160, gp120, gp55, p24, and p15 proteins. The results obtained with the corresponding antigens of HTLV-III were variable, and the pattern was very similar to the reaction of the serum of STLV-III-infected African green monkeys with the human and the simian lymphotropic viruses, respectively. A virus morphologically similar to HTLV-III was isolated which was not equally cytotoxic for T4+ T-cells (HTLV-IV). Based on serological studies and on the comparison of electrophoretic mobility, the viral proteins of HTLV-IV were indistinguishable from those of STLV-III. Thus the new African isolate appears to be much more related to STLV-III than to HTLV-III.

A second human retrovirus (LAV-2) (64), which fits into the gap between the highly related human (HTLV-IV) and simian (STLV-III) lymphotropic viruses recently isolated in West Africa and the widely distributed HTLV-III, has been isolated in Gambia. Isolates were obtained from two individuals not at risk for AIDS, but who developed the syndrome and were negative for HTLV-III/LAV. Serological studies, supported by hybridization data, suggest that this West African prototype is more closely related to STLV-IIImac than to HTLV-III/LAV, since it shares with the latter only some common epitopes and differs antigenically in the envelope proteins. Strong reactivity with the envelope antigen of the new strain was however observed with the serum of a macaque infected with STLV-III and affected by an immunodeficiency syndrome. The virus shows the same tropism as HTLV-III/LAV and grows only in fractions of PBL's enriched for T4+ T-cells. Seropositivity for the major antigens of LAV-2, gp120 and p26, was also observed with serum from other people of the same region, with or without AIDS, and with the serum of a healthy woman from Senegal, infected with the non pathogenic HTLV-IV.

A new strain of human T-lymphotropic virus (65), SBL 6669, has been recently isolated from a Gambian woman with early lymphoadenopathy who had a history of respiratory infections, low numbers of T4+ T-cells, and weight loss. The serum of the patient was positive for the gag and pol antigens of HTLV-III, as judged by the Western blot technique. Upon isolation of the virus, parallel studies were carried out to compare the properties of this isolate with other West African isolates and HTLV-III. With the exception of HTLV-IV, all the pathogenic viruses rapidly infected several human T-cell lines, induced cytopathic effects in tissue culture and generated, to various degrees, multinucleated cells. When the size of the viral proteins were compared, a similar value of approximately 120kd was calculated for the molecular weight of the major envelope protein, while the transmembrane protein was smaller in the West African isolates. Some distinct differences were noticed, however, in the size of the core proteins of the various isolates. Thus, the new strains of West African human lymphotropic viruses and the simian isolate could be distinguished from one another and from HTLV-III/LAV on the basis of the electrophoretic mobility of these proteins.

With the availability of probes specific for the constant (gag) and variable (env) regions of the STLV-III genome, the genetic relatedness of the human and simian T-lymphotropic viruses has been recently analyzed (66). The results of these studies have indicated that HTLV-IV and STLV-III are indistinguishable by Southern blot analysis, suggesting a high degree of homology in the genomic structure of these two viruses. When the DNA of cells infected with STLV-IIIagm and two isolates of HTLV-IV were digested with a battery of restriction enzymes and were analyzed by the Southern blot technique, the same restriction fragments were observed with all the isolates, with some degree of polymorphism between the two isolates of HTLV-IV. A very different electrophoretic pattern was obtained in the restriction analysis of DNA from cells infected with LAV-2. The pattern observed with DNA from SBL 6669-infected cells also differed from that of HTLV-IV-infected cells and was similar to that of LAV-2-infected cells, from which differences could be appreciated only by using a probe specific for the env region. In all cases a strong degree of hybridization was observed when the simian probe was annealed with the DNA of the newly isolated viruses. A restriction fragment common to all the strains of virus was detected with a probe specific for the gag region. These results confirm the relatedness observed among the simian STLV-III, HTLV-IV, LAV 2 and SBL 6669, with STLV-III and HTLV-IV being more closely related, LAV 2 being related to but genetically different from the latter two, and SBL 6669 being closely related to LAV-2 but still distinguishable from it in the env region, and with all of them being more closely related to one other than to HTLV-III.

Recently, the complete nucleotide sequence of the genome of STLV-III has been obtained and compared to that of the prototype HTLV-III/LAV (16). An overall homology of 55 % was observed with a very similar genetic organization. All the open reading frames present in HTLV-III were accounted for. Alignment of the two sequences disclosed an overall homology of 51 % between the _gag_ precursor regions of the two genomes. The highest degree of homology (66 %) was in the major core protein, p24, in agreement with other types of analysis. The _pol_ region was well conserved, with an amino acid homology of 56 % in the reverse transcriptase. In the open reading frames, located in the central region of the genome, major structural differences were observed in the second small open reading frame of the simian virus; two contiguous stretches of nucleotides with a strong homology for _R_ and _tat_-III coded for the corresponding proteins in the same open reading frame.

Some interesting observations were also made in the region of the _env_ gene: 1) the large and small _env_ proteins of the simian virus share 30 to 40 % homology, respectively, with the corresponding proteins of HTLV-III; 2) the pattern of cysteines is highly conserved; and 3) regions of high homology are interspersed with regions of lesser homology and correspond to the constant and variable regions common to the HTLV-III variants. It seems that the most conserved regions, the basic architecture of which is maintained in both viruses by invariant cysteine residues, are involved in the fulfillment of the same function, such as the binding to relevant components of a cellular receptor for these viruses.

10. CONCLUDING REMARKS

Helper T cell depletion is the recognized major cellular defect in AIDS. Because of the pivotal role of these cells, impairment of their function leads to the collapse of the immune system, rendered unable to combat or control the spread of opportunistic infections and/or some forms of cancer. The _in vitro_ infection of peripheral blood lymphocytes with HTLV-III offers a logical correlate to the situation _in vivo_.

Much has been learned about the structure and functions of the viral genes, but the mechanism of cell killing remains obscure. It is well established that transactivation is an essential prerequisite for virus replication, and that in the absence of the viral protein responsible for this function viral mRNAs are not translated into protein. Since _tat_-III is not associated with structural proteins inside the virion, the virus must initially utilize (with low efficiency) less specific cellular factors. Thus the success of a productive infection depends on the availability of these factors and the physiological state of the cell (resting or dividing). Once on control of its replicative system, the virus may exert cytotoxicity, altering specific cellular functions, by competitive mechanisms as well as transactivation of cellular genes. The impact will depend on the cell type and its metabolic state. It has been shown that cytotoxicity is not necessarily associated with HTLV-III replication, since the virus can replicate _in vitro_ in T4+ T-cells grown under reduced metabolic conditions and can replicate in different cell lines without killing them. Therefore, the effect of the virus on helper T-cells needs to be studied in the context of the immune system.

Replication of the virus in macrophages early in the course of the disease with continuous changes in the hypervariable regions of the _env_ gene has been proposed as a strategy by which the virus escapes from immunosurveillance and adapts to various cell receptors. In turn, continuous changes in the antigenicity of some (the env) but not all of the viral epitopes and death of infected helper cells upon strong antigenic stimulation is consistent with a continuous production of antibody to env proteins and the decline of antibodies against constant epitopes is explained in terms of the network between B and T cells for antibody production.

With regard to the new virus strains, studies have demonstrated that sera from individuals infected with the various strains of West African origin can identify the external envelope protein, the transmembrane and core proteins of all the West African isolates, and the core proteins but not the envelope protein of HTLV-III. On the other hand, sera positive for HTLV-III antigens, react only with the core proteins of West African isolates. In spite of the heterogeneity among the env genes of the HTLV-III isolates, the sera of individuals infected with HTLV-III can identify the gp120 protein of the widespread variants of HTLV-III/LAV. Major structural differences in the envelope of the West African strains and perhaps a lower rate of genetic variability may account for the presently limited diffusion of the West African strains.

In conclusion, it is hoped that by comparing the structures and different biological properties of these new virus strains more can be learned about the basis of their pathogenicity and that this information will be useful in the ongoing efforts to produce an effective vaccine.

REFERENCES

1. Wong-Staal F. in The Sympathoadrenal System (Christensen NJ, Henriksen O, Lassen NA, Eds). Munksgaard, Copenhagen, pp 1-11, 1986.
2. Wong-Staal F, Chandra PK, Ghrayeb J. Human T-Lymhotropic Virus Type III: The Eighth Gene, in press, 1987.
3. Starcich B, Ratner L, Josephs SF, Okamoto T, Gallo RC, Wong-Staal F. Science 227:538-540, 1985.
4. Rosen CA, Sodroski JG, Haseltine WA. Cell 41:813-823, 1985.
5. Jones KA, Kadonaga JT, Luciw PA, Tjian R. Science 232:755-759, 1986.
6. Josephs SF, Sadaie MR, Heisig V, Seigel LJ, Franza BR, Renan MJ, Wong-Staal F, in press, 1987.
7. Sodroski J, Patarca R, Rosen C, Wong-Staal F, Haseltine W. Science 229:74-77, 1985.
8. Guo H-G, Franchini G, Gurgo C, Beaver B, Colalti E, Gallo RC, Wong-Staal F, Reitz MS, in press, 1987.
9. Wong-Staal F, Gallo RC. Nature 317:395-403, 1985.
10. Rosen CA, Sodroski JG, Goh WC, Dayton AI, Lippke J, Haseltine WA. Nature 319:555-559, 1986.
11. Arya SK, Guo C, Josephs SF, Wong-Staal F. Science 229:69-73, 1985.
12. Fischer AG, Feinberg MB, Josephs SF, Harper ME, Marselle LM, Reyes G, Gonda MA, Aldovini A, Debouk C, Gallo RC, Wong-Staal F. Nature 320:367-371, 1986.
13. Dayton AI, Sodroski JG, Rosen CA, Goh WC, Haseltine WA. Cell 44:941-947, 1986.
14. Sodroski J, Goh WC, Rosen C, Dayton A, Terwilliger E, Haseltine WA. Nature 321:412-417, 1986.
15. Feinberg MB, Jarrett RF, Aldovini A, Gallo RC, Wong-Staal F. Cell 46:807-817, 1986.
16. Franchini G, Gurgo C, Gallo RC et al., in press, 1987.
17. Arya SK, Beaver B, Ensoli B, Kanki PJ, Biberfeld G, Zagury D, Wong-Staal F, Essex M, Norrby E, Gallo RC, in press, 1987.
18. Klatzmann D, Champagne E, Chamaret S, Gruest J, Guetard D, Hercend T, Gluckman J-C, Montagnier L. Nature 312:767-768, 1984.

19. Dalgleish AG, Beverley PCL, Clapham PR, Crawford DH, Greaves MF, Weiss RA. Nature 312:763-767, 1984.
20. McDougal JS, Mawle A, Cort SP, Nicholson JKA, Cross GD, Schleppler-Campbell JA, Hicks D, Sligh J. J. Immunol. 135:3151-3162, 1985.
21. McDougal JS, Kennedy MS, Sligh JM, Cort SP, Mawle A, Nicholson JKA. Science 231:382-385, 1986.
22. Sharer LR, Epstein LG, Cho E-S, Joshi VV, Meyenhofer MF, Rankin LF, Petito CK. Human Pathol. 17:271-284, 1986.
23. Shaw GM, Harper ME, Hahn BH, Epstein LG, Gajdusek DC, Price RW, Navia BA, Petito CK, O'Hara CJ, Groopman JE, Cho E-S, Oleske JM, Wong-Staal F, Gallo RC. Science 227:177-182, 1985.
24. Maddon PJ, Dalgleish AG, McDougal JS, Clapham PR, Weiss RA, Axel R. Cell 47:333-348, 1986.
25. Pert CB, Hill JM, Ruff MR, Berman RM, Robey WG, Arthur LO, Ruscetti FW, Farrar WL. Proc. Natl. Acad. Sci. USA, in press, 1987.
26. Levy JA, Cheng-Mayer C, Dina D, Luciw PA. Science 232:998-1001, 1986.
27. Tschachler E, Groh V, Popovic M, Mann DL, Konrad K, Safai B, Eron L, diMarzo Veronese F, Wolff K, Stingl G. J. Invest. Dermatol. 88:233-237, 1987.
28. Gartner S, Markovits P, Markovitz DM, Kaplan PH, Gallo RC, Popovic M. Science 233:215-219, 1986.
29. Biberfeld P, Chayt KJ, Marselle LM, Biberfeld G, Gallo RC, Harper ME. Am. J. Pathol. 125:436-442, 1986.
30. Gurgo C, Guroff MR, Gallo RC. Antibiot. Chemother., in press, 1986.
31. Starcich, BR, Hahn BH, Shaw GM, McNeely PD, Modrow S, Wolf H, Parks ES, Parks WP, Josephs SF, Gallo RC, Wong-Staal F. Cell 45:637-648, 1986.
32. Alizon M, Wain-Hobson S, Montagnier L, Sonigo P. Cell 46:63-74, 1986.
33. Gurgo C, Guo WH, Franchini G, et al., in press, 1987.
34. Hahn BH, Shaw GM, Taylor ME, Redfield RR, Markham PD, Salahuddin SZ, Wong-Staal F, Gallo RC, Parks ES, Parks WP. Science 232:1548-1553, 1986.
35. Modrow S, Hahn BH, Shaw GM, Gallo RC, Wong-Staal F, Wolf H. J. Virol. 61:570-578, 1987.
36. Sarngadharan MG, Popovic M, Bruch L, Schupbach J, Gallo RC. Science 224:506-508, 1984.
37. Blattner WA, Biggar RJ, Weiss SH, Melbye M, Goedert JJ. Ann. Intern. Med. 103:665-670, 1985.
38. Manca N, diMarzo Veronese F, Ho DD, et al. Eur. J. Epidemiol., in press, 1987.
39. Franchini G, Robert-Guroff M, Wong-Staal F, et al. Proc. Natl. Acad. Sci. USA, in press, 1987.
40. Robey WG, Safai B, Oroszlan S, Arthur LO, Gonda MA, Gallo RC, Fischinger PJ. Science 228:593-595, 1985.
41. diMarzo Veronese F, Copeland TD, DeVico AL, Rahman R, Oroszlan S, Gallo RC, Sarngadharan MG. Science 231:1289-1291, 1986.
42. Robert-Guroff M, Reitz MS Jr, Robey WG, Gallo RC. J. Immunol. 137: 3306-3309, 1986.
43. Robert-Guroff M, Brown M, Gallo RC. Nature 316:72-74, 1985.
44. Weiss RA, Clapham PR, Cheingsong-Popov R, Dalgleish AG, Carne CA, Weller IVD, Tedder RS. Nature 316:69-72, 1985.
45. Sodroski J, Goh WC, Rosen C, Tartar A, Portetelle D, Burny A, Haseltine W. Science 231: 1549-1553, 1986.
46. Sodroski J, Goh WC, Rosen C, Campbell K, Haseltine WA. Nature 322: 470-474, 1986.

47. Lifson JD, Reyes GR, McGrath MS, Stein BS, Engleman EG. Science 232: 1123-1127, 1986.
48. Zagury D, Bernard J, Leonard R, Cheynier R, Feldman M, Sarin PS, Gallo RC. Science 231:850-853, 1986.
49. Letvin N, Eaton KA, Aldrich WR, Sehgal PK, Blake BJ, Schlossman SF, King NW, Hunt RD. Proc. Natl. Acad. Sci. USA 80:2718-2722, 1983.
50. Henrickson RV, Maul DH, Osborn KG, Sever JL, Madden DL, Ellingsworth LR, Anderson JH, Lowensteine LJ, Gardner MB. Lancet i:388-390, 1983.
51. King NV, Hunt RD and Letvin NL. Am. J. Pathol. 113:382-388, 1983.
52. Stromberg K, Benveniste RL, Rabin H, Giddens WE Jr, Ochs HT, Morton WR, Tsai C-C. Science 224:289-292, 1984.
53. Guo H-G, Wong-Staal F, Gallo RC. Science 223:1195-1197, 1984.
54. Miyoshi I, Ohtsuki Y, Fujishita M, Yoshimoto S, Kubonishi I, Minezawa M. Gann 73:848-849, 1982.
55. Hayami M, Komuro A, Nozawa K, Shotake T, Ishikawa K-i, Yamamoto K, Ishida T, Honjo S, Hinuma Y. Int. J. Cancer 33:179-183, 1984.
56. Homma T, Kanki PJ, King NW Jr, Hunt RD, O'Connell MJ, Letvin NL, Daniel MD, Desrosiers RC, Yang CS, Essex M. Science 225:716-718, 1984.
57. Hunt RD, Blake BJ, Chalifoux LV, Sehgal PK, King NW, Letvin NL. Proc. Natl. Acad. Sci. USA 80:5085-5089, 1983.
58. Kanki PJ, McLane MF, King NW Jr, Letvin NL, Hunt RD, Sehgal P, Daniel MD, Desrosiers RC, Essex M. Science 228:1199-1201, 1985.
59. Daniel MD, Letvin NL, King NW, Kannagi M, Sehgal PK, Hunt RD, Kanki PJ, Essex M, Desrosiers RC. Science 228:1201-1204, 1985.
60. Saxinger WC, Levine PH, Dean AG, de Thé G, Lange-Wantzin G, Moghissi J, Laurent F, Hoh M, Sarngadharan MG, Gallo RC. Science 227:1036-1038, 1985.
61. Kanki PJ, Alroy J, Essex M. Science 230:951-954, 1985.
62. Hirsch V, Riedel N, Kornfeld H, Kanki PJ, Essex M, Mullins JI. Proc. Natl. Acad. Sci. USA 83:9754-9758, 1986.
63. Kanki PJ, Barin F, M'Boup S, Allan JS, Romet-Lemonne JL, Marlink R, McLane MF, Lee T-H, Arbeille B, Denis F, Essex M. Science 232:238-243, 1986.
64. Clavel F, Guétard D, Brun-Vézinet F, Chamaret S, Rey M-A, Santos-Ferreira MO, Laurent AG, Dauguet C, Katlama C, Rouzioux C, Klatzmann D, Champalimaud JL, Montagnier L. Science 233:343-346, 1986.
65. Albert J, Bredberg W, Chiodi F, et al., in press, 1987.
66. Franchini G, Collalti E, Arya SK, et al. AIDS Research, in press, 1987.

CONTRIBUTIONS FROM THE REGA INSTITUTE

I. RESEARCH UNIT OF PROF. H. VANDERHAEGHE

SYNTHESIS AND ACTIVITY OF SOME NEW CEPHALOSPORINS

M. VAN ROBAYS, P. CLAES and H. VANDERHAEGHE
Rega Institute, Katholieke Universiteit Leuven, B-3000 Leuven, Belgium

1. INTRODUCTION

Compared to the great number of parenteral cephalosporins, only a few orally active compounds of this group are available. Unlike the parenteral cephalosporins which possess a great variety of 7- and 3-substituents, the orally absorbed cephalosporins are very similar to one another. The 7-acyl group is an α-aminophenylacetyl group (or a dihydroxy derivative thereof); the 3-substituent is usually small and uncharged. These structural similarities point to the severe constraints on combining antibacterial activity with the capability of oral absorption. Orally absorbed cephalosporins are very similar in their antibacterial and pharmacokinetic properties; and as their antimicrobial activity is much lower than that of the injectable derivatives, further search for new compounds seems indicated (1).

2. SYNTHESIS

We have prepared a series of new phenylglycines, which could be considered as either benzimidazolyl-, indolyl-, indazolyl-, or -triazolylglycines (Fig. 1). These compounds were coupled with 7-aminodesacetoxycephalosporanic acid to give new cephalosporins.

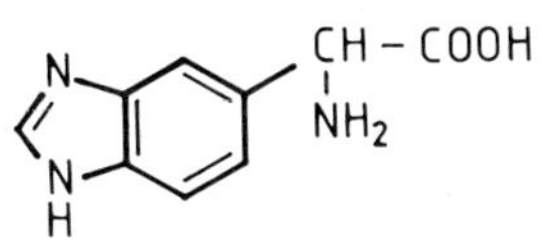

5-benzimidazolylglycine

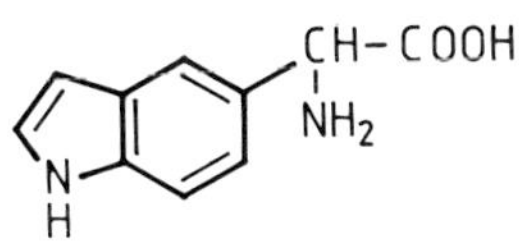

5-indolylglycine

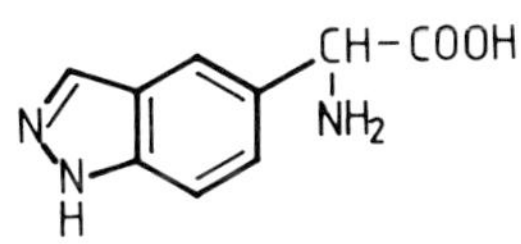

5-indazolylglycine

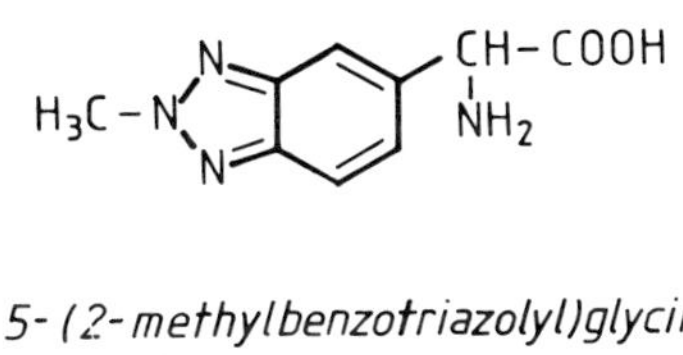

5-(2-methylbenzotriazolyl)glycine

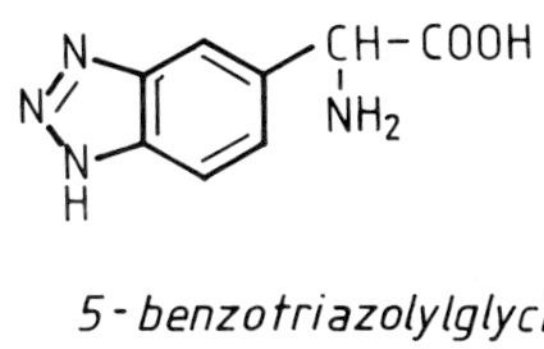

5-benzotriazolylglycine

FIGURE 1.

The Boc-protected amino acids were coupled with 7-aminodesacetoxycephalosporanic acid t-butyl ester via the mixed anhydride method, using EEDQ (2) or ethyl chloroformiate (3). Deprotection with trifluoroacetic acid afforded the desired D/L cephalosporines. The two diastereoisomers of 7-(5-benzimidazolylglycyl)aminodesacetoxycephalosporanic acid were separated on a XAD-2 column. The other racemic mixtures could not be separated because of the small difference in affinity for the XAD-2 resin.

3. BIOLOGICAL ACTIVITY

A biological assay was performed by diluting the cephalosporins in Mueller-Hinton agar and 5 % lysed horse blood. Inoculum : 10^5 cfu/spot by replicator. The incubation time was 18 h at 36°C.

TABLE 1. Biological activity against selected bacterial strains.

Strains	Minimum inhibitory concentration (mg/L)					
	A[a]	B	C	D	E	F
Str.pyogenes	0.5	0.5	1	0.5	0.25	2
S.aureus	2	2	8	4	16	8
S.epidermidis	1	1	4	1	8	4
Str.pyogenes	4	2	16	8	8	*
Sh.sonnei	8	4	*	16	16	*
S.typhimurium	8	8	*	16	16	*

[a]A : D 7-(5-benzimidazolylglycyl)aminodesacetoxycephalosporanic acid;
B : cephalexin;
C : D/L 7-(5-indolylglycyl)aminodesacetoxycephalosporanic acid;
D : D/L 7-(5-indazolylglycyl)aminodesacetoxycephalosporanic acid;
E : D/L 7-(5-benzotriazolylglycyl)aminodesacetoxycephalosporanic acid;
F : D/L 7-(5-(2-methylbenzotriazolyl)glycyl)aminodesacetoxycephalosporanic acid.

4. CONCLUSION

The antimicrobial spectrum of the new cephalosporins is similar to that of cephalexin. Their oral absorption is subject of further study.

REFERENCES

1. O'Callaghan CH. in Fortschritte der antimikrobiellen und antineoplastischen Chemotherapie (Adam D, Ed). Futuramed Verlag, München, vol 1, pp 1-6, 1982.
2. Belleau B, Malek G. J. Amer. Chem. Soc. 90:1651-1653, 1968.
3. Spencer JL, Flynn EH, Roeske RW, Siu FY, Chauvette RR. J. Med. Chem. 9:746-750, 1966.

INCORPORATION OF NUCLEOSIDE ANALOGUES INTO OLIGONUCLEOTIDES

A. VAN AERSCHOT and H. VANDERHAEGHE
Rega Institute, Katholieke Universiteit Leuven, B-3000 Leuven, Belgium

1. INTRODUCTION

Synthesis of deoxyoligonucleotides has become a routine procedure in nucleic acid biochemistry. Oligonucleotides can be used for several purposes. They can serve as linkers and adaptors, as primers for cDNA synthesis and DNA sequencing, and as hybridization probes. They can also be used for site-specific mutagenesis. Even whole genes can be synthesized from deoxyoligonucleotides.

The incorporation of nucleoside analogues into oligonucleotides may be instrumental in the elucidation of the mechanism of enzymatic reactions (1) as well as for conformational studies (2).

DNA probes are very useful for cloning complementary deoxyribonucleic acids (cDNAs). Since amino acid codons are apt to degeneration, DNA probes for a specific amino acid sequence should contain all possible codon sets. Of obvious importance is an ideal base analogue which would substitute for all four natural nucleosides without much destabilisation, thus obviating the need for mixed-sequence probes. Probes with the phenyl ring as the base substituent have failed (3), but deoxyinosine has been reported to be a suitable alternative (4).

We succeeded in incorporating deoxyinosine, deoxyxanthosine and deoxynebularine into oligonucleotides by the phosphotriester route on silica gel, as developed in our laboratory (5). Their incorporation has been demonstrated by enzymatic digestion and HPLC analysis.

2. PROCEDURE

2.1. Incorporation of analogues into oligonucleotides

Deoxynebularine (dN) (1) was made from deoxyadenosine (6); deoxyinosine (dI) (2) and deoxyxanthosine (dX) (3) were synthesized from deoxyguanosine and protected at the lactam functions with p-nitrophenylethyl blocking groups. Synthesis of the three building blocks has been described elsewhere (7). The protected monomers were used to synthesize heptamers. These heptamers had the following sequence : TITIITT. A bench top synthesizer (Omnifit) was used for assembly of the oligonucleotides.

2.2. Control for side reactions

After deprotection and purification by HPLC, heptamers were enzymatically degraded with nuclease P1, and the mixture was analyzed with ion exchange chromatography, giving 3 peaks : dT, pdT and 5'phosphate deoxy analogue. Subsequent treatment with alkaline phosphatase and reversed phase analysis gave dT and the deoxynucleoside analogue, which were then quantitatively determined.

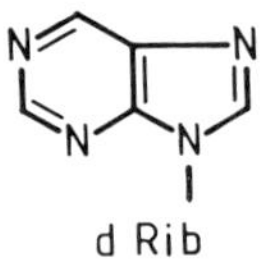

1

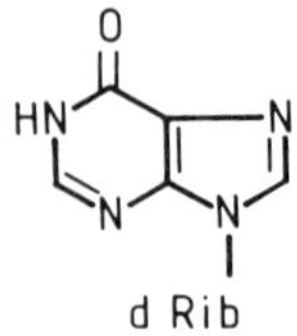

2

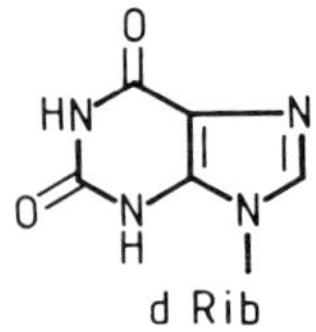

3

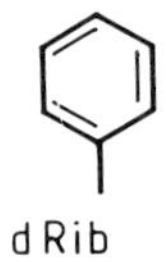

4

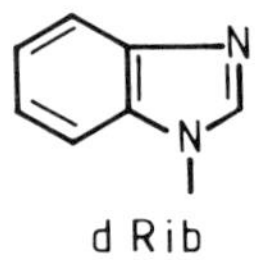

5

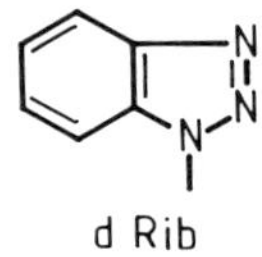

6

2.3. Biological evaluations

After the incorporation of the analogues was unambiguously proved, 23-mers were synthesized as probes for the isolation of the gene for the proline-rich protein of rats. The same sequence was synthesized three times, with either deoxyinosine, deoxyxanthosine or deoxynebularine at 6 ambiguous positions.

Nucleoside analogues may also be useful in mutagenesis studies. Experiments were performed to see which base was preferentially incorporated by DNA polymerase into DNA opposite an analogue. Therefore, nonamers were synthesized with 4 overlapping nucleotides at each side of a nucleoside analogue. In addition to dN, dX and dI, 1,2-dideoxy-1-phenyl-β-D-ribofuranose (4), 1,2-dideoxy-1-(N^1-benzimidazolyl)-β-D-ribofuranose (5) and 1,2-dideoxy-1-(N^1-benzotriazolyl)-β-D-ribofuranose (6) have also been used for mutagenesis primers.

REFERENCES

1. a) Brennan CA, Van Cleve MD, Gumport RI. J. Biol. Chem. 261:7270-7278, 1986; b) Seela F, Driller H, Nucleic Acids Res. 14:2319-2332, 1986.
2. a) Uesugi S, Miyashiro H, Tomita K, Ikehara M. Chem. Pharm. Bull. 34:51-60, 1986; b) Guy A, Molko D, Wagrez L, Téoule R. Helv. Chim. Acta 69:1034-1040, 1986.
3. Millican TA, Mock GA, Chauncey MA, Patel TP, Eaton MAW, Gunning J, Cutbush SD, Neidle S, Mann J. Nucleic Acids Res. 12:7435-7452, 1984.
4. Ohtsuka E, Matsuki S, Ikehara M, Takahashi Y, Matsubara K. J. Biol. Chem. 260:2605-2608, 1985.
5. Van Aerschot A, Heyndrickx L, Herdewijn P. Arch. Intern. Physiol. Biochim. 94:B159, 1986.
6. Nair V, Chamberlain SD. Synthesis, 401-403, 1984.
7. Van Aerschot A, Vanderhaeghe H. Nucleosides & Nucleotides, in press.

2'-5'-OLIGOADENYLATES (2-5A) AS MEDIATORS OF INTERFERON ACTION. SYNTHESIS AND BIOLOGICAL ACTIVITY OF NEW 2-5A ANALOGUES

P. HERDEWIJN, R. PAUWELS, E. DE CLERCQ, R. CHARUBALA[†] and W. PFLEIDERER[†]
Rega Institute, Katholieke Universiteit Leuven, B-3000 Leuven, Belgium
[†]Fakultät für Chemie, Universität Konstanz, D-7750 Konstanz, West Germany

1. INTRODUCTION

Double-stranded RNA (dsRNA) is a potent inhibitor of protein synthesis in extracts of interferon-treated cells. One of the mechanisms that has been proposed to explain this inhibition of protein synthesis is by the 2-5A pathway (1). Interferon induces the synthesis of an enzyme, 2-5A synthetase, which upon activation by dsRNA generates 2-5A from ATP. This 2-5A activates a pre-existing endonuclease for cleavage of single-stranded RNA. The biological activity of 2-5A is rapidly lost due to cleavage of the 2'-5' internucleotide bond by a specific 2'-5' phosphodiesterase starting at the 3'-end. This rapid cleavage and the poor uptake of 2-5A in intact cells, the latter because of its ionic character, limit the potential of 2-5A as a useful approach to the treatment of virus infections or cancer.

Although several modified 2-5A analogues have been synthesized to improve enzymatic stability, only few of them are resistant to degradation and still able to activate the 2-5A dependent endonuclease (2-4). On the other hand, relative drastic methodology such as calcium coprecipitation and microinjection has been used to introduce 2-5A into intact cells. In an effort to improve the enzymatic stability and the cellular uptake, the 2-5A molecule was chemically modified. We describe here the synthesis and biological activty of oligoadenylates in which one or more adenosine residues are replaced by 9-(3-azido-3-deoxy-β-D-xylofuranosyl)adenine or 9-(3-amino-3-deoxy-β-D-xylofuranosyl)adenine.

2. SYNTHESIS

From the three methods, now currently in use for the synthesis of oligonucleotides (phosphotriester method, phosphoramidite method and H-phosphonate method), the phosphotriester method is the most practical for the large scale synthesis in solution.

In recent years, a broad variety of new blocking groups has been developed for protecting the hydroxyl groups of the sugar part, the amino group of the aglycone, the internucleotide phosphate group and even the heterocyclic lactam function. One of these approaches makes use of p-nitrophenylethyl as an universal blocking group (6) (Fig. 1). This blocking group has the advantage of its chemical stability towards mild acid and base, its inertness in the condensation step, its cleavage method (β-elimination) which does not affect the P-O bonds, and its general application for the protection of all functions which can interfere with the internucleotide bond formation reaction. The monomethoxytrityl group and the 2,5-dichlorophenyl group were used for the transient protection of the 5'-hydroxyl group and the phosphate group. They can be selectively removed with p-toluenesulphonic acid and p-nitrobenzaldoximate, respectively. Triisopropylbenzenesulfonylchloride was used as condensing agent in the presence of N-methylimidazole.

The advantage of the use of a uniform blocking group is evident during the deprotection step which only requires DBU treatment eventually followed by acid cleavage to remove the monomethoxytrityl group.

The different building blocks which were used in the synthesis of the oligonucleotides were all fully identified by NMR, UV and elemental analysis.

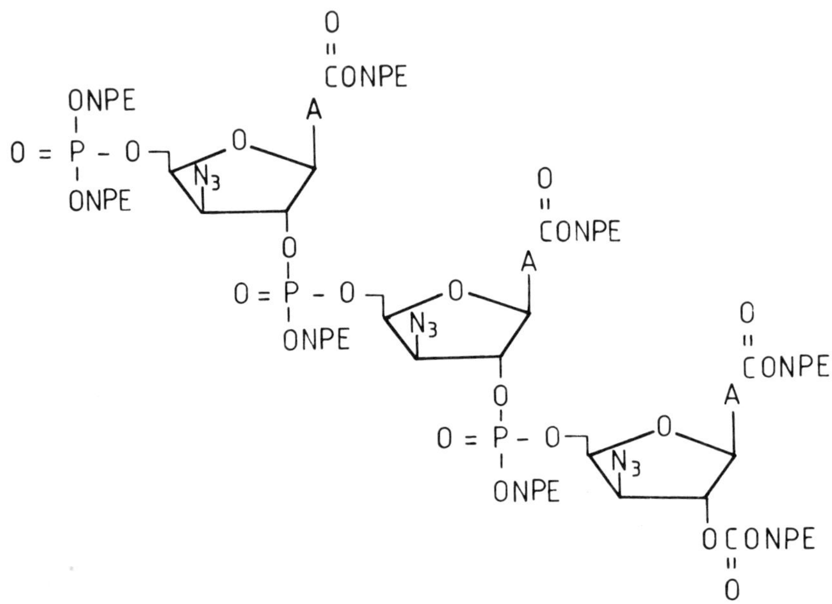

FIGURE 1. NPE = p-nitrophenylethyl.

3. BIOLOGICAL ACTIVITY

The antiproliferative activity of the 3'-substituted 2-5A analogues was determined by measuring the inhibition of murine leukemia L1210 cell growth. As shown in Table 1, all the oligonucleotides containing one or more azido-substituents show an activity comparable to that of 2-5A itself. The same activity was found for the parent 3'-azido-substituted monomer, whereas the amino analogue did not show any activity. Although it may be assumed that substitution of an azido or amino group for 3'-hydroxyl increases the resistance of 2-5A against degradation by intracellular 2',5'-phosphodiesterase, it did not lead to a marked increase in anticellular activity.

It has previously been suggested that the biological activity found for the cordycepin (7) and β-D-xylofuranosyladenine (8) trimers results from extracellular hydrolysis to the parent nucleosides which are then taken up by the cells and inhibit RNA and DNA synthesis probably after rephosphorylation.

It remains to be explored to what extent the azido- and amino-derivatives of 2-5A are degraded in the extra- or intracellular environment.

TABLE 1. Antiproliferative activity of 3'-substituted 2-5A analogues

Compound	Anticellular activity against L1210 cells ID_{50} (μM)
$A^{N_3}pA^{N_3}pA^{N_3}$	27
$pA^{N_3}pA^{N_3}pA^{N_3}$	29
$ApA^{N_3}pA^{N_3}$	45
$ApApA^{N_3}$	48
$A^{N_3}pA^{N_3}pA^{N_3}pA^{N_3}$	20
A^{N_3}	48
pA^{N_3}	122
$A^{NH_2}pA^{NH_2}pA^{NH_2}$	> 200
A^{NH_2}	> 200
ApApA	29

A^{N_3} = 9-(3-azido-3-deoxy-β-D-xylofuranosyl)adenine. A = adenine.

A^{NH_2} = 9-(3-amino-3-deoxy-β-D-xylofuranosyl)adenine.

REFERENCES

1. Kerr IM, Brown RE. Proc. Natl. Acad. Sci. USA 75:256-260, 1978.
2. Silverman RH, Wreschner DH, Gilbert CS, Kerr IM. Eur. J. Biochem. 115, 79-85, 1981.
3. Baglioni C, D'Alessandro SB, Nilsen JW, den Hartog JAJ, Crea R, van Boom JH. J. Biol. Chem. 256:3253-3257, 1981.
4. Imai J, Johnston MI, Torrence PF. J. Biol. Chem. 257:12739-12745, 1982.
5. Bayard B, Bisbal C, Silhol M, Cnockaert J, Huez G, Lebleu B. Eur. J. Biochem. 142:291-298, 1984.
6. Himmelsbach F, Schulz BS, Trichtinger T, Charubala R, Pfleiderer W. Tetrahedron 40:59-72, 1984.
7. Chapekar MS, Glazer RI. Biochem. Biophys. Res. Commun. 115:137-143, 1983.
8. Goswami BB, Gosselin G, Imbach JL, Sharma DK. Virology 137:403-407, 1984.

II. RESEARCH UNIT OF PROF. H. EYSSEN

CLONING AND EXPRESSION OF AN rRNA METHYLASE GENE IN STREPTOMYCES VENEZUELAE PHAGE VWB

J. ANNE, L. VAN MELLAERT and H. EYSSEN
Rega Institute, Katholieke Universiteit Leuven, B-3000 Leuven, Belgium

1. INTRODUCTION

Most work in molecular biology of Streptomycetes has concentrated on the genetically well characterized *Streptomyces coelicolor* and *S.lividans*, and less is known about industrially important species. Except for ΦC31 (1), only preliminary reports have been published on actinophage cloning systems. The availability of efficient transfection systems should be very useful to study the genetics and to elucidate the biological processes in those species.

To develop for *S.venezuelae* an efficient transfection system, we constructed a vector based on actinophage VWB. This phage, isolated from soil enriched with *S.venezuelae*, was chosen for the following reasons : VWB is a temperate phage and its DNA, estimated to be 47.3 kb, has cohesive ends (2). Because of its resistance to chelating agents this phage is supposed to have a special packaging mechanism. In addition, its DNA has an extremely high G+C content (77.2 mol %) and it is susceptible to several restriction endonucleases (3).

2. RESULTS AND DISCUSSION

2.1. Restriction endonuclease mapping of VWB and localization of the attachment site

The restriction map of phage VWB was constructed after single and double digestions of the DNA and measurements of the various fragments obtained. Restriction sites for 10 different restriction enzymes were localized. VWB DNA contains a single *Eco*RI recognition site, but this site is situated in an essential region. VWB DNA contains no sites for *Xba*I or *Xho*I.

The *att* site of VWB was localized on the restriction map after hybridization of [^{32}P]labeled VWB DNA to chromosomal DNA of a *S.venezuelae* strain that was lysogenized with phage VWB. The chromosomal DNA was digested with different restriction endonucleases and separated on an agarose gel prior to Southern transfer and hybridization. Its pattern was compared to that of VWB DNA treated in a similar way. In this manner the *att* site could be situated in a 2 kb fragment (Fig. 1).

2.2. Cloning in VWB of the tsr-gene and the XhoI site of pIJ702

The *tsr* gene was excised from pIJ702 by *Bcl*I digestion (Fig. 1) and, subsequently, ligated to VWB DNA partially digested with *Bgl*II (ca. 3 U/µg DNA, 2 min, 37°C). After transfection, phages containing the *tsr* insert were detected by selection of thiostrepton resistant lysogens. These lysogens spontaneously released phage particles which after infection produced again thiostrepton resistant lysogens. DNA of these phage particles contained the cloned fragment as proven by agarose gel electrophoresis.

To insert in VWB a single XhoI site usable for cloning, the BamHI/BclI fragment (i.e. 0.8 kb) of pIJ702 (Fig. 1) was ligated to the 1.09 kb BclI fragment having the tsr gene. This 1.89 kb fragment was ligated to VWB DNA partially digested with BglII. Phages containing the inserted DNA could be detected as described above and the 1.89 kb fragment was observed after restriction digestion of DNA isolated from phages released from the thiostrepton resistant lysogens. Although the insertion of the 1.09 kb and 1.89 kb fragment in VWB increased the amount of DNA to be packaged to 102.3 % and 104 %, respectively, the phages with the inserts were stable and remained resistant to chelating agents.

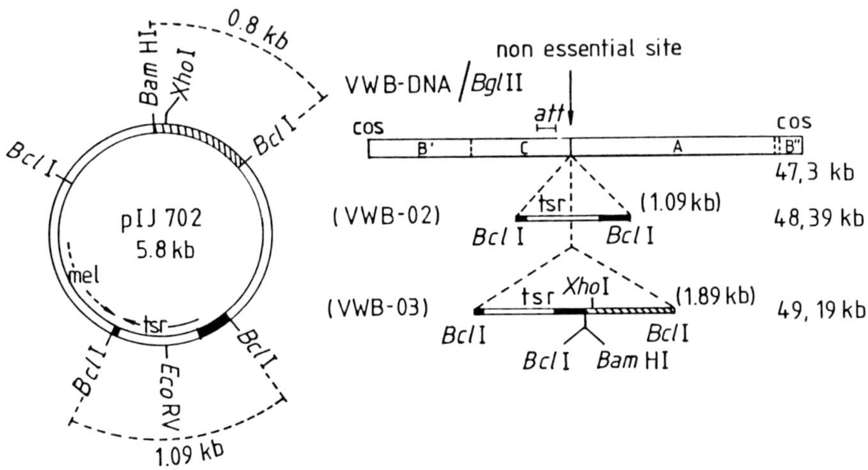

FIGURE 1. Construction of VWB vectors.

3. CONCLUSION

A gene coding for rRNA methylase and a DNA sequence containing a unique XhoI recognition site were cloned in S.venezuelae phage VWB. Although almost 2 kb of additional DNA were thus added to the 47.3 kb phage genome, the DNA could still be packaged, and the phages could lysogenize S.venezuelae. The rRNA methylase gene was expressed as revealed by the appearance of thiostrepton resistant lysogens.

The stable insertion of the tsr gene and of a single XhoI site in the VWB genome (47.3 kb) makes this phage appropriate for cloning provided still more DNA can be inserted. How much additional DNA can be packaged in VWB phage heads is currently under investigation, but its resistance to chelating agents - also for the phage with 4 pct (1.89 kb) extra DNA - suggests that VWB might have an unusual packaging mechanism capable to enclose still more DNA.

REFERENCES

1. King AA, Chater KF. J. Gen. Microbiol. 132:1739-1752, 1986.
2. Anné J, Verheyen P, Volckaert G, Eyssen H. Mol. Gen. Genet. 200:506-507, 1985.
3. Anné J, Wohlleben W, Burkardt HJ, Springer R, Pühler A. J. Gen. Microbiol. 130:2639-2649, 1984.

COPROSTANIC ACIDS AND A C_{29}-DICARBOXYLIC BILE ACID IN PEROXISOMAL DISEASES

G. PARMENTIER, G. JANSSEN, J. VAN ELDERE and H. EYSSEN
Rega Institute, Katholieke Universiteit Leuven, B-3000 Leuven, Belgium

1. INTRODUCTION

The biosynthesis of the two major bile acids cholic acid and chenodeoxycholic acid from cholesterol proceeds via trihydroxycoprostanic acid (THCA) and dihydroxycoprostanic acid (DHCA), respectively. These intermediates accumulate in inherited metabolic disorders characterized by impairment of the peroxisomal function. The Zellweger syndrome is the most typical example of this group of diseases and is characterized by a complete absence of peroxisomes.

In 1972, we first described two patients with markedly increased quantities of THCA in their bile (1). In 1979, we described three children with elevated amounts of DHCA and THCA in their serum, urine, bile and feces (2). Small amounts of 3α-OH-coprostanic acid, 3β-OH-5-cholesten-26-oic isomers of THCA, and varanic acid were also identified. Moreover, in serum and urine of these patients we detected appreciable amounts of an hitherto unknown bile acid, that contains 29 carbon atoms and has two carboxyl groups in the side chain. The structure of this new bile acid was subsequently established as 3α,7α,12α-trihydroxy-27-carboxymethyl-5β-cholestan-26-oic acid (3,4).

In this paper we report on the distribution of these bile acids in various forms of peroxisomal diseases.

2. RESULTS AND DISCUSSION

We have examined serum bile acids in more than 40 Zellweger patients. All had DHCA, THCA and the C_{29}-bile acid in their serum. Total amount of bile acids was always increased (8-90 μg/ml), as compared to normal value of 2-4 μg/ml. DHCA, THCA and the C_{29}-bile acid represented 35 to 90 % of the total bile acids; the percentage of the individual bile acid intermediates varied strongly from one patient to another.

We also analyzed serum bile acids in patients with other peroxisomal disorders (Table 1). Patients with the infantile form of Refsum disease and neonatal adrenoleukodystrophy consistently had increased levels of bile acid intermediates. Patients with X-linked adrenoleukodystrophy, classical Refsum disease, hyperpipecolic acidemia or rhizomelic chondrodysplasia punctata did not have increased bile acid intermediate levels. Total serum bile acid concentration (average : 41 μg/ml) and the percentage of bile acid precursors (80 %) were highest in typical Zellweger patients who died before the age of 1 year. Total serum bile acids and the relative amount of THCA, DHCA and C_{29}-bile acid decreased with age in Zellweger patients who lived for > 3 years.

No sufficient explanation as to the biosynthesis of the C_{29}-dicarboxylic acid has been provided yet. Experiments with labeled THCA seem to support the hypothesis that the C_{29}-dicarboxylic acid derivative is formed

TABLE 1. Serum bile acids in different peroxisomal disorders.

Diagnosis	Number of patients	Bile acid concentration (μg/ml) mean ± SD	Bile acid precursors (% of total bile acid concentration) mean ± SD
Zellweger (patients lived < 1 yr)	12	41 ± 20	80 ± 9
Zellweger (variants) (patients lived > 3 yrs)	11	22 ± 20	46 ± 27
Pseudo-Zellweger	1	10	18
Neonatal adrenoleukodystrophy	4	14 ± 13	70 ± 20
Infantile Refsum disease	6	12 ± 15	33 ± 21
X-linked adrenoleukodystrophy	5	3 ± 2	0
Classical Refsum disease	3	3 ± 3	0
Hiperpipecolic acidemia	4	4 ± 2	0
Chondrodysplasia punctata (rhizomelic type)	9	3 ± 3	0

by side chain elongation of THCA as small amounts of radioactivity were found in the C_{29}-dicarboxylic acid (5).

3. CONCLUSION

Screening for bile acid precursors, combined with the detection of very long chain fatty acids is, in our experience, an easy and reliable first line approach in the diagnosis of peroxisomal disorders.

REFERENCES

1. Eyssen H, Parmentier G, Compernolle F, Boon J, Eggermont, E. Biochim. Biophys. Acta 273:212-221, 1972.
2. Parmentier G, Janssen G, Eggermont E, Eyssen H. Eur. J. Biochem. 102: 173-183, 1979.
3. Janssen G, Toppet S, Parmentier G. J. Lipid Res. 23:456-465, 1982.
4. Eyssen H, Eggermont E, Van Eldere J, Jaeken J, Parmentier G, Janssen G. Acta Paediatr. Scand. 74:539-544, 1985.
5. Kase BF, Björkhem I, Hägä P, Pedersen JI. J. Clin. Invest. 75:427-435, 1985.

BILE ACID DESULFATION BY THE GUT MICROFLORA

J. ROBBEN, J. VAN ELDERE and H. EYSSEN
Rega Institute, Katholieke Universiteit Leuven, B-3000 Leuven, Belgium

1. INTRODUCTION

Sulfate esterification of bile acids in the liver decreases their entero-hepatic circulation and enhances their fecal excretion.However, the intestinal microflora is capable of hydrolyzing these sulfate esters and thus counteracting sulfate esterification process by the liver and the elimination of toxic bile acids such as lithocholate. We have isolated sulfatase-producing microorganisms from the rat and human fecal microflora, and we compared their substrate specificities *in vitro*. The influence of bile acid desulfation by *Clostridium* sp. S_2 on bile acid metabolism was studied *in vivo* in female gnotobiotic rats. In these rats the half-lifes of $[^{14}C]$-labeled sulfated and unsulfated taurolithocholate was determined following their intraperitoneal (i.p.) injection.

2. METHODS

Isolation and cultivation of the microorganisms was carried out in an anaerobic chamber (80 % N_2, 10 % H_2 and 10 % CO_2). Substrate specificities of sulfatase-producing microorganisms were determined by cultivation on bile acid sulfate containing media followed by extraction and gas-liquid chromatography of the free bile acids. *Clostridium* S_2 was associated with gnotobiotic rats previously infected with a limited microflora which normalized cecal size and intestinal transit. The half-life of i.p. injected bile acids was determined according to the Lindstedt and Norman procedure (1).

3. RESULTS

Three and five sulfatase-producing microorganisms were isolated from rat and human fecal flora, respectively (Table 1). The isolates showed limited substrate specificities. All isolates hydrolyzed sulfate esters of phenolphtaleine and aromatic steroids (aryl sulfates). Bile salt sulfatase activity was present only in *Clostridium* sp. strains S_1 and S_2 from the rat (2,3) and *Peptococcus niger* strain H_4 isolated from humans. Bile acid desulfation was limited to the 3-sulfate esters; 7- and 12-sulfates were not hydrolyzed. Desulfation of bile acid 3-sulfates was determined by the stereoconfiguration at C-3 and C-5, and appeared to be strain-specific : *Clostridium* S_1 hydrolyzed all bile acid 3-sulfates, except for the 3α-sulfates of the 5α-(allo)-bile acids; *Clostridium* S_2 hydrolyzed the 3α-sulfates of 5α- and 5β-bile acids; *Peptococcus niger* H_4 hydrolyzed bile acid 3β-sulfates. Desulfation of other bile acids by strain H_4 required induction by androstan 3-sulfates with the corresponding configuration at C-3 and C-5.

When associated with female gnotobiotic rats, *Clostridium* sp. S_2 desulfated nearly all large intestinal and fecal bile acid sulfates and cau-

TABLE 1. Substrate specifities of sulfatase-producing microorganisms isolated from rat and human fecal flora.

	Strains[a] isolated from							
	rat			human				
	S_1	S_2	R_9	H_1	H_2	H_4	H_{65}	H_{89}
3α,5β-bile acid sulfates								
lithocholate 3-sulfate	+	+	–	–	–	+[b]	–	–
chenodeoxycholate 3-sulfate	+	+	–	–	–	+[b]	–	–
deoxycholate 3-sulfate	+	+	–	–	–	+[b]	–	–
cholate 3-sulfate	+	+	–	–	–	+[b]	–	–
chenodeoxycholate 7-sulfate	–	–	–	–	–	–	–	–
cholate 7-sulfate	–	–	–	–	–	–	–	–
deoxycholate 12-sulfate	–	–	–	–	–	–	–	–
cholate 12-sulfate	–	–	–	–	–	–	–	–
3β,5β-bile acid sulfates								
iso-lithocholate 3-sulfate	+	–	–	–	–	+	–	–
iso-chenodeoxycholate 3-sulfate	+	–	–	–	–	–	–	–
3α,5α-bile acid sulfates								
allo-lithocholate 3-sulfate	–	+	–	–	–	+[b]	–	–
allo-chenodeoxycholate 3-sulfate	–	+	–	–	–	–	–	–
3β,5α-bile acid sulfates								
iso-allo-lithocholate 3-sulfate	+	–	–	–	–	+	–	–
iso-allo-chenodeoxycholate 3-sulfate	+	–	–	–	–	+	–	–
Aryl sulfates								
estron 3-sulfate	+	+	+	+	+	+	+	+
fenolfthaleine disulfate	+	+	+	+	+	+	+	+

[a]Strains : S_1, S_2: Clostridium sp.; H_1, H_2: Eubacterium cilindroides; H_4: Peptococcus niger; R_9, H_{65}, H_{89}: unidentified.
[b]Desulfation induced by androstan 3-sulfate with the corresponding configuration at C-3 and C-5.

sed a 25 % decrease of fecal bile acid excretion as compared to control rats. Desulfation of allochenodeoxycholate led to an enhanced reabsorption from the intestine and an increased 12α-hydroxylation of this compound by the liver, resulting in an increased faecal output of allocholate. As shown in Table 2, i.p. injected [^{14}C]lithocholate 3-sulfate was excreted considerably faster than the unsulfated analogue in the feces and urine of gnotobiotic rats not infected with the desulfating strain S_2. The cecal size reducing flora (CRF) which helped to normalize the small intestinal transit rate and cecal size, accelerated the excretion of both sulfated and unsulfated lithocholate. In rats associated with Clostridium S_2 (Cl_8+ S_2 rats and Cl_8+CRF+ S_2 rats) the excretion rates of the injected sulfated lithocholate fell to about the same level of those of the unsulfated lithocholate.

TABLE 2. 50 % Excretion times (in days) of label in feces and urine after i.p. injection of [^{14}C]-labeled bile acids in female gnotobiotic rats.

Intestinal microflora	[^{14}C]-labeled compound injected	
	Taurolithocholate 3-sulfate	Taurolithocholate
Cl_8 [a]	2.6 ± 0.3	8.3 ± 0.9
Cl_8+S_2 [b]	9.7 ± 0.3	7.3 ± 1.3
Cl_8+CRF [c]	0.7 ± 0.1	4.0 ± 0.1
Cl_8+S_2+CRF	4.8 ± 0.2	5.0 ± 0.6
CV [d]	1.8 ± 0.2	2.8 ± 0.2

[a] *Clostridium* Cl_8.
[b] S_2 : bile acid sulfatase-producing *Clostridium* S_2.
[c] CRF : cecal size reducing flora.
[d] CV : normal rat flora.

4. DISCUSSION AND CONCLUSION

Isolation of several sulfatase-producing intestinal microorganisms and subsequent investigation of the sulfatase substrate specificities revealed that sulfatase activity depended upon the stereoconfiguration at C-3 and C-5 of the bile acid molecule. Sulfatase activity was strain-specific, indicating that different microorganisms take part in the hydrolysis of intestinal bile acid sulfates. In gnotobiotic rats, desulfation of taurolithocholate 3-sulfate resulted in a 5-fold retardation of the fecal plus urinary excretion of the [^{14}C]-labeled isotope, pointing to the importance of intestinal microbial bile acid desulfation in the overall elimination of sulfated bile acids.

REFERENCES

1. Lindstedt S, Norman A. Acta fysiol. Scand. 38:121-128, 1956.
2. Huijghebaert SM, Mertens JA, Eyssen HJ. Appl. Environ. Microbiol. 43: 185-192, 1982.
3. Robben J, Parmentier G, Eyssen H. Appl. Environ. Microbiol. 51:32-38, 1986.

STEROID DESULFATION BY THE GUT MICROFLORA

J. VAN ELDERE, J. ROBBEN and H. EYSSEN
Rega Institute, Katholieke Universiteit Leuven, B-3000 Leuven, Belgium

1. INTRODUCTION

Most steroid hormones are eventually taken up by the liver cells, often reduced in ring A, always conjugated with glucuronic and/or sulfuric acid, and excreted into the bile. In the gut, the conjugates of steroid hormones are deconjugated by intestinal and bacterial glucuronidases and bacterial sulfatases and can be modified by other bacterial enzymes. Most of the steroids are reabsorbed and returned to the liver (enterohepatic circulation). Bacterial desulfation is necessary for reabsorption of sulfated steroids. Elimination of the desulfating intestinal bacterial strains in rats leads to a reduced reabsorption of estrone, and administration of antibiotics to pregnant women decreased plasma estriol levels. From the rat and human intestinal flora, we isolated a number of bacterial strains capable of steroid desulfation.

2. RESULTS

The substrate specificity of the steroid-desulfating bacterial strains is illustrated in Table 1. The broadest substrate specificity was found for the sulfatase of a strain identified as Peptococcus niger, isolated from human fecal material (Table 2). Apart from its sulfatase activity, this strain also developed C-17 oxido-reductase activity towards estrogens and androsta(e)nes and C-3 oxido-reductase activity towards androsta(e)nes and pregna(e)nes.

To demonstrate the importance of the intestinal microbial desulfation

TABLE 1. Steroid and bile salt sulfatase activity in bacterial strains isolated from the rat and the human intestinal microflora

	Rat			Man		
	Clostridium		Strain R_9	Eubacterium cylindroides		P.niger H_4
	sp. S1	sp. S2		H1	H2	
Bile salt sulfates	+	+	-	-	-	+
Arylsulfates (estrogen sulfates)	+	+	+	+	+	+
Alkylsulfates (androstane- & pregnane sulfates)	-	-	-	-	-	+

TABLE 2. Steroid-desulfating activity of the Peptococcus niger strain isolated from the human intestinal microflora

Bile salt sulfates	+ (20-80)[a]
Steroid sulfates	
1) Arylsulfates	
estrone sulfate	+ (100)
β-estradiol-3-sulfate	+ (50)
β-estradiol-17-sulfate	-
estriol-3-sulfate	trace
2) Alkylsulfates	
dehydroisoandrosterone sulfate	+ (100)
5α-androstane-17-one-3β-sulfate	+ (100)
5α-androstane-17-one-3α-sulfate	+ (100)
5β-androstane-17-one-3α-sulfate	+ (100)
5α-pregnane-20-one-3β-sulfate	+ (60)
Δ5-pregnene-20-one-3β-sulfate	+ (100)
Synthetic sulfate esters	
paranitrocatechol sulfate	+ (40)
paranitrophenyl sulfate	+ (70)
phenolphtaleine disulfate	+ (40)

[a]Between parentheses : percentage of desulfated product after 3 days of incubation on TSB + 0.1 % taurine and 50 µg/ml steroid sulfates or synthetic sulfate esters and 10 µg/ml of bile salt sulfates.

on the enterohepatic circulation of steroids, we studied the fecal and urinary excretion of orally administered [^{14}C]-labeled estrone-3-sulfate in germfree (GF) rats, conventional (CV) rats and gnotobiotic rats selectively associated with estrone-desulfating and/or cecal volume reducing microorganisms. The time required to excrete 50 % of the total label recovered ($t_{1/2}$) was 22 h in CV rats versus 32 h in GF rats (Fig. 1). Gnotobiotic rats selectively associated with a cecal volume-reducing flora[1] (CRF rats) excreted the label even faster ($t_{1/2}$ = 13 h) than did the CV rats. Association of GF rats as well as CRF rats with estrone-desulfating microorganisms (termed S1+S2+R9 rats and CRF+S1+S2+R9 rats, respectively) led to a slower excretion of labeled products ($t_{1/2}$ = 38 h in S1+S2+R9 rats and $t_{1/2}$ = 27 h in CRF+S1+S2+R9 rats). Intestinal microbial desulfation also increased the relative part of the urinary excretion from 4 % in GF rats to 8 % in S1+ S2+R9 rats and from 3 % in CRF rats to 9 % in CRF+S1+S2+R9 rats.

3. CONCLUSION

Intestinal microbial desulfation decreases fecal plus urinary excretion of estrone-3-sulfate and, consequently, plays an important role in the enterohepatic circulation of estrone sulfate. Several steroid-desulfa-

[1]In female CV rats, the cecal weight constituted 2.2 % of the total body weight, versus 13.3 % in female GF rats. Association of the female GF rats with the CRF-flora reduced the cecal weight to 4.7 % of total body weight. The CRF flora did not desulfate estrone sulfate nor in any other way modified the estrogens.

ting microorganisms have been isolated from the rat and human intestinal microflora; their substrate specificity differs from one strain to another.

TABLE 3. Percentages of the different types of labeled fecal estrogen conjugates after oral administration of [^{14}C]-labeled estrone sulfate in gnotobiotic and conventional rats

	Free estrogen (%)	Estrogen glucuronide (%)	Estrogen sulfate (%)	Residue (%)
GF rats	2.8 (± 0.5)	10.6 (± 2)	73.8 (± 5)	12.8 (± 4)
S1+S2+R9 rats	24.7 (± 3)	32.8 (± 11)	24.6 (± 9)	18.0 (± 2)
CRF rats	5.1 (± 0.5)	7.4 (± 1)	79.5 (± 1)	7.9 (± 0.5)
CRF+S1+S2+R9 rats	39.4 (± 5)	27.8 (± 7)	19.5 (± 10)	13.3 (± 0.5)
CV rats	81.5 (± 3)	1.7 (± 0.3)	2.7 (± 0.5)	14.2 (± 2)

[a]Feces collected between 7 h and 24 h after administration of label; values are expressed as percentage of the total amount of label recovered. For each of the columns (except for the Residue column) the differences between GF and S1+S2+R9 rats, between CRF and CRF+S1+S2+R9 rats and between CV rats and all the other groups were statistically significant at a $p < 0.01$ level.

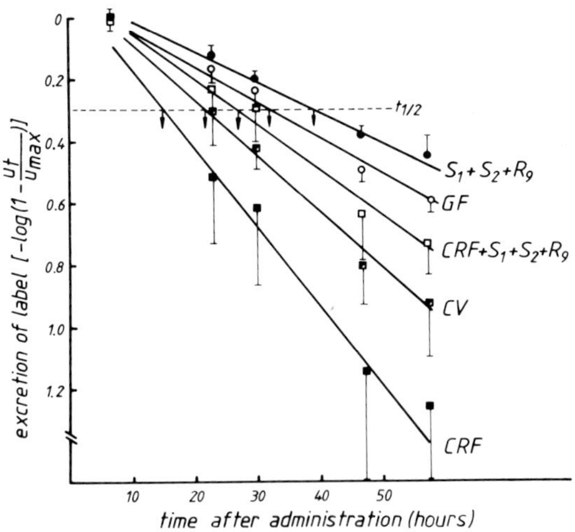

FIGURE 1. Semilogarithmic plot of the cumulative excretion curves after administration of labeled estrone sulfate. Transformation according to $-\log(1-u_t/u_{max})$. Fifty percent excretion ($t_{1/2}$) is reached at $-\log(1-u_t/u_{max}) = 0.301$.

u_t : total amount of label excreted at time t.

u_{max} : total amount of label recovered.

BIOHYDROGENATION OF FATTY ACIDS BY INTESTINAL BACTERIA

A. VERHULST, G. JANSSEN and H. EYSSEN
Rega Institute, Katholieke Universiteit Leuven, B-3000 Leuven, Belgium

1. INTRODUCTION

Fecal lipids are characterized by a high content of *trans*-monounsaturated and saturated fatty acids which is accounted for by preferential absorption and microbial hydrogenation. In earlier studies (1) *Eubacterium lentum* strain Rega was isolated which *in vitro* and *in vivo* (rats) hydrogenates linoleic acid to transvaccenic acid. Strain ATCC 25559 of human origin also performs this reaction. Presently, more than 40 different microorganisms isolated from mice with a limited anaerobic flora and more than 100 different microorganisms isolated from conventional rats have been examined for their ability to hydrogenate linoleic acid. We isolated 4 *Clostridium bifermentans* strains, 2 *Clostridium sporogenes* strains, 1 *Bacteroides* sp. and 1 *Propionibacterium acnes* strain which hydrogenates linoleic acid *in vitro*. We have now compared the linoleic acid-hydrogenating properties of these isolates with those of the type strains of *Eubacterium*, *Clostridium*, *Bacteroides* and *Propionibacterium*. Some of these bacterial strains were further examined *in vivo*.

2. METHODS

Biotransformation of unsaturated long-chain fatty acids was evaluated by cultivating the bacterial strains in medium containing the test substrate. Fatty acids were extracted from the culture medium, and the number and position of the double bonds in the reaction products were identified by combined gas-liquid chromatography and mass spectrometry. The geometry of the double bonds was determined by thin-layer chromatography of the monoenes obtained after partial reduction of the double bonds in the polyenes.

3. RESULTS

It was found that *E.lentum* (33 strains), *E.aerofaciens* (5 strains), *E.fissicatena* (2 strains), *E.contortum* (2 strains), *Cl.sporogenes* (7 strains), *Cl.bifermentans* (6 strains), *Cl.sordellii* (2 strains), *Bacteroides sp.* (1 strain), *P.freudenreichii* subsp. *freudenreichii* (2 strains), *P.freudenreichii* subsp. *shermanii* (2 strains), *P.acidi-propionici* (2 strains) and 1 strain of *P.technicum* hydrogenated linoleic acid (18:2; 9-*cis*,12-*cis*) to transvaccenic acid (18:1; 11-*trans*) *via* a 9-*cis*,11-*trans*-octadecadienoic acid. The *Eubacterium*, *Clostridium* and *Bacteroides* strains performed the total reaction for more than 80 % whereas the *Propionibacterium* strains isomerized linoleic acid for more than 50 % but reduced the isomer only for $\pm$ 10 %.

An *in vivo* study of *Cl.bifermentans*, *Cl.sporogenes* and *Cl.sordelii* showed that these species were also capable of hydrogenating linoleic acid in the rat. Gnotobiotic Fisher rats which were mono-associated with one of these three species excreted 15 to 20 % more octadecenoic acids in their feces than did the germfree control rats.

P.acnes (9 strains) were shown to isomerize in vitro linoleic acid for more than 80 % to 10-trans,12-cis-octadecadienoic acid. This isomer was reduced for ± 10 % to 10-trans-octadecenoic acid.

E.limosum (7 strains), E.alactolyticum (6 strains), E.contortum (3 strains), Cl.absonum (1 strain), Cl.perfringens (1 strain), Cl.hystolyticum (1 strain), Cl.butyricum (1 strain), Cl.tetanomorfum (1 strain), B.furcosus (1 strain), B.fragilis (1 strain), P.avidum (9 strains), P.granulosum (4 strains), P.jensenii (2 strains), P.thoenii (2 strains) and P.lympholphilum (1 strain) did not hydrogenate linoleic acid in vitro.

4. DISCUSSION AND CONCLUSION

Several common intestinal bacterial species were found to be capable of hydrogenating linoleic acid via the same reaction mechanism : isomerization of the 12-cis double bond followed by reduction of the 9-cis double bond. None of the type strains, except for P.acnes, saturated linoleic acid via another pathway. Since all strains of a positive species (except for E.contortum) were shown to hydrogenate linoleic acid, the hydrogenation pattern seems to be a species-specific marker. Our results further show that it is not a genus-specific marker of Bacteroides, Eubacterium and Clostridium. Since the positive species are all common intestinal microorganisms, and since the species we examined in vivo indeed hydrogenated linoleic acid in the rat, we may infer that at least part of the high fecal content of trans-monoenes results from microbial hydrogenation of polyunsaturated fatty acids.

REFERENCES

1. Eyssen H, Parmentier G. Zbl. Bakt., Suppl. 7: 39-43, 1979.

CLASSIFICATION OF THE GENUS *EUBACTERIUM*

A. VERHULST, F. SYMONS[†] and H. EYSSEN
Rega Institute and [†]Botanical Institute, Katholieke Universiteit Leuven, B-3000 Leuven, Belgium

1. INTRODUCTION

Gram-positive, non-sporeforming rods which produce butyric acid, acetic acid, formic acid or only trace amounts of fatty acids are classified in the genus *Eubacterium*, although this definition groups microorganisms which are not closely phylogenetically related. We determined cellular long-chain component patterns, mol % G+C of the DNA and the capability to hydrogenate linoleic acid to transvaccenic acid in 7 *Eubacterium* species, with the aim to establish their phylogenetic relationship.

2. METHODS

Cellular long-chain component patterns were identified according to Lambert and Moss (1). The G+C content of the DNA was estimated by determination of their thermal stability [melting temperature (Tm)]. Canonical correlation analysis (2) was used to graphically show the relationship between the test species.

3. RESULTS AND DISCUSSION (Figures 1 and 2).

The cellular long-chain component patterns enabled us to classify 33 *E.lentum* strains into 2 groups. In the first group the most abundant fatty acids were 15:0 and 14:0 branched-chain fatty acids. In the second group no branched-chain fatty acids were found. This group was characterized by 14:0 and 16:0 straight-chain saturated fatty acids. In the profiles of all 33 strains we found 14:0 and 16:0 fatty acid aldehydes. The cellular long-chain component patterns of 35 strains of the species *E.limosum*, *E.alactolyticum*, *E.aerofaciens*, *E.contortum*, *E.cylindroides* and *E.fissicatena* were species-specific; 14:0, 16:0, 16:1, 18:0 and 18:1 were the most abundant fatty acids. The G+C content of the DNA of all 68 strains was species-specific but varied from ± 30 mol % for the species *E.cylindroides* to ± 65 mol % for the species *E.lentum*. Biohydrogenation of linoleic acid was also a species-specific marker. Except for *E.contortum*, all strains of the positive species followed the same reaction mechanism to reduce linoleic acid: first, the 12-*cis*-double bond was isomerized to an 11-*trans*-double bond, then followed by a reduction of the 9-*cis*-double bond.

Our results show that microorganisms which according to the classical identification schemes belong to the *Eubacterium* species, are phylogenetically not very closely related. Determination of the patterns of cellular long chain components, the mol % G+C of the DNA and the ability to hydrogenate linoleic acid are helpful criteria in determinating species of the genus *Eubacterium*. According to their cellular long-chain component patterns and their mol % G+C values, the second group of *E.lentum* strains is more related to *E.aerofaciens* than to the first group of *E.lentum* strains.

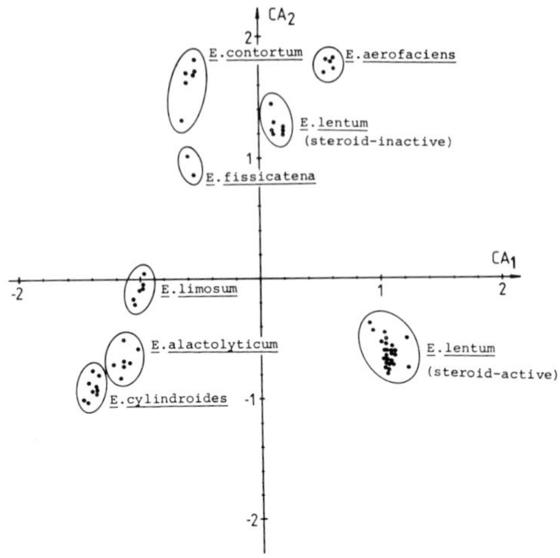

FIGURE 1. Canonical correlation analysis of 68 Eubacterium strains. Plot of the 1st canonical variable against the 2nd canonical variable. The 68 strains of the 8 different groups are indicated as dots.

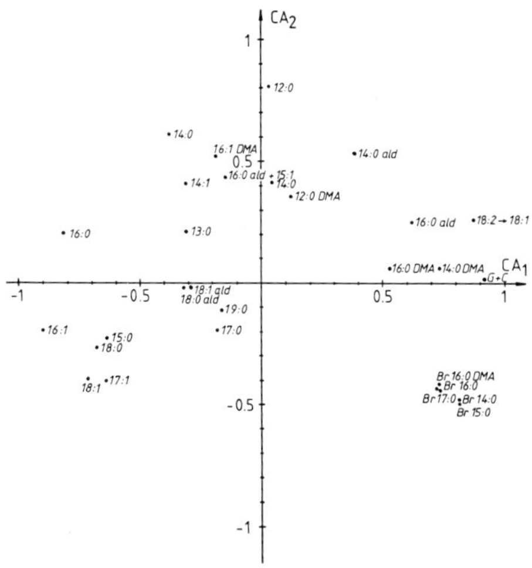

FIGURE 2. Canonical correlation analysis of 68 Eubacterium strains. Plot of the 1st canonical variable against the 2nd canonical variable. The cellular long chain components, the mol % G+C values and the capability to hydrogenate linoleic acid (original variables) are indicated as dots.

REFERENCES

1. Lambert MA, Moss CW. J. Clin. Microbiol. 18: 1370-1377, 1983.
2. Verhulst A, Symons F, Eyssen H. J. Gen. Microbiol. 133:275-282, 1987.

ISOLATION AND CHARACTERISTICS OF *CLOSTRIDIUM* HDCA-1, NOVA SPECIES

H. EYSSEN, G. DE PAUW and J. VAN ELDERE
Rega Institute, Katholieke Universiteit Leuven, B-3000 Leuven, Belgium

1. INTRODUCTION

Hyodeoxycholate is a major fecal bile acid in conventional rats, but does not occur in germfree rats, and arises via intestinal microbial 7β-dehydroxylation of β-muricholate which is a major bile acid in rat bile. We now isolated from rat feces pure cultures of a fastidious and unnamed microorganism that converts the muricholates into hyodeoxycholate.

2. ISOLATION

The hyodeoxycholate-producing microorganism was isolated from cultures of conventional rat feces heated for 20 minutes at 80°C. After repeated subculturing of this mixed culture under strict anaerobic conditions on a medium containing 125 mg of freeze-dried meat or liver particles, or 50 mg of freeze-dried feces from conventional rats, and supplemented with 300 µg of β-muricholate or hyocholate per ml medium, a pure culture (termed *Clostridium* HDCA-1) was isolated.

3. GROWTH REQUIREMENTS OF *CLOSTRIDIUM* HDCA-1

- Anaerobiosis : E_h below -250 mV.
- pH 7.00-7.75 : no growth below pH 6.8.
- Does not grow on any of the commercially available culture media.
- *Absolute requirement* for high levels of an unconjugated 3,6,7-trihydroxy bile acid, e.g. 200-800 µg/ml of β-muricholate or hyocholate.
- *Absolute requirement* for an unidentified heat-labile growth factor produced by certain other bacteria, e.g. *Peptostreptococcus productus*, and also present in filter-sterilized extract of conventional rat feces.

4. MORPHOLOGY OF *CLOSTRIDIUM* HDCA-1

- Small (± 1.5 µm), gram-labile, non-motile, tapered rods, in pairs or short chains.
- Spores formed *in vivo* in gnotobiotic rats.
- Colonies : - are small (0.05-0.1 mm), circular, entire, glistening.
  - consist of a monolayer of cells.

5. BILE ACID CONVERSIONS CARRIED OUT BY *CLOSTRIDIUM* HDCA-1

- β-Muricholate (3α,6β,7β-trihydroxy) converted to hyodeoxycholate (3α,6α-dihydroxy).
- α-Muricholate (3α,6β,7α-trihydroxy) converted to hyodeoxycholate (3α,6α-dihydroxy).
- ω-Muricholate (3α,6α,7β-trihydroxy) converted to hyodeoxycholate (3α,6α-dihydroxy).
- Hyocholate (3α,6α,7α-trihydroxy) converted to hyodeoxycholate (3α,6α-dihydroxy).

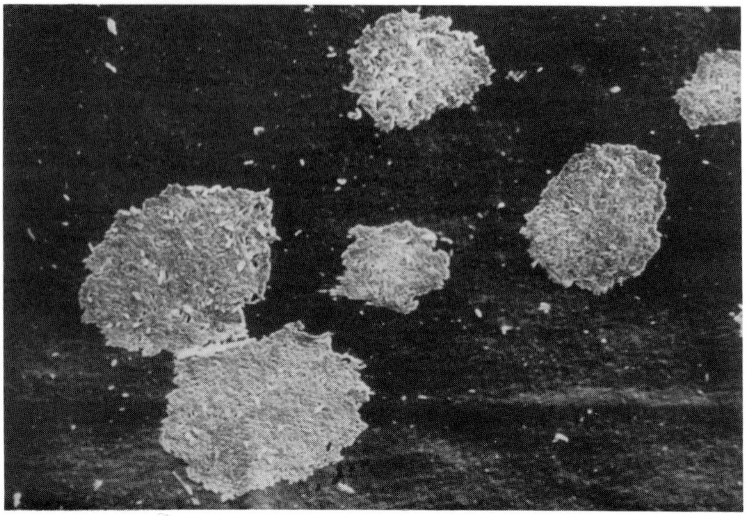

FIGURE 1. Colonies of Clostridium HDCA-1 consisting of a monolayer of cells. Colony diameter : 0.05-0.1 mm.

6. HYODEOXYCHOLATE FORMATION IN GNOTOBIOTIC RATS

To allow growth of Clostridium HDCA-1 the gnotobiotic rats had to be associated with Clostridium perfringens, so as to provide an anaerobic environment, and Peptostreptococcus productus as a source of the heat-labile unidentified growth factor.

TABLE 1. Production of hyodeoxycholate by Clostridium HDCA-1 in gnotobiotic rats. Per cent of total fecal bile acids.

	Male rats		Female rats	
	β-Muricholate	Hyodeoxy-cholate	β-Muricholate	Hyodeoxy-cholate
Before association with HDCA-1	41 %	0 %	29 %	0 %
After association with HDCA-1	8 %	30 %	7 %	23 %

7. CONCLUSION

Hyodeoxycholate can be formed from the muricholates by the fastidious bile acid-dependent Clostridium HDCA-1. The observation that growth of strain HDCA-1 is dependent on the presence of 3,6,7-tri-OH bile acids in the culture medium, may explain why previous attempts to isolate the microorganism have failed. The fact that strain HDCA-1 does not develop at pH values below pH 6.8 also may explain why the conversion of β-muricholate to hyodeoxycholate is suppressed in rats fed lactose-containing diets.

III. RESEARCH UNIT OF PROF. A. BILLIAU

HETEROGENEITY OF RECOMBINANT MURINE INTERFERON-γ (MuIFN-γ)

R. DIJKMANS, G. VOLCKAERT, J. VAN DAMME, M. DE LEY, H. HEREMANS and A. BILLIAU
Rega Institute, Katholieke Universiteit Leuven, B-3000 Leuven, Belgium

1. INTRODUCTION

Insertion of genes into DNA of heterologous prokaryotic or eukaryotic cells with the purpose to achieve high-yield expression of these genes, has become the method of choice for the production of proteins that, under natural conditions, are produced in small quantities amidst a majority of other "contaminating" proteins. We have followed this approach to develop a convenient and highly efficient production system for murine interferon-γ (MuIFN-γ).

2. ISOLATION OF THE MuIFN-γ cDNA

A cDNA library prepared from RNA rich in MuIFN-γ mRNA was screened for MuIFN-γ cDNA sequences using rat IFN-γ chromosomal gene fragments as probes. A complete cDNA clone of MuIFN-γ was obtained by recombining two appropriate segments from partial cDNA clones (1). Restriction mapping and sequence analysis of the MuIFN-γ cDNA allowed its comparison with other cloned interferon genes and the deduction of the amino acid sequence of MuIFN-γ (Fig. 1).

met asn ala thr his cys ile leu ala leu gln leu phe leu met ala val ser

gly cys tyr cys HIS GLY THR VAL ILE GLU SER LEU GLU SER LEU ASN ASN TYR

PHE ASN SER SER GLY ILE ASP VAL GLU GLU LYS SER LEU PHE LEU ASP ILE TRP

ARG ASN TRP GLN LYS ASP GLY ASP MET LYS ILE LEU GLN SER GLN ILE ILE SER

PHE TYR LEU ARG LEU PHE GLU VAL LEU LYS ASP ASN GLN ALA ILE SER ASN ASN

ILE SER VAL ILE GLU SER HIS LEU ILE THR THR PHE PHE SER ASN SER LYS ALA

LYS LYS ASP ALA PHE MET SER ILE ALA LYS PHE GLU VAL ASN ASN PRO GLN VAL

GLN ARG GLN ALA PHE ASN GLU LEU ILE ARG VAL VAL HIS GLN LEU LEU PRO GLU

SER SER LEU ARG LYS ARG LYS ARG SER ARG CYS

FIGURE 1. Amino acid sequence of MuIFN-γ as deduced from the nucleic acid sequence of MuIFN-γ cDNA. Residues of the mature MuIFN-γ protein are printed in capitals, amino acids of the signal peptide are in lower-case letters. Potential N-glycosylation sites are underlined. The single cysteine residue of the mature MuIFN-γ protein is indicated by dots.

3. TRANSIENT EXPRESSION OF MuIFN-γ IN COS CELLS

Further confirmation for the identity of the MuIFN-γ cDNA was based on the detection of MuIFN-γ as the expression product. To obtain such expression COS-1 cells were used. These monkey cells contain an integrated copy of the T-antigen coding sequences and allow the replication of exogenous plasmids carrying the SV40 replication origin. MuIFN-γ cDNA was inserted into such a plasmid (pKCR) (2) downstream the SV40 early promotor and upstream a rabbit β-globin intron and polyadenylation sequences. The resulting plasmid was transfected into COS-1 cells. A low and transient MuIFN-γ activity (10-100 units/ml) could be detected in the cell medium. The MuIFN-γ production could be increased about ten-fold by deleting the authentic MuIFN-γ polyadenylation signal sequence.

4. PERMANENT EXPRESSION OF MuIFN-γ IN CHINESE HAMSTER OVARY (CHO) CELLS

The COS cell expression system is not suitable for the large-scale production of MuIFN-γ. To establish a cell line that constitutively produces high levels of MuIFN-γ, a technique was used in which the exogenous DNA permanently co-replicated with the host cell DNA. To this end, the MuIFN-γ expression unit, as used for expression in COS cells, was linked to the dihydrofolate reductase (DHFR) cDNA as a marker gene and was subsequently integrated into the genome of $DHFR^-$ Chinese hamster ovary (CHO) cells. Selection of cells in nucleotide-depleted medium enabled us to isolate $DHFR^+$ cell clones that constitutively secreted MuIFN-γ (DICK cells). Some clones (e.g. DICK-234) were isolated which produced MuIFN-γ in excess of 10,000 units/ml/day.

The DHFR gene, which at first had been used for the selection of cell clones containing the MuIFN-γ cDNA, was then used as a selectable marker to increase the copy number of integrated MuIFN-γ sequences (3). Treatment of the high-producing DICK-234 cell line with cytotoxic concentrations of methotrexate, a specific inhibitor of the DHFR enzyme, resulted in the isolation of cell lines in which not only the DHFR sequences but also the MuIFN-γ cDNA sequences were amplified (Table 1). The MICK-3 cell line,

TABLE 1. MuIFN-γ production by methotrexate-resistant CHO cell lines

Cell line	Methotrexate resistance	MuIFN-γ production[a]	MuIFN-γ RNA[b]	MuIFN-γ DNA[c]
DICK-234	< 2 nM	17,000	1	1
MICK-1	60 nM	58,000	N.D.	8
MICK-2	500 nM	160,000	N.D.	30
MICK-3	3,000 nM	570,000	20	60

[a] Cells were grown to confluency in minimum essential medium-alpha with nucleotides and without methotrexate. Fresh medium was added and replaced after 48 hr for interferon assay. Interferon concentrations are expressed in laboratory units per ml.

[b] Poly(A)$^+$-RNA was isolated and analyzed by Northern blotting using MuIFN-γ cDNA as probe. Values are expressed in relative units. N.D. = not determined.

[c] High molecular weight DNA was isolated and the concentration of MuIFN-γ DNA was determined by Southern blotting. Values are expressed in relative units.

which was isolated after three consecutive amplification steps secreted approximately 30 times more MuIFN-γ than did the original line. Also, SDS-PAGE analysis of total [^{35}S]methionine labeled proteins secreted by the MICK-3 cells showed a major protein band of Mr 19,000 that co-migrated with MuIFN-γ biological activity. This band was not detected with the culture fluid of the original DICK-234 cells. The high interferon production level of MICK-3 cells (2-5 x 10^5 units per ml per day) remained constant during at least 60 passages in nucleotide-free medium, even if methotrexate was omitted from the culture medium.

5. CHARACTERIZATION OF CHO CELL-PRODUCED MuIFN-γ

The MICK-3 cell line was not only suitable for the production of large amounts of recombinant MuIFN-γ, but also proved to be an excellent tool for the molecular characterization of the MuIFN-γ proteins. Although [^{35}S]methionine-labeled MuIFN-γ proteins produced by MICK-3 cells could be detected on SDS-PAGE without any prior purification, the availability of monoclonal antibodies allowed us to examine these proteins more closely after immunoprecipitation. This analysis revealed a previously unsuspected molecular heterogeneity : in addition to 2 major protein bands (monomer : Mr 19,000 and dimer Mr 38,000) which together constituted about 90 % of the total immunoprecipitable material, many minor bands were observed. The electrophoretic profile for antiviral activity corresponded well with the band pattern, although the activity for the Mr 38,000 (dimeric) band was higher than expected.

Since natural MuIFN-γ is glycosylated (4), part of the heterogeneity may be due to variation in extent of glycosylation. We did indeed find that the pattern of immunoprecipitable bands was drastically altered when tunicamycin, an inhibitor of N-glycosylation, was included in the medium of MICK-3 cells during MuIFN-γ synthesis. The major components now appearing had Mr values of 27,000 (dimer) and 14,000 (monomer), probably representing the carbohydrate-depleted Mr 38,000 and Mr 19,000 components observed in the absence of the glycosylation inhibitor. Under these conditions, only two minor bands of Mr 12,000 and Mr 13,000 were observed. Assuming that the carbohydrate moiety would represent a Mr of 5,000 (as calculated for the major components), the N-glycosylated counterparts of the minor components would possess a Mr of 17,000 and 18,000, respectively. Indeed, protein bands corresponding to these Mr values were seen on SDS-PAGE gels with immunoprecipitates from MICK-3 cells that had not been treated with tunicamycin.

In contrast to what was observed for the major components, no bands corresponding to twice the molecular weight of the minor carbohydrate-deprived components were detected, suggesting that the latter were unable to form dimers. This, together with their smaller Mr value, suggested the possibility that they lacked a carboxyterminal fragment containing the unique and terminal cysteine residue (Fig. 1). This hypothesis was confirmed by the observation that the smaller carbohydrate-deprived components became undetectable when the cells were labeled with [^{35}S]cysteine instead of [^{35}S]methionine (Fig. 2). SDS-PAGE gels of proteins secreted by MICK-3 cells which had not been treated with tunicamycin showed a Mr 17,000 protein that could be labeled with [^{35}S]methionine but not with [^{35}S]cysteine, reinforcing the assumption that this protein is the glycosylated counterpart of the Mr 12,000 carbohydrate-deprived component.

Thus, MICK-3 cells produce three molecular forms of MuIFN-γ monomers, which, without N-glycosylation, have Mr values of 12,000, 13,000 and 14,000. Only the Mr 14,000 form is able to dimerize at its C-terminal cys-

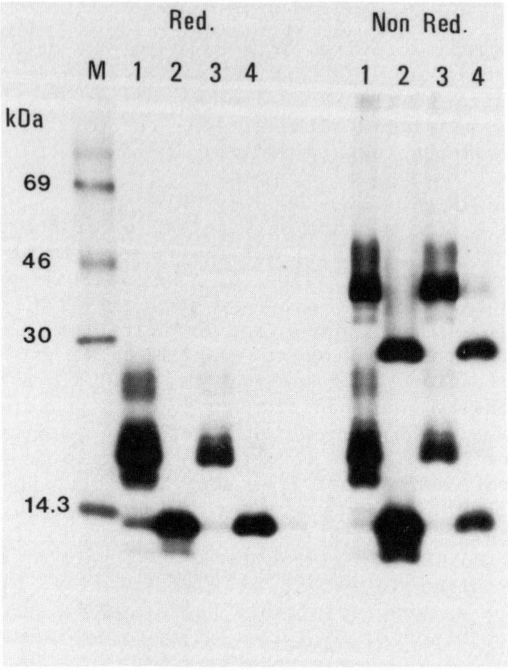

FIGURE 2. SDS-PAGE analysis of [^{35}S]methionine and [^{35}S]cysteine-labeled CHO cell-derived MuIFN-γ. Confluent monolayers of MICK-3 cells cultured in the presence or absence of 10 μg/ml tunicamycin were refed with fresh medium containing either [^{35}S]methionine or [^{35}S]cysteine. Secreted proteins were reacted with MuIFN-γ-specific monoclonal antibodies and immunoprecipitated proteins were submitted to SDS-PAGE analysis either in reducing (Red) or non-reducing (Non Red) conditions. Fluorography was for 5 days at -70°C. Lane M : molecular weight reference protein mixture. Lanes 1 : [^{35}S]methionine-labeled, without tunicamycin. Lanes 2 : [^{35}S]methionine-labeled with tunicamycin. Lanes 3 : [^{35}S]cysteine labeled without tunicamycin. Lanes 4 : [^{35}S]cysteine-labeled with tunicamycin.

teine residues. The two other forms lack this C-terminal fragment which must consist of several other amino acids in addition to the C-terminal cysteine. Other factors which contribute to the heterogeneity of MuIFN-γ are N-glycosylation of all three forms and dimerization of the Mr 14,000 form.

6. CONCLUSION

The MuIFN-γ cDNA was cloned and expressed in heterologous mammalian cells. In contrast to the COS cell system in which a transient expression was observed, the CHO cell expression system allowed constitutive production of high levels of MuIFN-γ. The CHO cell-derived MuIFN-γ was characterized by SDS-PAGE and found to be heterogeneous. At least three cumulating modifications account for this heterogeneity : carboxyterminal proteolytic processing, variations in N-glycosylation and dimerization.

REFERENCES

1. Dijkmans R, Volckaert G, Van Damme J, De Ley M, Billiau A, De Somer P. J. Interferon Res. 5:511-520, 1985.
2. O'Hare K, Benoist C, Breathnach R. Proc. Natl. Acad. Sci. USA 78: 1527-1531, 1981.
3. Kaufman RJ, Sharp PA. J. Mol. Biol. 159:601-621, 1982.
4. Havell FA, Spitalny GL. Arch. Virol. 80:195-207, 1984.

MONOCLONAL ANTIBODIES AGAINST MURINE INTERFERON-γ AND THEIR USE TO INVESTIGATE THE ROLE OF ENDOGENOUS INTERFERON-γ *IN VIVO*

H. HEREMANS, R. DIJKMANS and A. BILLIAU
Rega Institute, Katholieke Universiteit Leuven, B-3000 Leuven, Belgium

1. INTRODUCTION

Interferons are endowed with multiple regulatory effects on the immune system : killer cell activity, macrophage activity, T- and B-cell proliferation and differentiation. Inhibition as well as stimulation of many immune functions by interferon have been described. Interferons in general, and interferon-γ (IFN-γ) in particular, may therefore play an important role in the pathogenesis of diseases involving immune reactions or tissue rejection. One approach to study the possible contribution of IFN-γ in these reactions is by systemic administration of interferon. It is known that local inflammatory reactions are exceptionally complex and involve the participation of multiple endogenous mediators such as prostaglandins, histamine-like substances as well as cytokines. Therefore an alternative approach to evaluate the role of a particular cytokine in the pathogenesis of inflammation is the blockage of the action of that cytokine by administration of monoclonal antibodies. In our laboratory monoclonal antibodies directed against mouse IFN-γ were developed.

Here we describe experiments in which we have tested exogenous mouse IFN-γ as well as antibodies against IFN-γ for their modulatory effect on local inflammation elicited by bacterial lipopolysaccharide (LPS) and on rejection of allogeneic tumor cells *in vivo*.

2. MATERIALS AND METHODS

Six to 8 week-old female NMRI mice were obtained from the Experimental Animal Centre of the University of Leuven.

Natural mouse interferon-γ (MuIFN-γ) was prepared from splenocytes of NMRI mice exposed to killed *S.aureus* suspensions as described elsewhere (1) and purified by affinity chromatography on monoclonal antibody against MuIFN-γ. The preparations had a specific activity in the range of $10^{5.7}$ to $10^{7.4}$ units/mg protein. Recombinant MuIFN-γ was derived from Dick cells (2) and also purified by affinity chromatography on monoclonal antibody giving a preparation with a specific activity of approximately $10^{7.65}$ units/ mg protein. Monoclonal antibodies to MuIFN-γ were obtained by fusing spleen cells of a Lou/Ca rat immunized with MuIFN-γ with the non-secreting mouse myeloma cell line Sp2/0Ag14 using standard procedures. Three stable hybridomas secreting IgG2a antibodies directed against MuIFN-γ were selected. Two clones (F2 and F3) specifically bound and neutralized the antiviral activity of MuIFN-γ. One clone (F1) was weakly neutralizing. The hybridomas were propagated by *in vivo* passage of ascitic fluid in thymusless nude mice.

Lipopolysaccharide B of *Serratia marcescens* was obtained from Difco (Detroit, Mich., U.S.A.).

The TA3 tumor (provided by Dr. M. De Brabander, Janssen Pharmaceutica , Beerse, Belgium) was derived from a spontaneous mammary adenocarcinoma in an A/HeHa mouse (3). It was maintained as an ascitic tumor in the peritoneal cavity of NMRI mice. The length and width of the tumors were measured 3 times weekly with calipers. Tumor size was calculated as the square root of the tumor area. Footpad swelling was measured daily with the aid of micrometer-calipers. Footpad swelling was calculated as % increase in footpad thickness, by comparing thickness of the LPS-injected footpad (L) with that of the contralateral injected footpad (C) :

$$\% \text{ Increase} = \frac{L-C}{C} \times 100.$$

3. RESULTS AND DISCUSSION

3.1. Role of MuIFN-γ in rejection of allogeneic tumor cells

Groups of 6 NMRI mice were inoculated subcutaneously with 10^6 TA3 tumor cells and treated for 7 days, starting on the day of tumor cell inoculation, with daily injections of monoclonal antibodies to MuIFN-γ (clone F3), either intraperitoneally or subcutaneously at the tumor challenge site. Additional groups received daily intraperitoneal injections of monoclonal antibodies against HuIFN-β or daily subcutaneous injections of cyclosporin (0.4 mg/mouse), a potent immunosuppressive agent used in organ transplantation. The control group received PBS. Fig. 1 shows the evolution of tumor growth during an observation period of 40 days. Subcutaneous injection of TA3 tumor cells in allogeneic NMRI mice resulted in a temporary tumor growth at the site of injection, followed by a spontaneous and total rejection of the tumors in all mice. In contrast, seven daily intraperitoneal as well as subcutaneous injections of neutralizing monoclonal

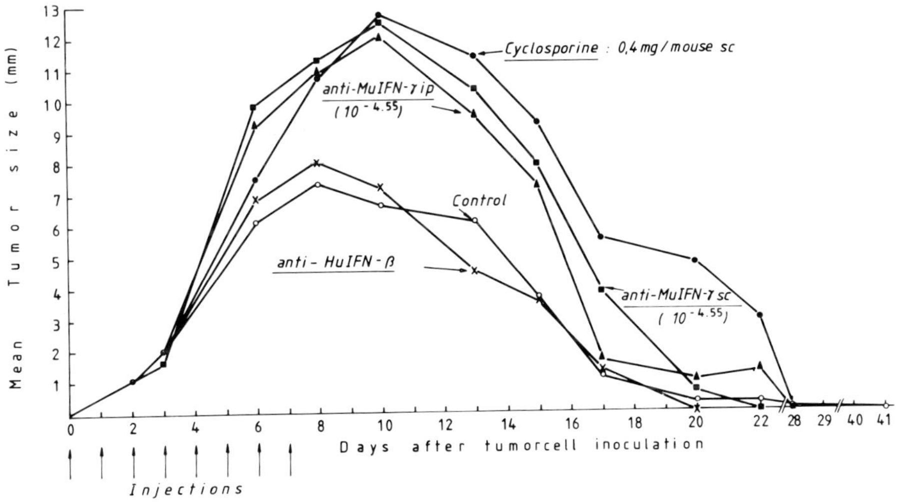

FIGURE 1. Effect of monoclonal antibodies to MuIFN-γ on the rejection of allogeneic tumor cells. Groups of mice were injected with 0.2 ml of monoclonal antibodies to MuIFN-γ (neutralizing titer: $10^{-5.55}$ against 30 units of MuIFN-γ); monoclonal antibodies to HuIFN-β (neutralizing titer: < 1/10 against 30 units/ml of MuIFN-γ); cyclosporine (15 mg/kg).

antibodies against MuIFN-γ stimulated tumor growth and caused a delay in rejection of the allogeneic tumor cells. Similar results were obtained with daily injections of cyclosporine (15 mg/kg). Injection of monoclonal antibodies to HuIFN-β had no effect.

These results show that monoclonal antibodies to MuIFN-γ can inhibit allograft rejection _in vivo_. This is in accordance with observations by Landolfo _et al_. (4) who studied rejection of a methylcholanthrene-induced fibrosarcoma (CE-2) of the BALB/C strain in CBA/J mice. Moreover, it has been shown that cyclosporine can completely inhibit the production of IFN-γ by human and mouse T-cells, which have been stimulated with mitogen or alloantigen (5).

These results indicate that endogenous IFN-γ plays an important role in the mechanism of allograft rejection and suggest that antibodies against or inhibitors of interferon-γ may be useful in controlling organ transplant rejections in the clinic.

2. _Role of MuIFN-γ in inflammatory skin reactions to bacterial lipopolysaccharide_

The role of IFN-γ in inflammatory processes was examined by studying footpad swelling in mice given local injections of bacterial lipopolysaccharide (LPS). For this purpose mice were injected into the right hind footpad with 5 μg of LPS (in 25 μl). Phosphate buffered saline (PBS) was injected as a control into the left hind footpad. Maximal footpad response (100 to 145 %) occurred on day 2 to 3 post LPS challenge (Fig. 2).

The effect of exogenous systemic MuIFN-γ on the footpad reaction was studied by treating groups of 4 to 8 mice with intraperitoneal injections of natural or recombinant MuIFN-γ. Interferon was given each day from day 0 till day 3 post LPS challenge. Results are summarized in Table 1. It can be seen that systemic administration of high doses ($10^{5.3}$ and $10^{5.0}$) of natural as well as recombinant MuIFN-γ depressed the footpad inflammatory reaction to LPS. The effect was dose-dependent. Using this model system, we previously described that MuIFN-α,β induced during viral infection or systemically administered natural MuIFN-α,β and recombinant MuIFN-α_1 were also found to cause inhibition of the LPS induced footpad reaction (6). The effect could not be demonstrated with recombinant MuIFN-β.

TABLE 1. Inhibition of LPS-induced footpad swelling by MuIFN-γ.

Interferon source	Dose ($\log_{10}$ units/mouse)	% Inhibition of footpad swelling on[a]:	
		day 2	day 3
Natural[b]	5.0	38	64
	4.3	9	16
	3.3	12	28
Recombinant[c]	5.3	30	41
	4.3	26	17
	3.3	-5	-26

[a] $\% \text{ Inhibition} = 100 - \frac{\text{mean swelling in experimental group}}{\text{mean swelling in control group}} \times 100.$

[b] Experiment done with 4 mice.

[c] Experiment done with 8 mice.

These findings suggest that systemic interferons may act as negative modulators in local inflammation caused by gram-negative bacteria and their lipopolysaccharides.

In order to further evaluate the role of interferons in LPS-induced inflammation we studied the effect of blockage of endogenous interferon by administration of antibodies. Groups of 4 mice were injected intraperitoneally, 24 h before LPS challenge (5 μg) with a single dose of monoclonal antibodies against MuIFN-γ (F1, F2 and F3), possessing different neutralizing potency. Control mice received saline. The results shown in Fig. 2 indicate that systemic administration of monoclonal antibodies to MuIFN-γ caused a strong and dose-dependent inhibition of the footpad reaction. The poorly neutralizing antibody, F1, failed to affect the footpad reaction.

The results indicate that endogenous IFN-γ is produced locally after LPS injection in the footpad. They also suggest that, although systemic interferons act as inhibitors, this local interferon acts as a positive mediator of the reaction, possibly as a trigger or link in the cytokine cascade which brings about the inflammation. Blockage of the action of this local IFN-γ by monoclonal antibody would then switch off the inflammatory response.

The influence of the time of administration of the antibody was also examined. Monoclonal antibody (F3) was given intraperitoneally in a single dose at different times before (-) or after (+) LPS challenge (5 μg) (Table 2). Antibodies given before or shortly after LPS injection nearly com-

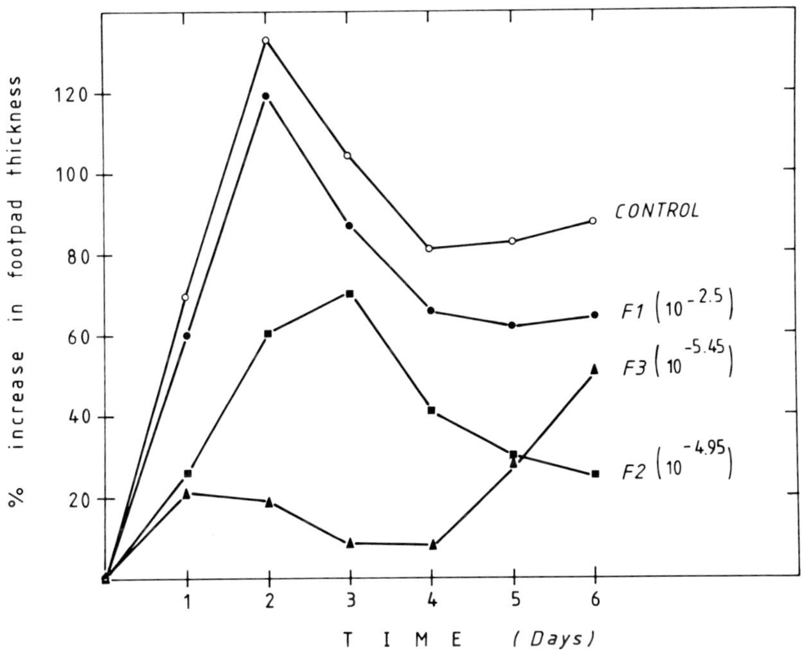

FIGURE 2. Effect of monoclonal antibodies against MuIFN-γ on LPS-induced footpad swelling in NMRI mice. F1, F2 and F3 possess different neutralizing potential (titers in brackets) for the in vitro antiviral effect of MuIFN-γ.

TABLE 2. Inhibition of LPS-induced footpad swelling by monoclonal antibody to MuIFN-γ : influence of time of administration.

Time of antibody administration[a] [days before (-) or after (+) LPS administration]	% Inhibition of footpad swelling[b] on	
	day 2	day 3
-5	84	84
-2	89	93
-1	82	73
0	86	90
+1	71	85
+2	10	32
+3	-1	3

[a]Antibody (F3) : 0.1 ml intraperitoneally (neutralizing potency : 5.75 $\log_{10}$ units/ml).
[b]Average values for 4 mice.

pletely inhibited the footpad swelling. Antibody given while the reaction was already underway, caused immediate arrest and regression of swelling.

In earlier work we have found that systemic administration of neutralizing polyclonal antibodies directed against MuIFN-α,β did not influence the footpad reaction (6) but could partially abrogate the inhibitory effect of MuIFN-α,β, suggesting that endogenous production of these interferons was insufficient to regulate inflammation.

In conclusion, the results described here clearly indicate that the use of monoclonal antibodies directed against interferons allows the analysis of the role of these mediators in tissue or organ reactions. Our findings also open new perspectives for therapy with those agents that specifically neutralize the action of IFN-γ in allograft rejection and certain inflammatory reactions to bacterial infections.

REFERENCES

1. Heremans H, De Ley M, Billiau A, De Somer P. Cell. Immunol. 71:353-364, 1982.
2. Dijkmans G, Volckaert G, Van Damme J, De Ley M, Billiau A, De Somer P. J. Interferon Res. 5:511-520, 1985.
3. Hauschka TS, Weiss L, Holdridge BA, Cudney TL, Zumpft M, Planinsck JA. J. Natl. Cancer Inst. 47:343-359, 1971.
4. Landolfo S, Cofano F, Giovarelli M, Prat M, Cavallo G, Forni G. Science 229:176-179, 1985.
5. Kalman VK, Klimpel GR. Cell. Immunol. 78:122-129, 1983.
6. Heremans H, Billiau A, Coutelier JP, De Somer P. Proc. Soc. Exp. Biol. Med., in press, 1987.

THE INTERRELATIONSHIP BETWEEN INTERLEUKIN-1, INTERFERON-ß AND HEMOPOIETIC GROWTH FACTORS

J. VAN DAMME, M. DE LEY[†], S. CAYPHAS[§], G. OPDENAKKER, J. VAN SNICK[§] and A. BILLIAU
Rega Institute and [†]Laboratory of Biochemistry, Department of Chemistry, Katholieke Universiteit Leuven, B-3000 Leuven; [§]Ludwig Institute for Cancer Research, Brussels Branch, B-1200 Brussels, Belgium

1. INTRODUCTION

Cytokines are humoral factors which play an important role in immunity. They are produced in extremely small quantities, but can be detected _in vitro_ due to their very high specific activity in various biological tests, such as growth assays, in which they are active at the ng level. By the use of advanced purification methods and recombinant techniques, research on identification and production of a number of cytokines has made a tremendous progress. Thus, it has become clear that several biologically defined factors correspond to the same molecular structure.

As a result of the multiple biological effects of a single cytokine it was found that a relationship exists between different cytokines in that one can induce the other. In clarifying the cytokine network, new activities were discovered and identified as separate molecules. At present a number of biologically defined factors still need to be isolated and others probably need to be discovered.

At our Institute we have studied various aspects of one of the first isolated cytokines, i.e. ß-interferon (IFN-ß). In doing so, we discovered a new cytokine, which could induce IFN-ß and identified it as an interleukin-1 (IL-1). Furthermore, we characterized another important cytokine induced by IL-1, namely a potent growth factor for B cell hybridomas and plasmacytomas (HPGF). These findings point to the existence of a cytokine network. The relationship between the three molecules, IFN-ß, IL-1 and HPGF, will be documented here.

2. RESULTS

2.1. Fibroblast interferon : large-scale production, purification and characterization

Induction of diploid cells with the double-stranded RNA, poly-rI:rC in combination with the metabolic inhibitors cycloheximide and actinomycin D (superinduction) resulted in the production of high-titered IFN-ß (20,000 U/ml, 20 U/1000 cells) (1). This prompted us to start a production unit in the Rega Institute, in order to produce large quantities of this antiviral substance for clinical trials. Because normal fibroblasts have a limited lifespan and are difficult to grow, production of IFN-ß using superinduction of large numbers of stationary cultures suffered from several inconveniences : (a) labor-intensive induction system; (b) high rate of contamination; (c) irregular interferon yields. In order to overcome these problems several steps were undertaken : (a) development of an immortal cell line, MG-63, which was easier to grow and to induce. Moreover, this cell line also grew to higher cell densities and produced more IFN-ß per cell than diploid fibroblasts (Table 1); (b) development of stringent sterility precautions in order to reduce cell product contamination; (c) introduction of techniques for large-scale culture of anchorage-dependent cells (roller bottles, bulk culture vessels, microcarriers), significantly reducing the number of cell cultures to be handled. The results obtained with these culture systems are summarized in Table 1. They illustrate that, although such large-scale production of natural IFN-β was a challenge, it was a feasible method on long term.

TABLE 1. IFN-β yields[a] on diploid fibroblasts and MG-63 grown in different systems for large-scale production.

Determination	Diploid fibroblasts			MG-63 cells	
	Roller bottle	Bulk vessel	Micro-carrier	Roller bottle	Micro-carrier
1. Surface area (dm^2) per culture	8.5	85	120	8.5	75
2. Number of IFN-β batches tested	40	33	22	32	10
3. Volumes of crude IFN-β					
Total (liter)	84	317	6.6	72	9.6
Per batch (liter)	2.1	9.6	0.3	2.3	1.0
Per culture vessel (ml)	27	1600	300	28	955
4. Average IFN-β yield ($\log_{10}$ units)					
Per 10^3 cells	1.3	1.2	1.0	1.7	1.7
Per cm^2 surface area	3.0	2.8	2.2	3.6	3.3
Per ml of culture fluid	4.5	3.6	3.8	5.0	4.2
Per culture vessel	5.9	6.8	6.3	6.5	7.2
Per batch	7.8	7.5	6.3	8.4	7.2

[a]Confluent monolayer cultures were induced at 37°C with poly-rI:rC (50 µg/ml) in the presence of cycloheximide (10 µg/ml) for 4 h. Actinomycin D (1 µg/ml) was then added for 2 h. Cell cultures were washed 6 h after induction, refed with serum-free medium and incubated overnight for IFN-β production.

Although all IFN-β could be produced under serum-free conditions it needed to be further purified before use in humans. Soon it appeared, however, that this fibroblast-derived type of IFN was quite unstable in pure form. Nevertheless, it was successfully purified with high recovery rates to a specific activity of $10^{8.0}$ U/mg by using a 2-step purification procedure, consisting of (a) adsorption to controlled pore glass beads (2) and (b) zinc chelate chromatography (3). After the first purification step partially purified IFN-β still contained other induced proteins, such as a 26kDa protein, which would later be characterized as the HPGF already mentioned. In fact, an antiserum directed against such impure IFN-β preparations was used for the cloning of IFN-β and led to the discovery of the 26kDa protein (4). In contrast, an antiserum directed against Zn-chelate-purified IFN-β was found to show no cross-reactivity with the 26kDa protein.

2.2. 22K factor : relation with IFN-β and identification as IL-1β

Since fibroblasts are known to be the main producer cells of IFN-β while mononuclear cells are producers of IFN-α and IFN-γ, it came somewhat as a surprise that peripheral blood leukocytes released a 22kDa protein (22K factor) with antiviral activity neutralisable by a highly specific antibody against IFN-β (5). The evidence that this antiviral activity was in fact different from classical fibroblast-derived IFN-β, is as follows : (a) different cell-specificity for antiviral activity; (b) different behaviour on zinc-chelate. The finding that 22K factor did not bind to an anti-IFN-β column and that a specific antibody against 22K factor failed to neutralize IFN-β then indicated that 22K factor is an inducer of IFN-β in the fibroblast cells used for antiviral assay (6). This hypothesis was further supported by demonstrating the existence of IFN-β mRNA in 22K factor-treated cells (7).

Further characterization of the 22K factor and determination of its relation with other cytokines required purification to homogeneity. This was achieved by isolating mononuclear cells from a large number of blood donations and by stimulating them with mitogen. The crude lymphokine preparation so-obtained was then purified and fractionated by a sequence of chromatographical methods (Table 2) including (a) adsorption to silicic acid and elution at neutral (for IFN-γ) or acid pH (for 22K factor), (b) immuno-affinity chromatography, (c) gel filtration (22kDa) and (d) fast protein liquid chromatography (FPLC). Homogeneity of the final product was demonstrated by SDS PAGE, showing a single band at Mr 17,000 containing the biological activity of the 22K factor (8).

When testing the pure protein in several cytokine assays, the factor was found to be highly active in a lymphocyte activating factor (LAF) assay, the method for measuring IL-1 in vitro (8). Moreover, the 22K factor was highly pyrogenic in rabbits and showed a dramatic effect on the number

TABLE 2. Procedure for purification of human leukocyte-derived IL-1 and fibroblast-derived HPGF to homogeneity.

Purification method (B = binding; E = elution)	Leukocyte IL-1		Fibroblast HPGF	
	Vol (ml)	Conc[a]	Vol (ml)	Conc[a]
1) Silicic acid (IL-1) or CPG (HPGF)				
B: in batch (10 g/l, neutral pH)	30,000	2.5	4500	4.9
E: 0.3 M glycine.HCl, pH 2.0	1,000	3.7	500	5.7
2) Antibody chromatography				
B: neutral pH	1,000	3.7	500	5.7
E: 0.1 M citrate, pH 2.0 + 0.5 M NaCl	30	5.0	20	6.7
3) Gel filtration (2.6 x 100 cm column)				
B: 1.55 M NaCl, pH 7.2 + 15 % (v/v) EG	10	5.5	10	7.0
E: Mr 20-25,000 kDa fractions	40	4.7	20	6.1
4) Fast protein liquid chromatography (mono S, cation exchanger)				
B: 50 mM formiate, pH 4.0	10	5.1	10	6.0
E: 50 mM formiate, pH 4.0 + 1 M NaCl	1	6.0	3	6.7

[a]Concentration as $\log_{10}$ units per ml.

of circulating neutrophils (8,9). These *in vivo* activities had already been assigned to a factor, called endogenous pyrogen and related to IL-1.

Final identification of 22K factor was obtained by N-terminal amino-acid sequence analysis of the homogeneous protein, using a gas-phase sequenator. The sequence obtained was highly homologous with the middle part of the cDNA-derived sequence of the IL-1ß precursor molecule (8). This finding was important for two reasons : identification of the 22K factor as IL-1ß, and determination of the cleavage site in the inactive IL-1 precursor.

2.3. Hemopoietic growth factors (HPGF, CSF-GM) : induction by IL-1

In addition to its antiviral effect on fibroblasts, pure 22K factor (i.e. IL-1) was also found to have growth promoting activity for bone marrow cultures (8). Like in the case of the antiviral effect, the colony growth promoting effect of IL-1 was also indirect, i.e. by induction of colony-stimulating factor for granulocytes and macrophages (CSF-GM) in monocytes present in the marrow cultures (10). Furthermore, we found IL-1 to be a potent inducer in various cells of the mRNA for the so-called 26K protein (11). Finally, we recently discovered that IL-1 was an excellent physiological inducer in fibroblasts of a growth factor (HPGF) for B-cell hybridoma and plasmacytoma cell lines (Table 3). The conditions for induction of HPGF and IFN-ß were different in that IL-1 was a better inducer for HPGF in diploid fibroblasts and MG-63 cells than for IFN-ß and that poly-rI:rC was a better inducer of IFN-ß than for HPGF in MG-63 cells. However, the conditions for induction of HPGF were strikingly similar to those for induction of the 26kDa protein (12).

TABLE 3. Induction of hybridoma growth activity (HGF) and IFN-ß in fibroblastoid cells by IL-1 and poly-rI:rC.

Cell treatment		Production ($\log_{10}$ U/ml) of HGF and IFN-ß[a] on			
Inducer	Dose	Diploid fibroblasts		MG-63 cells	
		HGF	IFN-ß	HGF	IFN-ß
IL-1	100	4.4	2.3	4.2	3.1
	10	4.0	-	4.1	-
	1	3.4	-	3.4	-
	0.1	2.7	-	2.7	-
Poly-rI:rC	100	4.7	3.9	3.5	4.6
	10	4.5	3.5	2.6	4.0
	1	2.0	2.3	2.5	3.7
Control	-	1.6	1.7	2.3	2.2

[a]Superinduction method (cycloheximide, actinomycin D) on stationary cultures as described in Table 1.

The relationship between HPGF and the 26kDa protein was demonstrated by binding mRNA for HPGF to immobilized 26K cDNA and concomitant enrichment in oocyte-translated HPGF activity (13). Final evidence for identity between HPGF and 26kDa protein was obtained by purification to homogeneity and sequence analysis of HPGF produced by IL-1-stimulated MG-63 cells. Crude supernatant from a large number of cultures was therefore subjected to a 5-step purification procedure : (a) adsorption to controlled pore glass; (b) immuno-affinity chromatography; (c) gel filtration, (d) FPLC and (e) HPLC (Table 2). N-terminal amino-acid analysis of the pure protein resulted a sequence that exactly matched that predicted from the known nucleic acid sequence of the 26kDa cDNA (13). This demonstrated that the biological function of the 26kDa protein is that of a growth factor for B-cells. This function could recently be extended to B-cell differentiation factor (BCDF, BSF-2) because the sequence of this factor is also identical to that of the 26K protein (14).

3. DISCUSSION

The data presented here demonstrate one aspect of the cytokine network, namely the fact that one particular cytokine can induce production of another and thereby mimick its action. Thus, our data demonstrate that the monokine, IL-1 (22K factor), induces several secondary factors possessing antiviral and/or cell growth controlling activities : IFN-ß, CSF-GM and the 26kDa protein, recently identified as the B-cell hybridoma plasmacytoma growth factor (HPGF) and the B-cell differentiation factor BSF-2. Other activities that could be demonstrated with the 22K factor include induction of prostaglandin E, caseinase and plasminogen activator in human articular chondrocytes (15) and plasminogen activator inhibitor in human endothelial cells (16).

These factor-inducing abilities of IL-1 provide a basis to explain the observation that IL-1, aside from its T-cell activating effect mediated by IL-2-induction, exerts a number of other biological activities, such as (a) interferon-like antiviral effect on cells, (b) promotion of bone marrow colony formation, (c) activation of B-cells.

The antiviral effect of IL-1 was found to be blocked by sera that are highly specific of IFN-ß, indicating that this cytokine is always implicated in the antiviral effect. It should be mentioned, however, that an interferon-like antiviral effect, which is neutralizable by highly specific anti-IFN-ß-antisera, has also been assigned to the 26kDa protein. Therefore, the possibility remains that both IFN-ß and the 26kDa protein act as mediators of the antiviral effect of IL-1. If so, they may be connected to IL-1 in either of the following ways

$$IL\text{-}1 \begin{array}{l} \nearrow 26kDa \rightarrow \text{antiviral effect (Ia)} \\ \searrow IFN\text{-}\beta \rightarrow \text{antiviral effect (Ib)} \end{array}$$

$$IL\text{-}1 \rightarrow 26kDa \rightarrow IFN\text{-}\beta \rightarrow \text{antiviral effect (II)}$$

$$IL\text{-}1 \rightarrow IFN\text{-}\beta \rightarrow 26kDa \rightarrow \text{antiviral effect (III).}$$

Pathway (I) postulates that 26kDa protein has a direct antiviral effect. Its neutralizability by highly specific anti-IFN-ß-antibody would derive from common epitopes between the 26kDa protein and IFN-ß, despite the fact that (a) their primary structures are entirely different, (b) anti-IFN-ß sera fail to precipitate or bind the 26kDa protein, and (c) anti-IFN-ß sera fail to neutralize the most prominent biological activity of the 26kDa protein, namely the HPGF-activity. Pathway (II) postulates that 26kDa protein acts as an inducer of IFN-ß. Pathway (III) postulates that IFN-ß induces formation of the 26kDa protein and that the two factors act additively or synergistically in bringing about their antiviral effect. Which of these pathways is the correct one is presently being investigated.

Our data also emphasize that ordinary fibroblasts are potent producers of cytokines (fibrokines) endowed with extremely diverse biological activities. Whereas it has been known for a long time that "artificial" stimulation of fibroblasts can lead to the production of IFN-β, our data demonstrate that exposure to an endogenous factor, IL-1, can achieve production of interferon and several other cytokines. These findings illustrate the diversity and complexity of processes taking place in connective tissues. Further clarification of this fibrokine network should help to elucidate the pathogenesis of connective tissue diseases as well as physiological reactions to tissue damage.

ACKNOWLEDGEMENTS

We thank the Belgian Red Cross (Blood Transfusion Centers of Leuven and Antwerp) for providing buffy coats. J. Van Damme is a research associate and G. Opdenakker a senior research assistant of the Belgian NFWO.

REFERENCES

1. Billiau A, Van Damme J, Van Leuven F, Edy VG, De Ley M, Cassiman JJ, Van Den Berghe H, De Somer P. Antimicrob. Agents Chemother. 16:49-55, 1979.
2. Edy VG, Braude IA, De Clercq E, Billiau A, De Somer P. J. Gen. Virol. 33:517-521, 1976.
3. Heine JW, Van Damme J, De Ley M, Billiau A, De Somer P. J. Gen. Virol. 54:47-66, 1981.
4. Content J, De Wit L, Pierard D, Derynck R, De Clercq E, Fiers W. Proc. Natl. Acad. Sci. USA 79:2768-2772, 1982.
5. Van Damme J, De Ley M, Claeys H, Billiau A, Vermylen C, De Somer P. Eur. J. Immunol. 11:937-942, 1981.
6. Van Damme J, Billiau A, De Ley M, De Somer P. J. Gen. Virol. 64:1819-1822, 1983.
7. Van Damme J, Opdenakker G, Billiau A, De Somer P, De Wit L, Poupart P, Content J. J. Gen. Virol. 66:693-700, 1985.
8. Van Damme J, De Ley M, Opdenakker G, Billiau A, De Somer P, Van Beeumen J. Nature 314:266-268, 1985.
9. Van Damme J, Opdenakker G, De Ley M, Heremans H, Billiau A. Clin. Exp. Immunol. 66:303-311, 1986.
10. Fibbe WE, Van Damme J, Billiau A, Voogt PJ, Duinkerken N, Kluck PMC, Falkenburg JHF. Blood 68:1316-1321, 1986.
11. Content J, De Wit L, Poupart P, Opdenakker G, Van Damme J, Billiau A. Eur. J. Biochem. 152:253-257, 1985.
12. Van Damme J, Cayphas S, Opdenakker G, Billliau A, Van Snick J. Eur. J. Immunol. 17:1-7, 1987.
13. Van Damme J, Opdenakker G, Simpson RJ, Rubira MR, Cayphas S, Vink A, Billiau A, Van Snick J. J. Exp. Med., in press, 1987.
14. Billiau A. Nature 324:415, 1986.
15. Bunning RAD, Van Damme J, Richardson HJ, Hughes DE, Opdenakker G, Billiau A, Russell RGG. Biochem. Biophys. Res. Commun. 139:1150-1157, 1986.
16. Gramse M, Breviario F, Pintucci G, Millet I, Dejana E, Van Damme J, Donati MB, Mussoni L. Biochem. Biophys. Res. Commun. 139:720-727, 1986.

STUDIES ON GENE EXPRESSION IN STREPTOMYCES LIVIDANS

K. NEESEN, R. AERT, A. DE SMEDT, W. DALEMANS AND G. VOLCKAERT
Rega Institute, Katholieke Universiteit Leuven, B-3000 Leuven, Belgium

1. INTRODUCTION

Streptomyces are gram-positive bacteria which grow as mycelia and may differentiate to spores. A wide range of different species have been characterized for antibiotic production and secretion of exoenzymes (1). Little is known about the gene expression mechanisms in *Streptomyces*.

In our laboratory a project has been initiated to study the expression of (heterologous) genes in *Streptomyces*, in particular the molecular signals responsible for transcription and translation. A human IFN-β gene expression unit was constructed and introduced in *Streptomyces lividans*.

2. STUDY OF TRANSCRIPTION SIGNALS IN *STREPTOMYCES LIVIDANS*

The extremely high GC-content of the *Streptomyces* genome (up to 73 %) raises the question of how transcriptional promoters look like since promoters in other prokaryotic organisms are rather rich in adenine and thymine (2).

Nucleotide sequences responsible for initiation of transcription in *Streptomyces* have been identified recently (3). Moreover, at least two kinds of RNA polymerase, recognizing different promoters, have been shown to exist in *Streptomyces* (4) and at least some *Escherichia coli* promoters are recognized by one of these RNA polymerases (5).

As a first step towards a systematic search and analysis of promoter sequences in *Streptomyces* we elaborated a promoter-probe shuttle plasmid integrated into a vector system for rapid chemical sequencing (6). To construct the shuttle vector we fused the plasmids pIJ702 and pGV462. pIJ702 is a 5650 base-pair (bp) *Streptomyces* recombinant plasmid carrying the thiostrepton resistance gene (*tsr*) of *S.azureus* and the tyrosinase gene (*mel*) of *S.antibioticus* (7). pGV462 is a 1617 bp *E.coli* plasmid conferring chloramphenicol resistance and bearing a rapid chemical sequencing module and an artificial replicon with a very high copy number (G. Volckaert *et al.*, in preparation). Expression of the *cat* gene in this plasmid is driven by the β-lactamase promoter. pIJ702 and pGV462 were cleaved with *Sph*I and *Sac*II which are both unique in each of these plasmids. This removed 430 bp from the tyrosinase gene. The vector fragments were ligated and transferred to *E.coli*, thereby selecting for chloramphenicol resistance. The shuttle plasmid was named pSKN01 and conferred not only thiostrepton resistance, but also chloramphenicol resistance to *S.lividans*. This may corroborate the earlier observation that the β-lactamase promoter (controlling the *cat* gene in pGV462) is active in *S.lividans* (5). However, chloramphenicol resistance might also arise from the promoter of the *tsr* gene which lies further upstream (Fig. 1).

Rapid chemical sequencing with pSKN01 was not possible because of the presence of two additional *Tth*111I sites outside the sequencing configura-

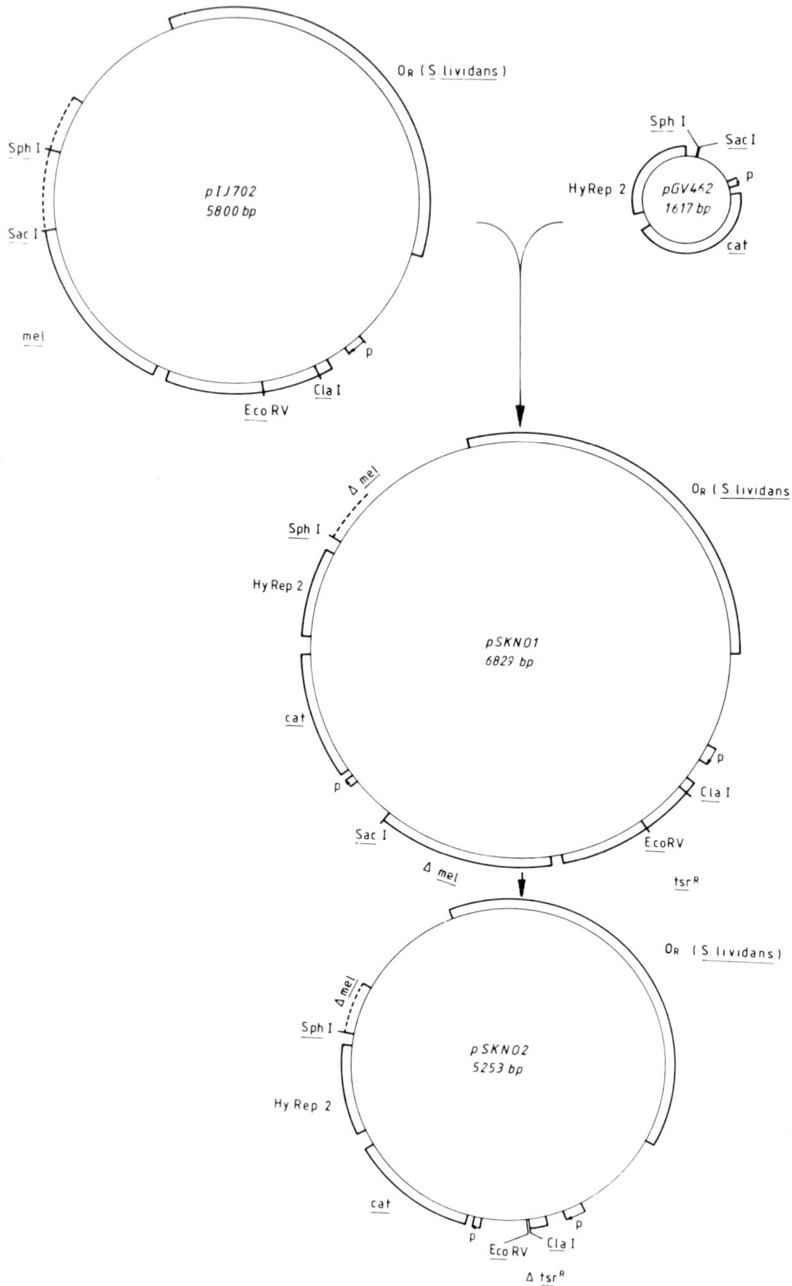

FIGURE 1. Construction of pSKN01.

tion. This problem could be solved by deletion of (part of) the *tsr* gene, which encloses these *Tth*111I sequences. pSKN01 was cleaved at its unique *Cla*I and *Sac*I sites and religated by insertion of an adaptor oligonucleotide which restores both *Cla*I and *Sac*I and also provides a unique *Eco*RV recognition sequence. Thus, the *tsr* gene was inactivated. The expression of the *cat* gene in *E.coli* as well as *S.lividans* provides for the essential selection criterion for transfer of the shuttle vector into both hosts. The new plasmid was named pSKN02.

In pSKN02, the chemical sequencing configuration is flanked by a unique *Bgl*II site. A portable promoter-less neomycin resistance gene (*aph*) of *S.fradiae* has been engineered so as to allow its transfer into this position. Insertion of the fragment reconstitutes a unique *Bgl*II site at the 5' end of the *aph* coding sequence. Hence, this will allow shotgun-cloning of *Sau*3AI-generated chromosomal fragments of *S.lividans* to select promoters which can be directly subjected to sequence analysis.

3. STUDY OF TRANSLATION SIGNALS IN *STREPTOMYCES LIVIDANS*

The major steps in mRNA-directed translation are formation of the initiation complex, elongation of the polypeptide chain and termination. Neither one of these steps has been studied in detail in *Streptomyces*. By understanding the mechanisms which underly translation in these microorganisms one may be able to design approaches to manipulate protein levels in a directed fashion. We have developed a two-step procedure to study translation signals in *Streptomyces* by reverse genetics. The *S.fradiae aph* gene (9) encoding neomycin resistance, was transferred from pIJ211 (9) to the *E.coli* vector pGV462 (G. Volckaert *et al*., in preparation) in such a way that the nucleotide sequences upstream and neighbouring the ATG initiation codon were readily accessible for manipulation. The recombinant plasmid was named pSRA07. It contained a unique *Nco*I site, covering the ATG initiation triplet and a unique *Sst*II site at an upstream distance of 124 bp. Hence, the potential ribosome binding site region was excisable and could be replaced after mutagenesis (Fig. 2). Concomitantly with the construction of pSRA07, site specific mutations were introduced inside the *aph* gene to remove unwanted restriction recognition sequences. These mutations were chosen as third letter alterations which did not affect the primary sequence of the neomycin phosphotransferase, but resulted in the presence of codons which are rarely found in the *Streptomyces* genes sequenced so far (3). We thus envisaged to obtain data on the effect of codon usage at the expression level of the *aph* gene.

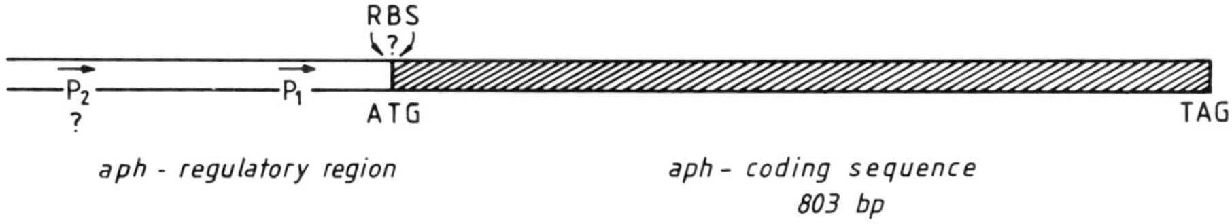

FIGURE 2. The *aph* expression module.

The *aph* expression module can be recloned in *Streptomyces* by ligation into either the shuttle plasmid pSKN01 (K. Neesen and G. Volckaert, submitted) or the *Streptomyces* plasmid pIJ702 (7). With both vector systems, neomycin resistance could be conferred to *S.lividans*. The presence of neomycin phosphotransferase in the transformed cells was also confirmed by the in vitro assay described by Reiss et al. (10). Mycelium, grown in the presence of different neomycin concentrations, was disrupted by sonication in a cup-horn device. The cell extracts were cleared by centrifugation and fractionated on polyacrylamide gel. Following blotting to P81 phosphocellulose ion exchange paper, enzyme activity was localized by autoradiography after incubation with kanamycin and γ-^{32}P-ATP. S.lividans harboring the engineered aph gene expressed neomycin phosphotransferase only when exposed to the antibiotic. The kinetics of this induction remain to be studied as well as the nucleotide sequences responsible for it.

4. CONSTRUCTION OF A HUMAN IFN-β GENE EXPRESSION UNIT FOR STREPTOMYCES LIVIDANS

Cloning of the MG63 HuIFN-β gene has been reported previously (11). The gene was manipulated to create a Sau3AI restriction site at the junction of the DNA sequences encoding signal peptide and mature interferon-β, respectively (12). Thus, a derivative was obtained from which the almost entire HuIFN-β coding sequence could be excised on a single Sau3AI or DpnI fragment. This fragment lacks the codon for the first amino acid, which is methionine. The interferon gene fragment was cloned in the unique Sau3AI site of pGV473 (a derivative of pGV451 : ref. 13), thereby causing duplication of the Sau3AI sequence. The site flanking the termination codon of the gene was removed by partial cleavage with Sau3AI and treatment with nuclease S1. Thus, Sau3AI was made unique again. This plasmid was designated pGVIFN06.

Subsequently, a Sau3AI fragment of pSRA02 (14), encompassing the 5' regulatory region and part of the coding sequence of the aph gene (3) was cloned in pGVIFN06. For this purpose, the vector was cleaved with Sau3AI and XmnI. Ligation of the fragment to the vector resulted in reconstruction of one Sau3AI site, whereas forced ligation of the blunt (XmnI) end to the second 5' protruding G-A-T-C eliminated the overhang (G. Volckaert, unpublished observations). Hence, Sau3AI remained unique. This construction led to an out-of-frame fusion of the aph and HuIFN-β genes and was designated pGVIFN09 after identification of the appropriate orientation. The ATG initiation codon of the aph gene was embedded in an NcoI site (3), which is unique in pGVIFN09.

Finally, cleavage of pGVIFN09 with NcoI and DpnI (a blunt-cleaving isoschizomer of Sau3AI) led to excision of the intervening aph coding sequences and brought the interferon DNA sequence immediately downstream of the Streptomyces expression signals. Moreover, filling-in of the NcoI site with Klenow polymerase, prior to ligation, re-created both the initiation triplet and the amino-terminal codeword for mature HuIFN-β (Fig. 3).

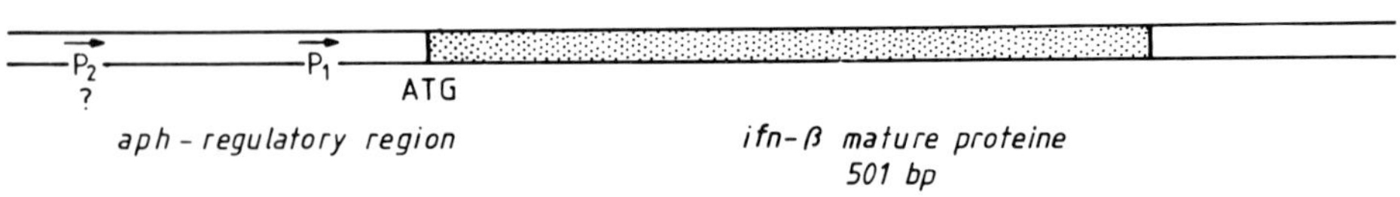

FIGURE 3. The interferon-β expression unit.

Two routes were followed to transfer the interferon-β expression unit into *S.lividans*. Firstly, the hybrid gene was cloned into the shuttle plasmid pSKN01 (15) and transformed to *S.lividans*. A second strategy involved direct transfer into the *Streptomyces* chromosome by means of a novel approach developed recently by Neesen and Volckaert (manuscript in preparation). Expression of the chloramphenicol acetyl transferase (*cat*) gene of the vector in *S.lividans* (15) proved that transfer and/or integration had been successful. The integrity and stability of the HuIFN-β gene in *S.lividans*, as well as the level of its expression, are being analyzed.

REFERENCES

1. Chater KF, Hopwood DA, Kieser T, Thompson CJ. Current Topics in Microbiol. Immunol. 96:69-93, 1982.
2. Hawley D, McClure W. Nucleic Acids Res. 11:2237-2255, 1983.
3. Bibb MJ, Ward JM, Cohen SN. Mol. Gen. Genet. 199:26-36, 1985.
4. Westpheling J, Ranes M, Losik R. Nature 313:22-27, 1985.
5. Jaurin B, Cohen S. Gene 28:83-91, 1984.
6. Volckaert G, De Vleeschouwer E, Blöcker H, Frank R. Gene Anal. Techn. 1:52-59, 1984.
7. Katz E, Thompson CJ, Hopwood DA. J. Gen. Microbiol. 129:2703-2714, 1983.
8. Thompson CJ, Gray GS. Proc. Natl. Acad. Sci. USA 80:5190-5194, 1983.
9. Kieser T, Hopwood DA, Wright HA, Thompson CJ. Mol. Gen. Genet. 185: 223-238, 1982.
10. Reiss B, Sprengel R, Will H, Schaller H. Gene 30:211-218, 1984.
11. Opdenakker G, Dijkmans R, Volckaert G, Billiau A, De Somer P. Arch. Int. Physiol. Biochim. 92:B49, 1986.
12. Winderickx J. Graduate Thesis, K.U.Leuven, 1984.
13. Volckaert G. Methods in Enzymol., in press, 1986.
14. Aert R, Dalemans W, Volckaert G, Eyssen H. Arch. Int. Physiol. Biochim. 94:B12, 1986.
15. Neesen K, Van Aerschot A, Volckaert G. Arch. Int. Physiol. Biochim. 94:B24, 1986.

IV. RESEARCH UNIT OF PROF. M. VANDEPUTTE

THE INFLUENCE OF LOCAL IMMUNE SUPPRESSION ON POLYOMA VIRUS-INDUCED VASCULAR TUMORS*

L. LU, L. VAN HOVE, H. SOBIS and M. VANDEPUTTE
Rega Institute, Katholieke Universiteit Leuven, B-3000 Leuven, Belgium

1. INTRODUCTION

The induction of tumors by polyoma virus is very much dependent upon the cell-mediated reaction of the host. In rats and mice this oncogenic virus only induces tumors when inoculated at birth or into immunosuppressed animals (1,2). Normal adult rodents are totally resistant to polyoma oncogenesis. Yet, the injection of polyoma (Py) virus into the placenta of fetectomized rats induces vascular tumors of yolk sac origin (3). The development of vascular tumors in these adult rats may result from : (a) the non-immunogenicity of these vascular tumors; (b) the special immunological mother-fetus relationship that exists during pregnancy; and (c) a particular sensitivity of the visceral yolk sac to neoplastic transformation that may eventually be linked to the growth and differentiation potential of the cells in this extra-embryonic membrane.

In previous experiments we could show that the Py-induced visceral yolk sac tumors were immunogenic and displayed the Py-specific tumor transplantation antigens (TSTA). The presence of these antigens was shown both in hemangiosarcomas induced in fetectomized rats (3) and in Py-virus-induced hemangiomas in grafts of visceral yolk sac (4). The possibility that the special mother-fetus immunological relationship might explain the outgrowth of these immunogenic vascular tumors could be discarded by our finding that grafts of visceral yolk sac infected *in vitro* with the virus led to the development of hemangiomas in virgin as well as pregnant rats (4). The particular sensitivity of the endothelial cells of the visceral yolk sac to the oncogenic potential of Py-virus was demonstrated by ectopic grafts of this membrane in adult syngeneic rats and by morphological studies on organ cultures (5). In both cases the endothelial cells were the only cells found to be transformed by the Py-virus. Since the transformed endothelial cells were always found to possess the Py-specific tumor antigens and to be immunogenic for the host it is likely that their outgrowth in an adult host is favored by a local or systemic immunosuppression due to the surgical procedure used in these experiments. In the present experiment this possibility was investigated in both pregnant fetectomized and virgin laporatomized rats.

2. MATERIALS AND METHODS

2.1. Surgical procedure

Rats of the inbred strain R were used. The R females mated with R males (day 0) were operated on at day 12 of pregnancy. The fetuses were removed and the visceral yolk sac was pulled through the incision and left

*This study was supported by the A.S.L.K. Cancer Fund.

outside the uterine horns (group 1). In virgin female rats (group 2) laparatomy was done and the uterus and peritoneal cavity were exposed for the same period of time as was necessary to perform the fetectomy in the previous (pregnant) group. In a third group of (virgin) female rats a large ($\pm$ 3 cm) longitudinal incision was made in the right flank. Twelve day old pregnant animals of the same age and weight but not operated upon were included as control group. All operations were performed under ether anesthesia.

2.2. Immunological tests

Three rats of each group were bled by heart punction at 3, 5 and 10 days after the surgical procedures. The peritoneal cavity was rinsed with MEM containing 50 U/ml heparin. Lymph node and spleen cells were teased apart with forceps and filtered through a stainless steel sieve. The lymphocytes were collected after centrifugation on Ficoll and washed with PBS. Cell viability was verified with the trypan blue dye exclusion test and in all experiments exceeded 98 %.

2.3. NK cytotoxicity

10^6 Effector lymphoid cells in 100 µl of RPMI-1640 medium containing 10 % fetal calf serum (FCS) were incubated for 4 hours at 37°C with 100 µl medium containing 10^4 YAC cells labeled with ^{51}Cr (Amersham) in round-bottom 96-well plastic microtiter plates. Three to four replicates were used for each group. After incubation, the plates were centrifuged and 100 µl of the supernatant were measured for radioactivity in a gamma counter. The spontaneous release of ^{51}Cr was determined in wells containing target cells and medium. Maximum release of ^{51}Cr was determined by addition of 10 % saponin. The results were expressed as percentage of cytotoxicity according to the formula : % cytotoxicity = 100 x [(experimental release (cpm) - spontaneous release (cpm)]/[maximum release (cpm) - spontaneous release (cpm)].

2.4. Macrophage cytotoxicity

Macrophages were separated from the lymphoid cells by incubation for 1 hr at 37°C in FCS-coated plastic petri dishes. After removing the non-adherent cells by washing, the adherent cells were collected by vigorously pipetting with PBS containing 0.2 % EDTA and 5 % FCS. The adherent cells (mostly macrophages) were then washed with PBS and resuspended in RPMI-1640 medium containing 10 % FCS. The macrophage cytotoxicity for ^{51}Cr-labeled F3/1(p) cells (6) was measured at an effector to target cell ratio of 100:1 for 18 hrs at 37°. The further procedure and the calculation of cytotoxicity level were similar as for the NK cell assay.

3. RESULTS

3.1. NK activity and macrophage cytotoxicity

As shown in Table 1 a marked decrease in NK activity was observed after laparatomy in both the virgin and pregnant rats. This decrease was restricted to the peritoneal cells and was not found in the spleen cells and lymph node cells. It was recorded at 3 and 5 days after surgery. Ten days after surgery the NK activity reached normal levels. As illustrated in Table 2 the macrophage cytotoxicity followed a similar pattern. Only the macrophages isolated from the peritoneal cavity displayed a much lower cytotoxicity in the groups of rats which underwent laparatomy whether or not combined with fetectomy (groups 1 and 2).

TABLE 1. NK[a] activity determined at 3, 5 and 10 days after surgery

Groups of rats	3 days after surgery			5 days after surgery		10 days after surgery	
	PEC[b]	LNC[b]	SP[b]	PEC	SP	PEC	SP
1. Pregnant fetectomy-laparatomy	18.7[c]±2.8	7.2±0.8	57±1.2	20±1.2	48±2.2	43±3.2	51±1.8
2. Virgin laparatomy	16±2.8	6.1±0.6	51±2.1	18±1.6	50±1.6	45±4.2	N.T.
3. Incision in flank	46.5±1.9	6.5±0.8	53±2	50±1.8	51±4.2	46±1.8	N.T.
4. Non-operated pregnant	48±4.8	7.1±0.6	51±1.2	52±2.4	50±1.2	47±2.8	50±1.2

[a]Effector to target cell (YAC) ratio 100:1.
[b]PEC = peritoneal cells, LNC : lymph node cells, SP : spleen cells.
[c]Percentage cytotoxicity in a ^{51}Cr 4 hour-release assay ± S.E.
═ : Statistically significant ($p < 0.05$).

TABLE 2. Macrophage[a] cytotoxicity at 3, 5 and 10 days after surgery

Groups of rats	3 days after surgery		5 days after surgery		10 days after surgery	
	PEC	SP	PEC	SP	PEC	SP
1. Pregnant laparatomy + fetectomy	1.7[b]±0.2	61.2±2.4	3.2±0.2	58±1.8	24±0.6	59±1.4
2. Virgin laparatomy	2.3±0.6	63.2±1.8	3.6±0.4	N.T.	23±1.2	N.T.
3. Incision in flank	26±1.8	N.T.	24±1.2	60±1.2	25±1.0	62±1.2
4. Non-operated pregnant	28±2.4	66±2.8	26±1.4	59±1.6	22±0.8	59±1.2

[a]Effector to target cells [F3/1(p)] ratio 100:1.
[b]Percentage cytotoxicity in a ^{51}Cr 18 hour-release assay ± S.E.
═ : Statistically significant ($p < 0.05$).

4. DISCUSSION

We previously demonstrated that the vascular tumors derived from the rat visceral yolk sac and induced by polyoma virus are immunogenic in the host since they possess the Py-tumor specific transplantation antigens (TSTA). The outgrowth of these immunogenic tumors seems to be favored by the local immune suppression induced by the surgical procedure required for the fetectomy or for the ectopic grafting of the Py-infected visceral yolk sac. Indeed, both the NK cell activity and the macrophage cytotoxicity are markedly depressed after laparatomy whether or not combined with fetectomy. However, this immune suppression remained restricted to the peritoneal cavity since it was absent in the peripheral lymphoid organs (lymph nodes and spleens). Moreover, PHA-stimulation data on lymphocytes from the peripheral blood and from the lymph nodes did not show differences between the various groups (results not shown), also indicating that the surgery induces mainly a local immune suppression. This suppression is no longer detectable ten days after surgery. Although the time during which the local immune response remains depressed may seem rather short to allow neoplastic transformation and outgrowth, morphological data indicate that the vascular tumors do indeed appear very rapidly. In a sequential morphological study of rats inoculated with Py-virus after fetectomy, proliferation of the endothelial cells was regularly observed as early as seven days after infection and three days later the development of hemangiomas was frequently recorded. Such rapid development of vascular tumors may be due to the particular sensitivity of fetal tissue to virus transformation (4,5). Hence, the local immune suppression induced by surgery together with the high sensitivity of fetal endothelial cells for Py virus neoplastic transformation may explain the outgrowth of immunogenic vascular tumors.

5. SUMMARY

The surgical procedure (laparatomy) used for the induction of vascular tumors by polyoma virus was shown to depress markedly the immune response in the peritoneal cavity. This immune depression involved both the natural killer activity and macrophage cytotoxicity. The immune response in the peripheral lymphoid tissues was found to be normal. The high sensitivity of the endothelial cells of the rat visceral yolk sac for the oncogenic activity of polyoma virus together with the local inhibition of the cellular immunity explain the outgrowth of immunogenic vascular tumors in adult rats.

REFERENCES

1. Vandeputte M, Denys P, Leyten R, De Somer P. Life Sci. 2:475-478, 1963.
2. Allison AC, Monga JN, Hammond V. Nature 252:746-748, 1974.
3. Vandeputte M, Meyer G, Sobis H. J. Natl. Cancer Inst. 56:517-521, 1976.
4. Lu YL, Sobis H, Van Hove L, Vandeputte M. Eur. J. Cancer Clin. Oncol. 21:631-636, 1985.
5. Lu YL, Sobis H, Vandeputte M. Eur. J. Cancer Clin. Oncol., in press.
6. Sobis H, Van Hove L, Heremans H, De Ley M, Billiau A, Vandeputte M. Int. J. Cancer 26:93-99, 1980.

INDUCTION OF TRANSPLANTATION TOLERANCE AFTER TOTAL LYMPHOID IRRADIATION : INFLUENCE OF THYMUS TRANSPLANTATION OR BLOOD TRANSFUSIONS

M. WAER, E. VAN DER SCHUEREN[†] and M. VANDEPUTTE
Rega Institute, Katholieke Universiteit Leuven and [†]Department of Radiotherapy, University Hospital, B-3000 Leuven, Belgium

1. INTRODUCTION

Total lymphoid irradiation (TLI), a radiotherapy regimen which has been successfully used for several decades in the treatment of Hodgkin's disease, was recently shown to result in a long lasting state of immunosuppression (1). In animal experiments, TLI enabled the induction of transplantation tolerance when allogeneic solid organs were transplanted together with bone marrow from the same donor (1). These experimental results prompted several clinical transplantation teams, including our own (2,3), to use TLI for clinical organ transplantation. Preliminary results of these trials are promising.

The exact mechanism of action of TLI is not known yet. Our and others' experiments (1,4,5) point to the pivotal role of suppressor cells which are temporarily induced after TLI. These suppressor cells are very similar to neonatal suppressor cells. Since in the neonatal period transplantation tolerance can be induced under certain circumstances, the immunological status created post-TLI is presently compared with the neonatal one (6). Because the thymus plays a predominant role in the development of the fetal and neonatal immune system, the present experiments were designed to determine whether an allogeneic thymus could instruct the immune system, which is thought to develop again after TLI, to become tolerant to the transplantation antigens exposed by this allogeneic thymus.

As mentioned before, transplantation tolerance, without the need for concomitant immunosuppression, can be induced after TLI only when also donor bone marrow is infused at the time of transplantation (1). Another reason why we were interested in the present experiments was to see whether the pivotal role of bone marrow to induce tolerance after TLI could be taken over by the thymus or by blood transfusions originating from the donor. Blood transfusions have indeed been found at many occasions to result in or to facilitate the induction of tolerance (7).

2. MATERIALS AND METHODS

2.1. Mice

Specific pathogen-free inbred BALB/c ($H2^d$), C3H ($H2^k$) and C57BL/Ka ($H2^b$) mice were used in all experiments. From the time of irradiation all animals were kept in conventional housing conditions, and a broad spectrum antibiotic (Tylan) was added to the drinking water.

2.2. Irradiation

Total lymphoid irradiation was delivered in daily fractions of 0.2 Gy each up to a total dose of 34 to 44 Gy, as described previously (4).

2.3. Thymus transplantation

One thymus lobe was aseptically removed on day 1 after birth from the donors and transplanted with a pipette under the kidney capsule of the recipient using the transrenal way. The success of the technique was verified using control animals receiving syngeneic thymus transplantation.

2.4. Skin grafting

Skin grafting was performed, as described previously (8) on day 30 after TLI.

2.5. Cell mediated lympholysis (CML)

CML was done with spleen lymphocytes, as described in detail previously (9).

2.6. Blood transfusion

0.5 ml of peripheral blood was collected on heparin from the retro-orbital veins of the donors and injected *via* the tail vein on day 1, 4 and 5 after TLI.

3. RESULTS

3.1. Skin graft experiments

From Table 1 it is clear that TLI results in a significant prolongation of allogeneic skin grafting when performed immediately after TLI. When transplantation is delayed until day 30 after TLI, no skin graft prolongation is oberved. These results are similar to those previously obtained by Strober *et al*. (1).

TABLE 1. Skin graft survival

Recipients (number)	Graft survival (days, mean ± SD)
- Control mice (n = 14)	11.2 (± 1.5)
- TLI-treated; skin graft on day 1 post TLI (n = 13)	33.0 (± 9)
- TLI-treated; skin graft on day 30 post TLI (n = 6)	11.5 (± 1.2)
- TLI-treatment, allogeneic thymus graft on day 1, allogeneic skin graft on day 30 post TLI (n = 8)	12.0 (± 0.5)

3.2. Cell-mediated lympholysis assay (CML)

The results of the CML tests performed between 2 and 5 months after TLI show that TLI-treated animals partially recovered in their capacity to develop *in vitro* cytotoxic cells against allogeneic stimulator cells (Table 2). These results are compatible with the findings of Strober *et al*. (1) showing a recovery of the mixed lymphocyte reaction from the second month post-TLI onwards. The presence of syngeneic (unirradiated) neonatal thymus under the kidney capsule does not accelerate but rather retards the recovery of CML reactivity. The presence of an allogeneic neonatal thymus under the kidney capsule does not seem to result in the development of transplantation tolerance. On the contrary, CML reactivity seems to be stimulated by the presence of the allogeneic antigens of the neonatal thymus. Finally, blood transfusions administered in the early post-TLI period are unable to induce CML non-responsiveness against the relevant antigens.

TABLE 2. CML reactivity

Mice (number experiments)	CML reactivity[a] against stimulator A[b]	CML reactivity against stimulator B
- TLI (n = 6)	70 % (± 12)	56 % (± 8)
- TLI + syng. thymus[c] (n = 4)	48 % (± 15)	38 % (± 17)
- TIL + allog. thymus[d] (n = 9)	87 % (± 21)	62 % (± 23)
- TLI + BTFS[e] (n = 4)	68 % (± 19)	N.D.

[a]Expressed as % of CML reactivity of unirradiated, untransplanted, untransfused control mice.
[b]Stimulator A : same strain as allogeneic thymus donor or donor of allogeneic blood transfusion.
Stimulator B : third party control (different from A).
[c]These animals received a syngeneic thymus transplant on day 1 after TLI.
[d]These animals received an allogeneic thymus transplant from donor A on day 1 after TLI.
[e]These animals received allogeneic blood transfusions from donor A on day 1, 4 and 5 after TLI.

4. DISCUSSION

The present study shows that, when administered after TLI, allogeneic blood or allogeneic neonatal thymus, in contrast to allogeneic bone marrow, are unable to induce permanent transplantation tolerance. This was ascertained by following the survival of allogeneic skin grafts originating from the same donor and evaluating CML non-responsiveness against the donor-antigens.

The negative thymus transplant experiments could be accounted for by a number of reasons. The possibility that the thymus transplants would not grow can be excluded since syngeneic control thymus transplants were uniformly successful as shown by microscopic examination at the time the spleens were removed for the CML experiments. Another explanation is that the newly developing lymphocytes of the TLI-treated mice are preferentially instructed by their own thymus, before the allogeneic thymus becomes functional, to reject the latter. Experiments to verify this hypothesis in thymectomized TLI-treated animals are in progress. A third explanation could be that, whereas the suppressor cells observed after TLI, are similar to neonatal suppressor cells, not the whole situation of the early post-TLI immune system can be regarded as "neonatal" and, hence, the post-TLI lymphocytes may no longer be susceptible to tolerance induction by the thymus. A fourth explanation could be that the "self-tolerance" created in the thymus relies mainly on bone marrow-derived cells. It is thus conceivable that after TLI the neonatal allogeneic thymus is very rapidly repopulated by recipient-type bone marrow-derived cells, which thus induce tolerance towards recipient-type, and not donor-type, transplantation antigens.

The absence of transplantation tolerance in the blood transfusion experiments probably indicates that blood transfusions, after TLI, are unable to interfere with suppressor cells, anti-idiotypic antibodies or clonal deletion, which is in contrast to their capacity to do so in other experimental as well as clinical situations (7).

The present findings illustrate that the exact immunological status following TLI is still puzzling and that more experiments are needed to understand the basic mechanisms underlying the post-TLI induction of tolerance.

REFERENCES

1. Strober S, Slavin S, Gottlieb M, Zan-Bar I, King DP, Hoppe RT, Fuks Z, Grumet FC, Kaplan HS. Immunol. Rev 46:87-112, 1979.
2. Levin B, Collins G, Waer M, Gerinsky T, Hoppe R, Miller E, Bieber C, Strober S. Lancet ii:1321-1326, 1985.
3. Waer M, Vanrenterghem Y, Roels L, Ang KK, Bouillon R, Lerut T, Gruwez J, van der Schueren E, Vandeputte M, Michielsen P. Transplantation, in press, 1986.
4. Waer M, Ang KK, van der Schueren E, Vandeputte M. J. Immunol. 132: 985-990, 1984.
5. Waer M, Ang KK, van der Schueren E, Vandeputte M. J. Immunol. 132: 991-996, 1984.
6. Strober S. Ann. Rev. Immunol. 2:97-115, 1984.
7. Van Rood JJ. Transplantation Proc. 15:915-916, 1983.
8. Waer M, Ang KK, Michielsen P, van der Schueren E, Vandeputte M. Transplantation Proc. 18:1362-1364, 1986.
9. Van Vlasselaer P, Vandeputte M. Transplantation 39:541-547, 1985.

V. RESEARCH UNIT OF PROF. E. DE CLERCQ

MECHANISM OF ANTIVIRAL ACTION OF 5-SUBSTITUTED 2'-DEOXYURIDINES: (E)-5-(2-IODOVINYL)-2'-DEOXYURIDINE (IVDU) AS COMPARED TO ITS CARBOCYCLIC ANALOGUE (C-IVDU)

R. BERNAERTS, A. VERBRUGGEN[†] and E. DE CLERCQ
Rega Institute and [†]Laboratory of Nuclear Medicine and Radiopharmacy, Katholieke Universiteit Leuven, B-3000 Leuven, Belgium

1. INTRODUCTION

The 5-substituted 2'-deoxyuridines (E)-5-(2-bromovinyl)-dUrd (BVDU) and (E)-5-(2-iodovinyl)-dUrd (IVDU) are highly potent and selective inhibitors of the replication of herpesviruses, i.e. herpes simplex virus (HSV).

A possible limitation to the therapeutic usefulness of these compounds is their rapid degradation by pyrimidine nucleoside phosphorylases. As a consequence, IVDU and BVDU are rapidly cleared from the bloodstream (1). In an attempt to circumvent this problem, carbocyclic analogues have been synthesized; i.e. the carbocyclic analogues of 5-iodo- and 5-fluoro-2'-deoxyuridine (2,3), carbocyclic IVDU (C-IVDU) and carbocyclic BVDU (C-BVDU) (4).

BVDU and IVDU are preferentially phosphorylated by the viral thymidine kinase to their 5'-mono- and 5'-diphosphate, which, in turn, is further phosphorylated by cellular kinases to their 5'-triphosphate. The 5'-triphosphates of BVDU and IVDU are inhibitory to the viral DNA polymerase and/or incorporated into DNA. If incorporated, they may render the DNA less functional as a template for replication and/or transcription.

As has been described (5), the carbocyclic analogues of IVDU (Fig. 1) and BVDU are, like IVDU and BVDU themselves, preferentially phosphorylated by HSV-1-infected cells, and eventually incorporated into DNA of the infected cells. The present study was designed to establish whether there was a direct correlation between the antiviral effects of IVDU and C-IVDU and the extent of their incorporation into DNA.

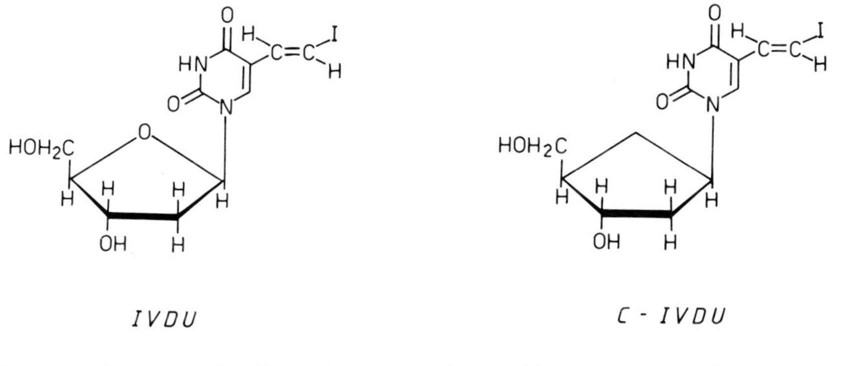

FIGURE 1. Structural formulae of (E)-5-(2-iodovinyl)-2'-deoxyuridine (IVDU) and its carbocyclic analogue (C-IVDU).

2. MATERIALS AND METHODS

2.1. Virus yield reduction

Vero cells were grown to confluency in 60-mm Falcon Petri dishes and infected with HSV-1 (strain KOS) at 0.1 PFU/cell. After a 1-hour adsorption period residual virus was removed and the cells were further incubated for 24 hours in the presence of varying concentrations of either IVDU or C-IVDU. Cells were freeze-thawed and scraped off with a rubber policeman. Cell debris was removed by centrifugation and virus yield of the supernatants was determined by plaque formation in Vero cells.

2.2. Cesium chloride gradient analysis

Confluent Vero cells, grown in 60-mm Falcon Petri dishes, were infected with HSV-1 (strain KOS) at 0.1 PFU/cell, and after 1-hour virus adsorption period, further incubated for 24 hours in the presence of either $[^{125}I]$IVDU or C-$[^{125}I]$IVDU at the indicated concentrations. The cells were lysed and analyzed on CsCl gradients as described previously (5). Fractions containing viral DNA were pooled, desalted on Sephadex G-50 columns and further examined by sucrose gradient analysis.

2.3. Sucrose gradient analysis

Fractions containing viral DNA were layered on top of 4 ml of a continuous sucrose gradient (from 10 to 30 % w/v) and centrifuged for 2.5 hours at 15°C in a SW56 rotor. Five-drop fractions were collected from the top of the tube, and DNA was precipitated on Whatman GF/C paper disks with ice-cold trichloroacetic acid 5 %. Filters were dried with ethanol and radioactivity was determined.

3. RESULTS

3.1. Virus yield reduction

IVDU and C-IVDU differ markedly in the dose-response of their antiviral effects (Fig. 2). For IVDU, the virus yield decreases proportionally with increasing concentrations of the compound (in the range of 0.1 to 5 μM), whereas for C-IVDU virus yield reduction levels off from a compound concentration of 5 μM.

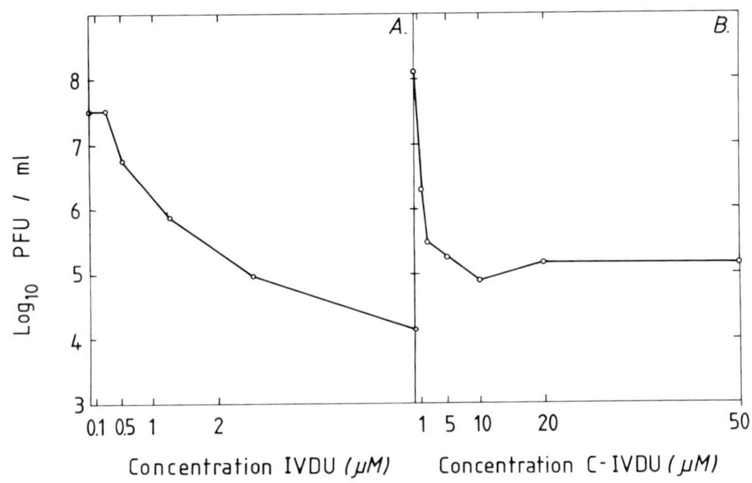

FIGURE 2. Dose-related inhibitory effects of IVDU (panel A) and C-IVDU (panel B) on HSV-1 (strain KOS) yield in Vero cells.

3.2. Interaction of IVDU and C-IVDU at the DNA level

The incorporation of [^{125}I]IVDU into viral DNA, as measured on CsCl gradients, increases with increasing concentrations of [^{125}I]IVDU (Fig. 3-A,B,C). On the contrary, the incorporation of C-[^{125}I]IVDU into viral DNA first increases, but then decreases, with increasing concentrations of C-[^{125}I]IVDU (Fig. 3-D,E,F).

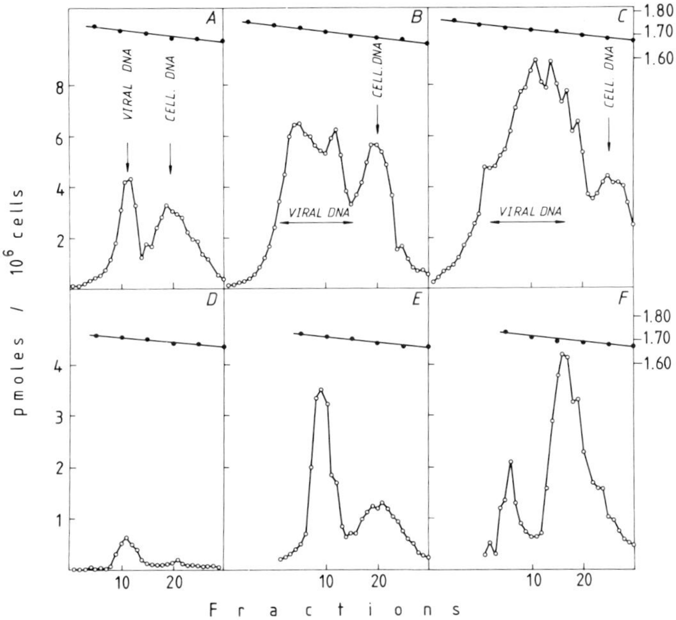

FIGURE 3. CsCl equilibrium gradient analysis of DNA from HSV-1 (strain KOS)-infected Vero cells, incubated for 24 hours in the presence of either [^{125}I]IVDU at 0.2 µM (panel A), 0.5 µM (panel B) and 2 µM (panel C), or C-[^{125}I]IVDU at 0.5 µM (panel D), 5 µM (panel E), and 20 µM (panel F).

At concentrations achieving an equivalent antiviral effect, [^{125}I]-IVDU is incorporated into DNA to a much greater extent than C-[^{125}I]IVDU. Moreover, the [^{125}I]IVDU-substituted DNA is clearly heterogeneous as it is dispersed over a large part of the gradient. The latter effect probably results from a shift in density of the viral DNA upon substitution of [^{125}I]IVDU for dThd. Alternatively or additionally, incorporation of [^{125}I]IVDU may distort the DNA helix structure to such an extent that strand breaks occur readily. This would give rise to smaller fragments with more varying density and difference in molecular weight. Changes in molecular weight can easily be visualized by sucrose gradient analysis. [^{125}I]IVDU-substituted DNA clearly showed an heterogenous DNA population, apparently pointing to double-strand breakage (Fig. 4). With C-[^{125}I]IVDU, when used at an equivalent antiviral concentration as [^{125}I]IVDU, no evidence of strand breakage could be detected when the C-[^{125}I]IVDU-substituted DNA was analyzed on sucrose gradient.

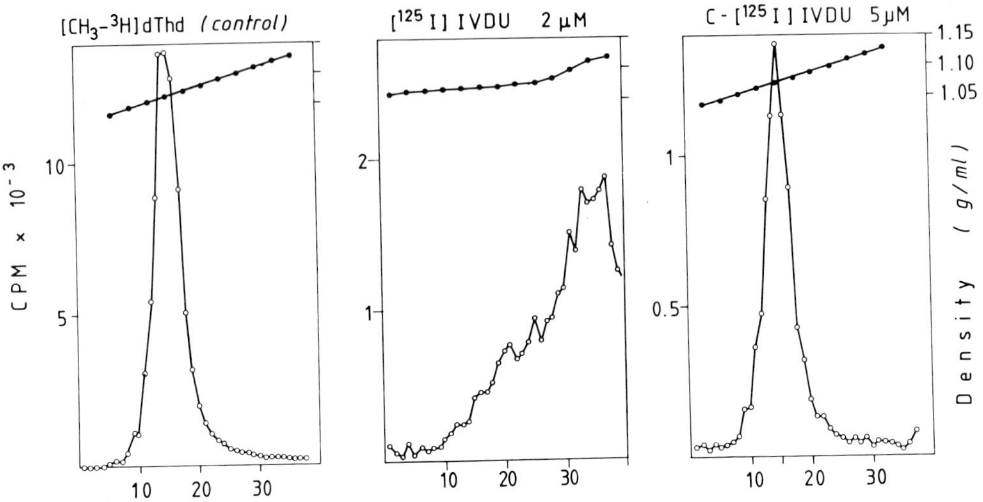

FIGURE 4. Sucrose gradient analysis of viral DNAs isolated from CsCl gradients (Figure 3).

4. CONCLUSION

From CsCl gradient analysis it is clear that both [^{125}I]IVDU and C-[^{125}I]IVDU are incorporated into DNA. For [^{125}I]IVDU, incorporation into viral DNA is clearly related to, and probably responsible for, the antiviral activity of the compound, as the DNA becomes more labile when increasing amounts of IVDU are incorporated.

For C-[^{125}I]IVDU, however, there is no direct correlation between antiviral activity and the amount of C-IVDU incorporated. C-IVDU most probably owes its antiviral action to an inhibitory effect on the viral DNA polymerase, exerted either directly by the 5'-triphosphate or indirectly, as the consequence of an impaired functioning of the C-[^{125}I]IVDU-substituted viral DNA template. Thus, the antiviral effect of C-IVDU may be due to an inhibitory effect on viral DNA synthesis during its first round. IVDU, however, would not impair viral DNA synthesis during its first round but rather interfere with the subsequent replication and transcription steps.

REFERENCES

1. Desgranges C, Razaka G, Drouillet F, Bricaud H, Herdewijn P, De Clercq E. Nucleic Acids Res. 12:2081-2090, 1984.
2. Shealy YF, Frye JL, DuBois NF, Shaddix SC, Brockman RW. J. Med. Chem. 24:1083-1086, 1981.
3. Shealy YF, O'Dell CA, Shannon WM, Arnett G. J. Med. Chem. 26:156-161, 1983.
4. Herdewijn P, De Clercq E, Balzarini J, Vanderhaeghe H. J. Med. Chem. 28:550-555, 1985.
5. De Clercq E, Bernaerts R, Balzarini J, Herdewijn P, Verbruggen A. J. Biol. Chem. 260:10621-10628, 1985.

THERAPEUTIC IMPLICATIONS OF THE UNUSUALLY LONG HALF-LIFE OF (*E*)-5-(2-BROMOVINYL)URACIL (BVUra) IN VIVO

C. DESGRANGES[†], G. RAZAKA[†] and E. DE CLERCQ
[†]Unité 8 de l'INSERM, Avenue du Haut-Lévêque, F-33600 Pessac, France and Rega Institute, Katholieke Universiteit Leuven, B-3000 Leuven, Belgium

1. INTRODUCTION

The *in vivo* degradation of those pyrimidine analogues that are endowed with antiviral or antitumor activity, may obviously affect the therapeutic usefulness of these drugs. Indeed, pyrimidine nucleosides or bases are generally subjected to rapid catabolism leading to products devoid of therapeutic activity. Thus, several 5-substituted-2'-deoxyuridine (dUrd) analogues are substrates for pyrimidine nucleoside phosphorylases (PNPases) which degrade them to their corresponding bases (1,2). Likewise, several 5-substituted uracil (Ura) analogues, and in particular 5-halogenated Ura analogues, are subject to a reductive catabolic pathway with, as the first step, the reduction of the 5:6 double bond of the pyrimidine ring by a dihydropyrimidine (H_2Thy) dehydrogenase (3). Advantage can be taken from the long half-life of BVUra in the plasma, (*i*) to regenerate the potent antiviral drug (*E*)-5-(2-bromovinyl)-2'-deoxyuridine (BVdUrd) (4) from BVUra by a pentosyl transfer and (*ii*) to enhance the therapeutic activity of 5-fluoro-Ura (FUra) by inhibition of its degradation by BVUra. The present study is focussed on the pharmacokinetics of BVUra and BVdUrd in rat plasma and the modalities of the (re)generation of BVdUrd from BVUra.

2. RESULTS AND DISCUSSION

2.1. Plasma pharmacokinetics of BVUra and BVdUrd in the rat

In vitro studies have previously demonstrated that BVUra is not a substrate but an inhibitor of H_2Thy dehydrogenase (5), the first enzyme of the reductive pathway for pyrimidine degradation. Consequently, BVUra is cleared from the blood stream at a much slower rate than are the other pyrimidine bases (5-7).

Thus, after intraperitoneal (i.p.) administration at 200 µmol/kg (Fig. 1b), BVUra appears rapidly in the plasma where it maintains a concentration of 50-70 µM for at least 6 h. BVUra is then cleared from the plasma according to an apparent first order reaction with a mean half-life of about 8 h (Table 1). Even at 24 h after BVUra administration, there is still a plasma level of 5-20 µM. Following intravenous (i.v.) administration of the same dose (Fig. 1b), BVUra attains a plasma concentration of 100 µM at 20 min after injection; the half-life derived from a first order elimination is about 7.5 h (Table 1).

BVdUrd is a good substrate for pyrimidine nucleoside phosphorylases (PNPases) (2,8). Consequently, this dUrd analogue is rapidly cleared from the plasma after i.p. or i.v. administration to rats (9). When BVdUrd is administered i.p. at 200 µmol/kg, it achieves a maximum plasma concentration (50 to 90 µM) at 20 min after injection. At 4 h, it is completely cleared from the plasma (Fig. 1a). Its half-life is similar to that of other dUrd analogues (Table 2).

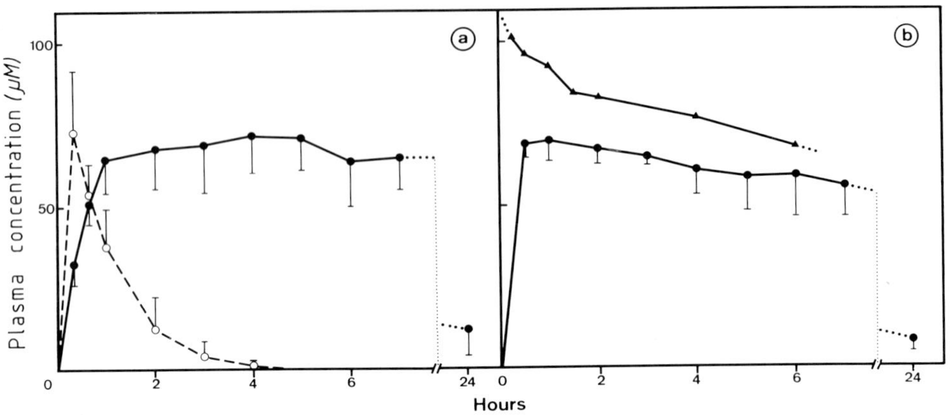

FIGURE 1. Plasma concentration of BVUra (●, ▲) and BVdUrd (O) in rat plasma : a) after i.p. BVdUrd administration (200 μmol/kg; 5 rats); b) after i.p. (●) or i.v. (▲) BVUra administration (200 μmol/kg; 3 rats and 1 rat, respectively).

TABLE 1. Mean half-lifes ($t_{1/2}$) and areas under the curve (AUC) of 5-substituted Ura analogues in the plasma of rats following i.p. or i.v. administration at 200 μmol/kg

Compound		After base administration		After nucleoside administration	
		$t_{1/2}$ (min)	AUC (μMxhr)	$t_{1/2}$ (min)	AUC (μMxhr)
Ura	i.p.	25	160	24	122
Thy	i.p.	22	171	14	50
Thy	i.v.	24	222	15	32
BVUra	i.p.	480	1000	450	1075
BVUra	i.v.	450	1175	426	782
FUra	i.p.	14	66	21	44
FUra	i.v.	15	144	24	31
IUra	i.p.	45	334	33	109
CF_3Ura	i.p.	73	559	30	291

The free base BVUra resulting from the phosphorolysis of BVdUrd attains a concentration of 50-80 μM within 1 h following i.p. administration of BVdUrd (200 μmol/kg) and is then eliminated at the same rate as occurring after BVUra administration, that is with an half-life of 7.5 h. Regardless of its mode of administration (i.v. or i.p., as nucleoside or base), BVUra has a much longer half-life and higher AUC than the other 5-substituted pyrimidine analogues (Table 1), probably because it is not a substrate for dihydropyrimidine dehydrogenase. This remarkable property has enabled us to develop two therapeutic strategies, the first leading to

TABLE 2. Mean half-life of 5-substituted dUrd analogues in the plasma of rats following i.p. administration at 200 µmol/kg.

Compound	Number of experiments	Mean $t_{1/2}$ (min)
dUrd	6	15.6 ± 5.2
dThd	3	25 ± 1.4
BVdUrd	5	28.4 ± 8.4
EtdUrd	3	27 ± 9.5
ClEtdUrd	3	24.3 ± 7.4
FdUrd	4	17 ± 3.6
IdUrd	3	14.2 ± 1.1

the (re)generation of the antiviral agent BVdUrd from BVUra *via* a pentosyl transfer reaction (9-11), and the second leading to the enhancement of the antitumor properties of FUra *via* inhibition of its degradation by BVUra (5-7,12,13).

2.2 De novo formation and regeneration of BVdUrd from BVUra in vivo

As stated above, BVdUrd is rapidly transformed *in vivo* to BVUra by PNPases, and this may obviously affect the therapeutic properties of BVdUrd. However, BVdUrd can be generated from BVUra by a pentosyl transfer reaction catalyzed by PNPases (7), according to the following scheme : BVUra + 5-X-dUrd ⇌ BVdUrd + 5-X-Ura, where X is a 5-substituent on the pyrimidine ring. This process has allowed us to (re)generate BVdUrd from BVUra upon the administration of an appropriate donor of the deoxyribosyl group. First, thymidine (dThd) has been used as the donor (9); however, since this nucleoside by itself counteracts the antiviral activity of BVdUrd and is therefore not therapeutically useful, other pyrimidine deoxynucleosides, some of which are by themselves antivirally active, i.e. 5-ethyl-dUrd (EtdUrd), have been found effective in regenerating BVdUrd (11), and here we report that purine deoxyribonucleosides such as deoxyadenosine (dAdo), deoxyinosine (dIno) or deoxyguanosine (dGuo) are effective deoxyribosyl donors for the regeneration of BVdUrd from BVUra *in vivo* (Fig. 2).

Some of the pentosyl donors appear useful for association with BVUra to regenerate BVdUrd, because (*i*) they do not counteract the antiherpetic capacity of BVdUrd: for example, dUrd; (*ii*) they have a low toxicity and an important antiviral activity by themselves: for example, EtdUrd; or (*iii*) they are natural nucleosides which do not directly interfere with the properties of BVdUrd: for example, the purine deoxynucleosides dGuo, dIno and dAdo.

3. CONCLUSION

The unusually long half-life of BVUra *in vivo* has two important therapeutic implications : (*i*) it permits the (re)generation of the antiherpetic drug BVdUrd through the combined use with an appropriate deoxyribosyl donor, and (*ii*) it leads to an enhancement of the antitumor activity of FUra through protection of the latter against premature degradation.

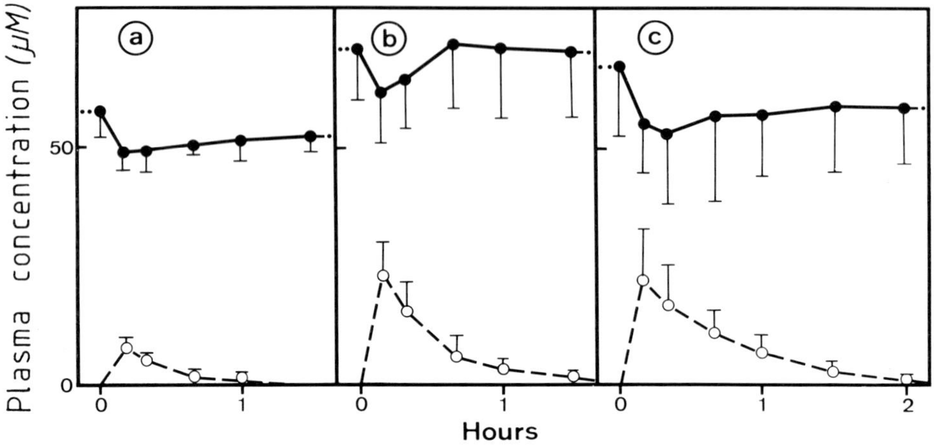

FIGURE 2. Regeneration of BVdUrd from BVUra in rats by deoxyribosyl transfer from purine deoxyribonucleosides. Three hours after a previous i.p. administration of BVdUrd (200 μmol/kg) to rats, this nucleoside was entirely cleared from the plasma; only BVUra was present at that time (T = 0). At T = 0 Guo (a, 3 rats), dIno (b, 4 rats) or dAdo (c, 5 rats) were administered i.p. to these rats (at 200 μmol/kg). Under these conditions, BVdUrd (O) was regenerated from BVUra (●).

REFERENCES

1. Nakayama C, Wataya Y, Meyer Jr RB, Santi DV, Saneyoshi M, Ueda T. J. Med. Chem. 23:962-964, 1980.
2. Desgranges C, Razaka G, Rabaud M, Bricaud H, Balzarini J, De Clercq E. Biochem. Pharmacol. 32:3583-3590, 1983.
3. Wasternak C. Pharmacol. Ther. 8:629-651, 1980.
4. De Clercq E, Descamps J, De Somer P, Barr PJ, Jones AS, Walker RT. Proc. Natl. Acad. Sci. USA 76:2947-2951, 1979.
5. Desgranges C, Razaka G, De Clercq E, Herdewijn P, Balzarini J, Drouillet F, Bricaud H. Cancer Res. 46:1094-1101, 1986.
6. Desgranges C, Razaka G, Bricaud H, De Clercq E. Biochem. Pharmacol. 34:403-404, 1985.
7. Desgranges C, Razaka G, Belloc I, Drouillet F, De Clercq E, Bricaud H. J. Pharm. Clin. 4:97-105, 1985.
8. Liermann B, Hermann G, Biomed. Biochim. Acta 42: K35-K38, 1983.
9. Desgranges C, Razaka C, Drouillet F, Bricaud H, Herdewijn P, De Clercq E. Nucleic Acids Res. 12:2081-2090, 1984.
10. Desgranges C, Razaka G, Bricaud H, De Clercq E. Biochem. Pharmacol. 34:405-406, 1985.
11. Desgranges C, De Clercq E, Razaka G, Drouillet F, Belloc I, Bricaud H. Biochem. Pharmacol. 35:1647-1653, 1986.
12. Ben-Efraim S, Shoval S, De Clercq E. Brit. J. Cancer 54:847-852, 1986.
13. Iigo M, Nakajima Y, Hoshi A, De Clercq E. Eur. J. Cancer Clin. Oncol., in press, 1987.

S-ADENOSYL-L-HOMOCYSTEINE HYDROLASE AS TARGET ENZYME FOR ANTIVIRAL AGENTS: STUDIES WITH ADENOSINE KINASE-DEFICIENT RAT CELLS

M. COOLS, J.C. DRACH[†] and E. DE CLERCQ
Rega Institute, Katholieke Universiteit Leuven, B-3000 Leuven, Belgium and [†]Department of Oral Biology, School of Dentistry, The University of Michigan, Ann Arbor, Michigan 48109, U.S.A.

1. INTRODUCTION

During the last decade several adenosine analogues with potent antiviral activities have been developed. Their mode of action mostly remains to be elucidated : they may be targeted at a number of enzymes including S-adenosylhomocysteine (SAH) hydrolase, adenosine kinase, adenosine deaminase, ribonucleotide reductase and DNA polymerase.

SAH hydrolase can be considered as a mediator of methyltransferases; methylation reactions are involved in the synthesis of different classes of compounds including the 5'-cap formation of viral mRNA, which is essential for the maturation and, hence, translation, of viral mRNA to viral proteins. Methylation reactions yield as reaction product SAH, a negative feedback inhibitor of methyltransferases. SAH is further hydrolyzed by the enzyme SAH hydrolase to adenosine and homocysteine. Inhibition of this enzyme leads to an accumulation of SAH and a concomitant inhibition of methyltransferases.

To assess the necessity of phosphorylation of the adenosine analogues by adenosine kinase, we evaluated their antiviral potencies in a paired cell system existing of adenosine kinase-positive (AK^+) rat cells and an adenosine kinase-negative (AK^-) mutant cell line thereof; both cell lines being devoid of adenosine deaminase activity.

2. MATERIALS AND METHODS

2.1. Test compounds

The origin of the compounds was as follows : 9-β-D-arabinofuranosyladenine (Ara-A) (Sigma), 3-deazaadenosine (c^3Ado) (L.M. Kedda, NCI, Bethesda, Maryland), carbocyclic 3-deazaadenosine (C-c^3Ado) (J.A. Montgomery, Southern Research Institute, Birmingham, Alabama), (S)-9-(2,3-dihydroxypropyl)adenine [(S)-DHPA] and (RS)-3-adenin-9-yl-2-hydroxypropanoic acid [(RS)-AHPA] (A. Holý, Czechoslovak Academy of Sciences, Prague, Czechoslovakia), neplanocin A (J. Murase, Toyo Yozo Co., Shizuoka-Ken, Japan), ribavirin (ICN Nutritional Biochemicals, Cleveland, Ohio), tubercidin (Sigma), sangivamycin and toyocamycin (M.J. Robins, University of Alberta, Edmonton, Alberta, Canada), β-xyloadenosine and α-lyxoadenosine (J.-L. Imbach, Université des Sciences et Techniques du Languedoc, Montpellier, France).

2.2. Viruses and cells

The source of viruses was as follows : vaccinia virus, vesicular stomatitis virus and Sindbis virus (American Type Culture Collection, Rockville, Maryland). The AK^+ cells (B-mix K-44/6 Rous sarcoma virus-transformed rat cells) were a gift from Dr. J. Svoboda (Institute of Experimental Biology and Genetics, Prague, Czechoslovakia). The AK^- cell line was derived thereof by selecting a population of cells stably resistant to a concentration of Ara-A 100-fold greater than that tolerated by the AK^+ cells (1).

2.3. Inhibition of virus-induced cytopathogenicity in vitro

Confluent cell cultures in microtiter trays were inoculated with 100 $CCID_{50}$ (1 $CCID_{50}$ corresponding to the virus stock solution that proved infective for 50 % of the cell cultures). After 1 h of virus adsorption to the cells, residual virus was removed and replaced by cell culture medium (Eagle's minimum essential medium) containing 3 % horse serum and various concentrations of the test compounds. Viral cytopathogenicity was recorded as soon as it reached completion in the untreated virus-infected cell cultures, i.e. at day 1 to 2 for vesicular stomatitis virus and at 2 to 3 days for vaccinia and Sindbis virus. The antiviral activity of the compounds is expressed as the concentration required to inhibit viral cytopathogenicity by 50 %.

2.4. Cytotoxicity

Cytotoxicity measurements were based on the alteration of normal cell morphology. Confluent cell cultures, which had not been infected, but were treated with various concentrations of the test compounds were incubated in parallel with the virus-infected cell cultures and examined microscopically at the same time as viral cytopathogenicity was recorded for the virus-infected cell cultures. A disruption of the cell monolayer was considered as evidence for cytotoxicity.

3. RESULTS

According to their MIC for virus replication, the adenosine analogues could be clearly divided in two classes (Table 1) : I, those compounds that inhibited virus replication in AK^+ cells at a 500- to 10,000-fold lower MIC_{50} than virus replication in AK^- cells; and II, those compounds that were only slightly (1- to 50-fold) more antivirally active in AK^+ cells than in AK^- cells. To class I compounds belong tubercidin, β-xyloadenosine and α-lyxoadenosine. To class II compounds belong c^3Ado, C-c^3Ado, (S)-DHPA, (RS)-AHPA and neplanocin A.

4. DISCUSSION

The antiviral activity of the class I compounds clearly depends on phosphorylation by the host cell adenosine kinase, which means that these compounds are antivirally active primarily as their phosphorylated products. Class II compounds, however, do not seem to require phosphorylation by the host cell adenosine kinase to express their antiviral activity. Yet, the cytotoxic action of neplanocin A in murine leukemic cells may be mediated by an activation by adenosine kinase (2). Striking is that all these compounds have been recognized as inhibitors of SAH hydrolase (3,4) and that their antiviral activity is correlated with the rate of inhibition of the enzyme (4). Thus, the present findings with the paired AK^-/AK^+ cell system are consistent with an action of the class II compounds at the level of SAH hydrolase.

REFERENCES

1. Shipman C Jr, Tong S-L, Smith SH, Katlama NB, Drach JC. Antimicrob. Agents Chemother. 24:947-949, 1983.
2. Inaba M, Nagashima K, Tsukagoshi S, Sakurai Y. Cancer Res. 46:1063-1067, 1986.
3. Chiang PK, Richards HH, Cantoni GL. Mol. Pharmacol. 13:939-947, 1977.
4. De Clercq E, Cools M. Biochem. Biophys. Res. Commun. 129:306-311, 1985.

TABLE 1. Antiviral effects of various compounds in adenosine kinase-positive (AK^+) and adenosine kinase-negative (AK^-) cells

	Minimum antiviral concentration[a] (μg/ml)						Minimum cytotoxic concentration (μg/ml)	
	Vaccinia		VSV		Sindbis			
	AK^+	AK^-	AK^+	AK^-	AK^+	AK^-	AK^+	AK^-
Ara-A	0.1	7	7	70	> 4	> 40	10	40
c^3Ado	2.5	25	25	> 100	> 4	> 40	≥ 4	100
C-c^3Ado	0.25	2.5	0.7	7	> 0.4	> 10	1	40
(S)-DHPA	7	70	7	300	> 40	> 400	100	> 400
(RS)-AHPA	4	200	25	> 400	> 4	> 400	100	> 400
Neplanocin A	> 0.01	0.007	> 0.01	> 0.01	> 0.01	> 0.01	0.04	0.04
Ribavirin	7	300	25	> 400	25	> 400	100	> 400
Tubercidin	> 0.004	7	0.007	10	0.007	70	0.01	≥ 10
Sangivamycin	> 0.1	0.7	> 0.1	> 10	> 0.04	> 1	0.4	10
Toyocamycin	> 0.1	0.7	> 0.4	> 4	> 0.04	> 4	0.4	10
α-lyxoadenosine	> 0.4	> 400	> 0.1	> 400	> 1	> 400	4	> 400
β-xyloadenosine	> 0.04	> 10	> 0.1	> 40	> 0.04	> 100	0.1	≥ 40

[a]When preceded by >, more accurate MIC_{50} values could not be determined because of cytotoxicity.

ANTIVIRAL ACTIVITY OF (S)-9-(3-HYDROXY-2-PHOSPHONYLMETHOXYPROPYL)ADENINE [(S)-HPMPA]

T. SAKUMA, E. DE CLERCQ, R. BERNAERTS, I. VOTRUBA[†] and A. HOLÝ[†]
Rega Institute, Katholieke Universiteit Leuven, B-3000 Leuven, Belgium and [†]Institute of Organic Chemistry and Biochemistry, Czechoslovak Academy of Sciences, 16610 Praha 6, Czechoslovakia.

1. INTRODUCTION

In recent years several new antiherpes compounds have been developed, i.e. BVDU and ACV, which owe their selective antiviral activity to a preferential phosphorylation by the virus-encoded thymidine kinase (TK). However, herpesviruses such as cytomegalovirus (CMV), which do not code for a specific viral TK, and thymidine kinase-deficient (TK$^-$) mutants of herpes simplex virus (HSV) or varicella-zoster virus (VZV), show little, if any, sensitivity to these antiherpetic compounds (1).

The search for a synthetic antiviral agent that may be effective against CMV and TK$^-$ mutant strains of HSV and VZV has led to the discovery of a new class of acyclic nucleoside analogues which exhibit a broad spectrum activity against a wide variety of DNA viruses (2). As the most potent congener of this class of compounds emerged (S)-9-(3-hydroxy-2-phosphonylmethoxypropyl)adenine [(S)-HPMPA], which can be considered as the 2-phosphonylmethyl derivative of (S)-9-(2,3-dihydroxypropyl)adenine [(S)-DHPA] (Fig. 1).

2. MATERIALS AND METHODS

2.1. Viruses

The origin of the HSV-1, CMV and VZV strains, including the TK$^-$ mutants, has been described previously (2).

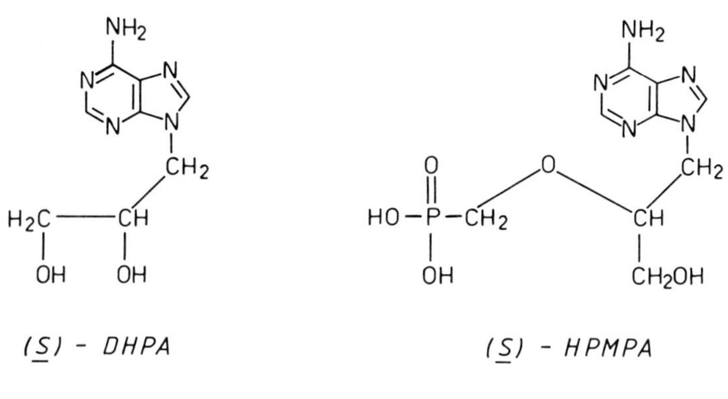

FIGURE 1. Structural formulae of (S)-DHPA and (S)-HPMPA.

2.2. Compounds

(S)-HPMPA, [U-^{14}C]-labeled (S)-HPMPA, (S)-HPMPA-DP and (S)-DHPA were synthesized at the Institute of Organic Chemistry and Biochemistry, Czechoslovak Academy of Sciences, Prague, Czechoslovakia.

2.3. Antiviral assay

Human embryonic lung fibroblast (HEL) cells grown in 96-well microplates were infected with 20 PFU of virus. After 2 h of incubation at 37°C, the infected cells were replenished with 0.1 ml of medium containing serial dilutions of the test compound. The antiviral activity is expressed as ID_{50} (i.e. the minimum concentration required to inhibit virus-induced plaque formation by 50 percent).

2.4. Cytotoxicity

HEL cells were seeded at a rate of 3 x 10^3 cells per cup in 96-well microplates and allowed to proliferate for 24 hours in Eagle's minimum essential medium (MEM) supplemented with 10 % fetal calf serum. The medium was then replaced by MEM containing various concentrations of test compound. After three days incubation at 37°C (70 % monolayer), the cell number was determined with a Coulter counter. The cytotoxic activity is expressed as IC_{50} (i.e. the minimum concentration required to inhibit cell growth by 50 percent).

2.5. DNA synthesis

Confluent HEL cell monolayers in 60-mm Falcon Petri dishes were infected with HSV-1 at a multiplicity of 0.1 plaque-forming units/cell. After 1 h incubation at 37°C, the cells were further incubated with 25 μCi [^{32}P]orthophosphate in phosphate-free medium, in the presence of (S)-HPMPA at different concentrations. The procedure for DNA synthesis analysis has been described previously (3).

2.6. DNA polymerase assay

HSV-1 infected HEL cells were sonicated with a MSE sonicator (150 W) in 0.25 M KH_2PO_4 (pH 7.5), 1 mM DTT, 1 mM EDTA, 1 mM PMSF and 0.5 % Triton X-100. After centrifugation at 10,000 g for 25 min, the supernatant was dialyzed overnight against 50 mM Tris-HCl (pH 7), 1 mM EDTA, 1 mM DTT and 20 % glycerol. The supernatant was then used as a crude enzyme material. The DNA polymerase assay was performed as described previously (4).

2.7. Acid soluble pool analysis of (S)-HPMPA

Confluent HEL cell monolayers (1.5 x 10^6 cells) in 60 mm Falcon Petri dishes were infected with HSV-1 at a multiplicity of 0.1 plaque-forming units/ cell. After 6 h incubation at 37°C, the cells were further incubated with [U-^{14}C]-(S)-HPMPA (15 nmoles per dish in 1 ml of medium) for 7 h. TCA soluble pool analysis of mock-infected and HSV-1-infected HEL cells was performed by thin layer chromatography (TLC).

3. RESULTS AND DISCUSSION

As shown in Table 1, (S)-HPMPA inhibited plaque formation of human CMV at a concentration of 0.3 μg/ml. Its potency against CMV was slightly higher than that of DHPG, a well-established anti-CMV compound (5). Plaque formation of both TK^+ VZV strains and TK^- VZV strains was inhibited by (S)-HPMPA at 0.004 μg/ml. Its potency was higher than that of various thymidine analogues which were active against TK^+ VZV strains but not active against TK^- VZV strains (Table 1).

(S)-HPMPA brought about a significant inhibition of HSV-1 DNA synthesis even at a concentration of 0.01 μg/ml (Fig. 2), while host cell DNA was not inhibited at concentrations up to 10 μg/ ml.

Neither deoxyadenosine (dAdo), nor deoxyguanosine (dGuo), nor deoxythymidine (dThd), nor deoxycytidine (dCyd), nor adenosine (Ado) were ef-

Table 1. Antiviral activity of (S)-HPMPA as compared to various other antiherpes compounds in HEL cells

Compound	IC_{50} for HEL cell growth (µg/ml)	ID_{50} (µg/ml)			
		VZV (20 PFU)		CMV (20 PFU)	
		YS (TK^+)	YSR (TK^-)	AD-169	Davis
(S)-HPMPA	5	0.004	0.004	0.3	0.3
(S)-DHPA	> 400	10	200	200	70
PFA	200	20	20	15	10
Ara-T	100	0.1	20	200	200
Ara-A	6	1.2	2	10	7
FIAC	10	0.07	20	0.8	0.6
FMAU	1	0.01	1	0.7	0.7
EDU	10	10	200	100	100
BVDU	200	0.006	> 400	200	300
ACV	250	0.2	20	30	20
BCV	400	7	> 400	> 400	> 400
DHPG	> 50	0.1	4	0.7	0.5
2'-nor-cGMP	2	0.2	0.2	0.8	1

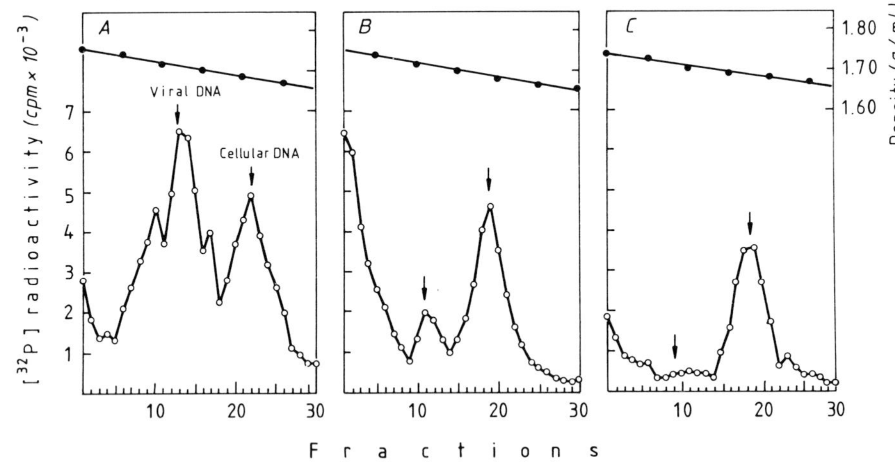

FIGURE 2. Effect of (S)-HPMPA on HSV-1 (KOS) and host cell DNA synthesis. CsCl equilibrium gradient analysis of DNA from HSV-1 (KOS)-infected HEL cells incubated for 24 h in the presence of [^{32}P]orthophosphate (25 µCi/2 ml of phosphate-free MEM supplemented with 2 % fetal calf serum/1.3 x 10^6 HEL cells) without (S)-HPMPA (panel A), with (S)-HPMPA at 0.01 µg/ml (panel B), or with (S)-HPMPA at 0.1 µg/ml (panel C).

Table 2. Effect of nucleosides on the antiviral activity of (S)-HPMPA

Compound	IC_{50} for HEL cell growth (μg/ml)	ID_{50} (μg/ml)			
		VZV (20 PFU)		CMV (20 PFU)	
		YS (TK^+)	YSR (TK^-)	AD-169	Davis
(S)-HPMPA	5	0.004	0.004	0.2	0.2
(S)-HPMPA plus 30 μg/ml dAdo	5	0.015	0.012	0.4	0.4
" 100 " "	5	0.08	0.06	0.4	0.4
(S)-HPMPA plus 10 μg/ml dGuo	5	0.003	0.003	0.2	0.2
" 30 " "	4	0.003	0.003	0.2	0.2
(S)-HPMPA plus 30 μg/ml dThd	6	0.01	0.005	0.2	0.2
" 300 " "	5	0.008	0.007	0.5	0.5
(S)-HPMPA plus 30 μg/ml dCyd	3	0.004	0.004	0.2	0.15
" 300 " "	6	0.004	0.004	0.5	0.5
(S)-HPMPA plus 30 μg/ml Ado	5	0.013	0.01	0.4	0.3
" 100 " "	5	0.045	0.04	0.6	0.6

fective in reversing the antiviral activity of (S)-HPMPA (Table 2).

Phosphorylation of [U-^{14}C]-(S)-HPMPA by HSV-1 (KOS)- and mock-infected HEL cells was analyzed by TLC (Fig. 3). One major and one minor radioactivity peak were detected on TLC of the acid-soluble fractions of the cell lysates. The major peak was identified as [U-^{14}C]-(S)-HPMPA and the minor peak as the diphosphate thereof ([U-^{14}C]-(S)-HPMPA-DP). The radioactivity peaks detected in HSV-1-infected HEL cells were higher than those obtained for mock-infected HEL cells. But the ratio of the major peak to the minor peak was essentially the same.

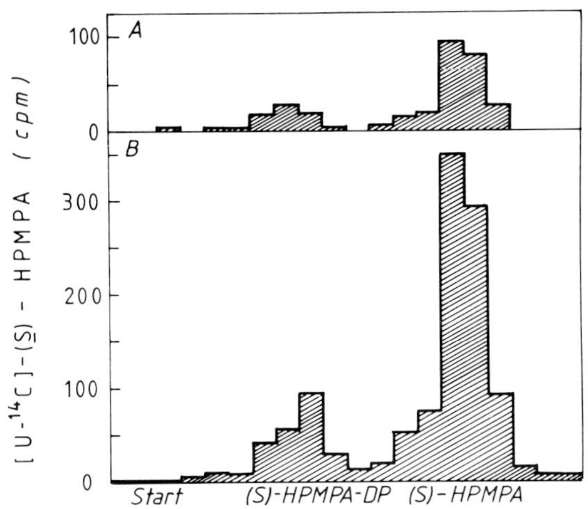

FIGURE 3. TLC analysis of the acid-soluble fraction of mock-infected (panel A) or HSV-1 (KOS)-infected (panel B) HEL cells exposed to 15 nM [U-^{14}C]-(S)-HPMPA (10 nmole/10^6 cells) .

(S)-HPMPA-DP inhibited HSV-1-induced DNA polymerase activity, measured with [8-^{3}H]dATP as substrate (Fig. 4A). (S)-HPMPA-DP did not serve as an alternate substrate for dATP for the DNA polymerase reaction (Fig. 4B).

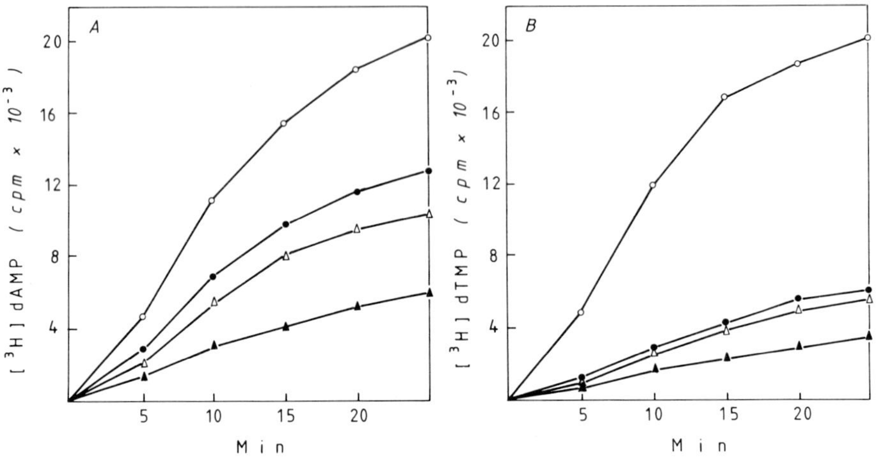

FIGURE 4. The standard mixture for the viral DNA polymerase assay contained 50 mM Tris-HCl (pH 8.0), 7.5 mM MgCl, 0.5 mM DTT, 100 mM ammonium sulfate, 0.1 mM each of dGTP, dCTP, dTTP, and [8-^{3}H]dATP (29 µCi/nmol), 25 µg of bovine serum albumin, 10 µg of activated calf thymus DNA, and enzyme in a total volume of 100 µl.

A. Effect of (S)-HPMPA-DP on HSV-1 DNA polymerase activity, as based upon [8-^{3}H]dAMP incorporation. [8-^{3}H]dATP was used at 0.005 mM in the absence (O) or presence of (S)-HPMPA-DP at 0.005 mM (●), 0.01 mM (△), or 0.04 mM (▲).

B. Effect of (S)-HPMPA-DP on HSV-1 DNA polymerase activity, as based upon [8-^{3}H]dTMP incorporation. [8-^{3}H]dTTP was used at 0.01 mM in the presence of dATP at 0.1 mM (O), (S)-HPMPA-DP at 0.04 mM and absence of dATP (△), (S)-HPMPA-DP at 0.005 mM and absence of dATP (▲), or absence of both dATP and (S)-HPMPA-DP (●).

4. CONCLUSION

(S)-HPMPA is better taken up by virus-infected cells than mock-infected cells. It is phosphorylated by cellular kinases. The active form of (S)-HPMPA may be its diphosphoryl derivative, and in this form, it may be targeted at the viral DNA polymerase.

REFERENCES

1. De Clercq E. Biochem. Pharmacol. 33:2159-2169, 1984.
2. De Clercq E, Holý A, Rosenberg I, Sakuma T, Balzarini J, Maudgal PC. Nature 323:464-467, 1986.
3. De Clercq E, Bernaerts R, Balzarini J, Herdewijn P, Verbruggen A. J. Biol. Chem. 260:10621-10628, 1985.
4. Knopf KW. Eur. J. Biochem. 98:231-244, 1979.
5. Mar E-C, Cheng Y-C, Huang E-S. Antimicrob. Agents Chemother. 24:518-521, 1983.

SELECTIVE INHIBITORY EFFECTS OF (S)-9-(3-HYDROXY-2-PHOSPHONYLMETHOXYPROPYL)ADENINE AND 2'-NOR-cGMP ON THE REPLICATION OF ADENOVIRUS *IN VITRO*

M. BABA and E. DE CLERCQ
Rega Institute, Katholieke Universiteit Leuven, B-3000 Leuven, Belgium

1. INTRODUCTION

Adenoviruses (AV) are medium-sized viruses (70-80 nm) containing linear DNA enclosed in a capsid with 252 capsomers. At least 40 serotypes of human AV have been identified and these can be divided into 7 subgroups (A-H). AV infect epithelial cells of mucous membrane, respiratory tract, and other organs. Most of the clinical manifestations caused by AV infections are not severe and heal spontaneously. However, in recent years it has been reported that severe and occasionally fatal AV infections may occur in immunocompromised patients, i.e. leukemic patients and recipients of kidney or bone marrow allografts (1-3). Also, epidemic keratoconjunctivitis (EKC) is a frequently occurring and highly infectious disease.

2. MATERIALS AND METHODS

2.1. Virus

Eight strains of AV divided into 4 subgroups were used for the antiviral assays. They were typed by cross-neutralization, passaged in human embryonic fibroblast (HEF) cells 3-4 times, and titrated.

2.2. Cells

HEF cell cultures were used throughout all assays.

2.3. Compound

The names and abbreviations of compounds evaluated for anti-AV activity are as follows : Ribavirin; (S)-9-(2,3-dihydroxypropyl)adenine [(S)-DHPA]; carbocyclic 3-deazaadenosine (C-c^{3}Ado); (S)-9-(3-hydroxy-2-phosphonylmethoxypropyl)adenine [(S)-HPMPA]; 9-[(2-hydroxy-1,3,2-dioxaphosphorinan-5-yl)oxymethyl]guanine P-oxide (2'-nor-cGMP); acyclovir (ACV); adenine arabinoside (AraA); (E)-5-(2-bromovinyl)-2'-deoxyuridine (BVDU); phosphonoformate (PFA).

2.4. Antiviral assay

The antiviral assays were carried out in confluent HEF monolayers in Falcon multiwell plates with 96 flat bottom wells of 5 mm diameter. The HEF monolayers were infected with 0.1 ml virus suspension containing approximately 100 $CCID_{50}$ (50 % cell culture infective dose) of AV. After adsorption of virus for 90 min at 37°C, each well received 0.1 ml maintenance medium containing various concentrations of the test compounds (250, 50, 10, 2, 0.4 and 0.08 µg/ml). Eight wells were used per compound concentration per virus strain. The monolayers were incubated at 37°C for 7 days, and then the cytopathic effect (CPE) of AV was recorded microscopically. The concentration of compound required to protect 50 % of the wells completely against viral CPE was determined as the 50 % inhibitory dose (ID_{50}). The ID_{50} endpoints were calculated by the Reed-Muench method.

2.5. Cytotoxicity

Cytotoxicity of the compounds for the host cells was examined by inhibition of host cell DNA synthesis. The amount of [^{3}H-methyl]thymidine incorporated into acid-insoluble materials was determined after 24 h.

3. RESULTS

The cytopathic effect of AV on HEF cells and uninfected control cells are shown in Fig. 1.

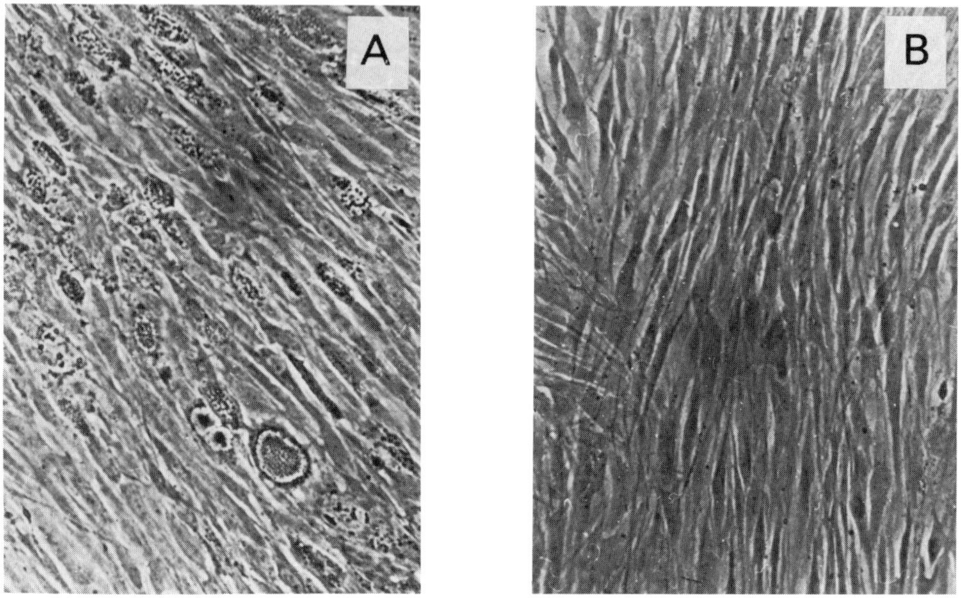

FIGURE 1. Cytopathic effect of AV on HEF cells (A) and uninfected control cells (B).

Only two compounds, (S)-HPMPA and 2'-nor-cGMP proved effective as inhibitors of AV replication at non-toxic concentrations (Table 1). They were effective against all eight types of AV. However, (S)-HPMPA was, at an average, 3 to 4 times more effective than 2'-nor-cGMP.

4. DISCUSSION

Antiherpes drugs, i.e. ACV, AraA, BVDU, PFA, were not effective against AV replication in HEF cell cultures. Broad spectrum anti-RNA virus compounds, i.e. ribavirin, (S)-DHPA, C-c^{3}Ado, were also ineffective. Tolman et al. reported that a cyclic GMP analog, 2'-nor-cGMP, inhibits the replication of several DNA viruses, including AV type 2, cytomegalovirus (CMV) and thymidine kinase-deficient (TK^{-}) herpes simplex virus (HSV) mutants (4). Recently, De Clercq et al. introduced a new compound, (S)-HPMPA, as a highly potent and selective agent against a broad spectrum of DNA viruses, including HSV, varicella-zoster virus (VZV), CMV, vaccinia virus, and TK^{-} mutants of HSV and VZV (5). The antiviral effects of 2'-nor-cGMP and (S)-HPMPA are not dependent on the help of a virus-specified thymidine kinase. Their exact mechanism of action remains to be elucidated. A putative target for their action against AV replication may be the viral

TABLE 1. Inhibitory effect of selected antiviral compounds on the replication of adenovirus (types 1-8) in human embryonic fibroblast (HEF) cell cultures

Compound	50 % Inhibitory dose (μg/ml)			S.I.[a]
	range of values	mean	HEF	
Ribavirin	> 125	> 125	N.D.[b]	-
(S)-DHPA	> 125	> 125	N.D.	-
C-c^{3}Ado	> 125	> 125	N.D.	-
(S)-HPMPA	0.6-2.5	1.1	52	47
2'-nor-cGMP	1.8-13.5	4.1	62	15
ACV	> 125	> 125	N.D.	-
AraA	> 125	> 125	N.D.	-
BVDU	> 125	> 125	N.D.	-
PFA	> 125	> 125	N.D.	-

[a]Selectivity index.
[b]Not determined.

DNA polymerase. Thus, (S)-HPMPA and 2'-nor-cGMP offer promise as broad-spectrum anti-DNA virus agents, and present results indicate that they should be further explored for their potential as anti-adenovirus agents.

REFERENCES

1. Shields AF, Hackman RC, Fife KH, Corey L, Meyers JD. New Eng. J. Med. 312:529-533, 1985.
2. Stalder H, Hierholzer JC, Oxman MN. J. Clin. Microbiol. 6:257-265, 1977.
3. Zahradnik JM, Spencer MJ, Porter DD. Am. J. Med. 68:725-732, 1980.
4. Tolman RL, Field AK, Karkas JD, Wagner AF, Germershausen J, Crumpacker C, Scolnick EM. Biochem. Biophys. Res. Commun. 128:1329-1335, 1985.
5. De Clercq E, Holý A, Rosenberg I, Sakuma T, Balzarini J, Maudgal PC. Nature 323:464-467, 1986.

IN SEARCH OF ANTIVIRAL COMPOUNDS AGAINST THE AIDS VIRUS

R. PAUWELS, P. HERDEWIJN, J. BALZARINI, M. BABA, E. DE CLERCQ, M. VANDEPUTTE and J. DESMYTER
Rega Institute, Katholieke Universiteit Leuven, B-3000 Leuven, Belgium

1. INTRODUCTION

The causative agent of AIDS has been identified as a human retrovirus, designated human T-cell lymphotropic virus type III (HTLV-III) (1), lymphadenopathy-associated virus (LAV) (2), AIDS-associated retrovirus (ARV) (3) and, more recently, human immunodeficiency virus (HIV).

The diameter of this enveloped RNA virus is 100 to 120 nm; it contains a cylindrical core and it is released from infected cells by budding at the cell membrane. A hallmark feature in the pathogenesis of AIDS is the marked tropism of HIV for the lymphocyte helper inducer subset bearing the T4 receptor molecule which leads to destruction of the target cells and ultimately results in inadequate host immune responses.

As there is an urgent need for a therapeutically active agent against AIDS, our efforts have been mainly directed toward the development of inhibitors of the replication of HIV.

Considering the importance of HIV replication in the pathogenesis of AIDS and the stringent dependence of HIV replication on the virus-associated reverse transcriptase (RT), this enzyme, which is needed by the virus for the "reverse" transcription of its RNA genome to (proviral) DNA, can be considered as the prime target in the design of chemotherapeutic agents against AIDS (4).

A simple, adequate and reproducible screening procedure is proposed which allows rapid (i.e. 5 days) *in vitro* evaluation of large numbers of compounds for anti-HIV activity. In additional experiments quantitative information as regards activity, toxicity and selectivity is obtained by more elaborate *in vitro* assays based upon cell viability and viral antigen expression.

2. MATERIALS AND METHODS

2.1. Compounds

2',3'-Dideoxycytidinene (ddeCyd) was synthesized in our laboratory according to the method of Horwitz *et al.* (5). All stock solutions were made in phosphate-buffered saline (PBS) and stored at 4°C until used. Concentrations of the compounds were determined by spectrophotometry.

2.2. Virus

HTLV-III_B (designated HIV) is derived from a pool of American patients with AIDS. It was obtained from the culture supernatant of a persistently HIV-infected H9 cell line (H9/HTLV-III_B) (6). Virus samples were stored in aliquots at -70°C until used. The titer was determined by the $CCID_{50}$ (50 % cell culture infective dose) method.

2.3. Cells

MT-4 cells, which is a human T4 cell line carrying HTLV-I (7) was used throughout all assays. The cells were grown in RPMI 1640 medium supplemented with 10 % heat inactivated foetal calf serum and antibiotics.

2.4. Antiviral assays

Rapid determination of anti-HIV activity during the initial screening procedure was based on the inhibition of HIV-induced alteration of MT-4 cell cluster morphology and 'reclustering' properties (8). Briefly, MT-4 cells were adjusted at 5 x 10^5 cells/ml with RPMI 1640 medium supplemented with 10 % fetal calf serum (FCS) and 2 ml portions of the cell suspension were infected with 200 µl of virus stock (final titer: 100 $CCID_{50}$/well). Control MT-4 cells were mock-infected with 200 µl of cell culture medium. After 90 minutes incubation at 37°C, 1.8 ml of culture medium was added to each tube. Various dilutions (100 µl) of the test compound in cell culture medium were prepared previously in flat bottomed 96-well plastic microtiter trays. Then, HIV-infected or mock-infected MT-4 cells (in 100 µl) were added to the wells, and the cell cultures were incubated at 37°C. After five days, the cluster morphology was evaluated microscopically. The cell aggregates were then broken up to a single cell suspension by pipetting. The cell suspensions were incubated at 37°C for 4 hours ans assessed microscopically for their 'reclustering' properties.

Viral cytopathogenicity assay : Viability of the MT-4 cells was determined 5 days after HIV-infection by trypan blue exclusion as previously described (8-10).

2.5. Inhibition of viral antigen expression in HIV-infected MT-4 cells

The inhibitory effects of the compounds on viral antigen expression in infected MT-4 cells were both determined by indirect immunofluorescence (IF) microscopy and quantitated by laser flow cytofluorography using a polyclonal antibody as probe, as previously described (8).

3. RESULTS

When infected MT-4 cells were grown in plastic microtrays under the conditions described earlier, they formed after 5 days incubation large clusters, each containing 1 x 10^4 to 2 x 10^4 cells (Fig. 1a). However, when the cells had been infected with HIV (100 CCID /well), cell cluster formation was observed but the morphology of these clusters gradually altered so that 4 or 5 days after infection HIV-infected cell clusters had a more dense appearance with irregularly shaped edges (Fig. 1b), whereas the mock-infected cell clusters had regular borders (Fig. 1a). When the mock-infected cell clusters were then dispersed to a single cell suspension, 'reclustering' took place within a few hours (Fig. 1c), whereas HIV-infected cell clusters did not show such an effect (Fig. 1d). These microscopic findings correlated in time and intensity with a gradual decrease in viability and a progressive expression of HIV antigens by the HIV-infected MT-4 cells (data not shown).

When the HIV-infected MT-4 cells were treated with 0.2 µM ddeCyd normal cell morphology and 'reclustering' was observed (Table 1). At this concentration ddeCyd afforded 61 % protection against HIV cytopathogenicity (Table 1) whereas viral antigen expression was significantly suppressed (81 % protective effect). Full protection against HIV infection was achieved at 1 µM ddeCyd (Fig. 2 and 3).

When assayed for cytotoxicity in mock-infected MT-4 cells, ddeCyd effected a 30 % reduction in cell viability at a concentration of 5 µM; whereas 100 % toxicity was noted at 25 µM ddeCyd (Table 1, Fig. 2). At this concentration 'reclustering' was absent and the dimensions of the cell clusters were significantly reduced.

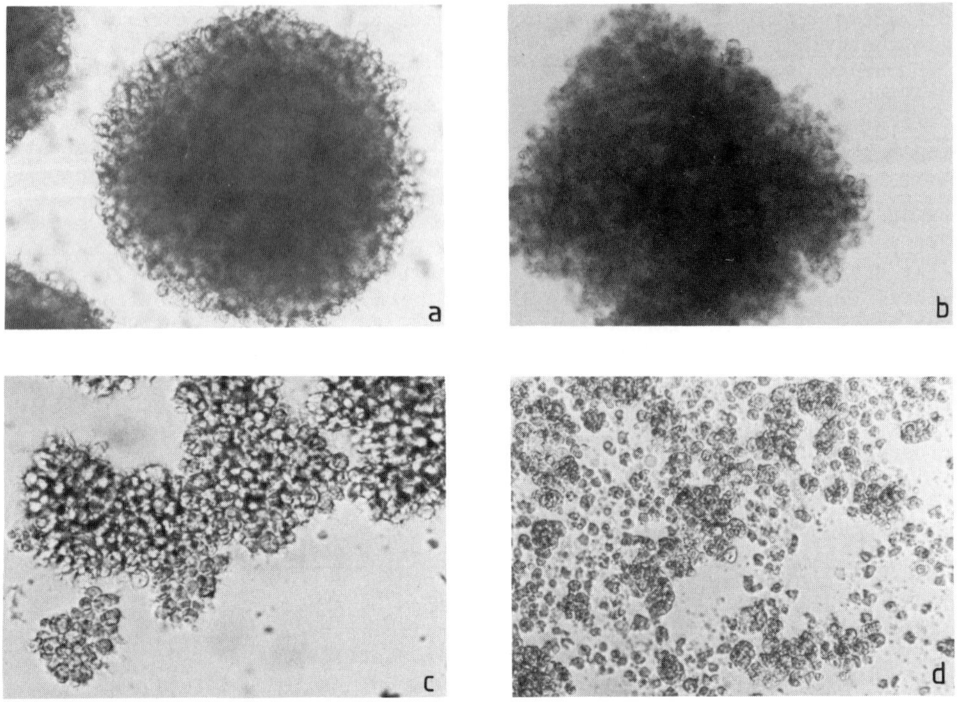

FIGURE 1. Cell cluster formation.
a : Mock-infected MT-4 cell cluster. b : MT-4 cell cluster 5 days after HIV infection. c : 'Reclustering' of mock-infected MT-4 cells. d : Absence of 'reclustering' of HIV-infected MT-4 cells.

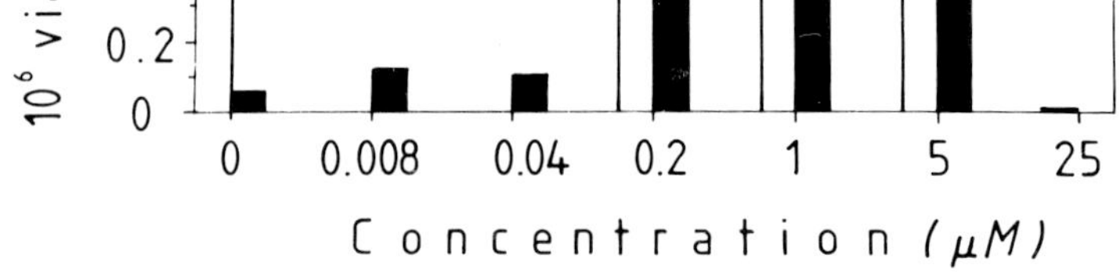

FIGURE 2. Inhibition of HIV-induced cytopathogenicity for MT-4 cells by 2',3'-dideoxycytidinene. Viability of the cells was measured by the trypan blue exclusion method on the 5th day after infection. □ : mock-infected cells; ■ : HIV-infected cells.

TABLE 1. Reclustering, viability and viral antigen expression of HIV-infected or mock-infected MT-4 cells in the presence of varying concentrations of 2',3'-dideoxycytidinene (ddeCyd).

Compound concentration (μM)	HIV-infected cells					Mock-infected cells	
	Reclustering[a]	Viability		Viral antigen expression		Reclustering	Viability
		Number of viable cells[b] (x 10^4)	Protection[c] (%)	Number of fluorescent cells[d] (%)	Protection[e] (%)		Number of viable cells (x 10^4)
0	-	6	0	69.0 ± 0.6	0	+	102
0.008	-	13	7	43.3 ± 5.5	38	N.D.	N.D.
0.04	-	11	5	43.7 ± 1.3	38	N.D.	N.D.
0.2	+	65	61	13.8 ± 1.8	81	+	107
1	+	99	97	2.2 ± 0.05	98	+	101
5	+	66	62	2.0 ± 0.2	98	+	71
25	N.D.	N.D.	N.D.	N.D.	N.D.	-	0

[a] As described in Materials and Methods.
[b] As determined by trypan blue exclusion, 5 days after infection.
[c] Percentage of protection was calculated by the following formula : 100 x [(number of viable cells exposed to HIV and incubated in the presence of ddeCyd) - (number of viable cells exposed to HIV and incubated in the absence of ddeCyd)] / [(number of viable mock-infected cells grown in the absence of ddeCyd) - (number of viable cells exposed to HIV and grown in the absence of ddeCyd)].
[d] As measured by indirect immunofluorescence and laser flow cytofluorometrhy 4 days after infection using polyclonal antibodies from an AIDS patient. A control sample of mock-infected MT-4 cells contained 1 % fluorescent cells.
[e] Percentage of protection was calculated by the following formula : 100 - 100 x [(number of positive cells in the presence of inhibitor) - (number of positive cells in a mock-infected control sample)] / [(number of positive cells in the absence of inhibitor) - (number of positive cells in a mock-infected control sample)].

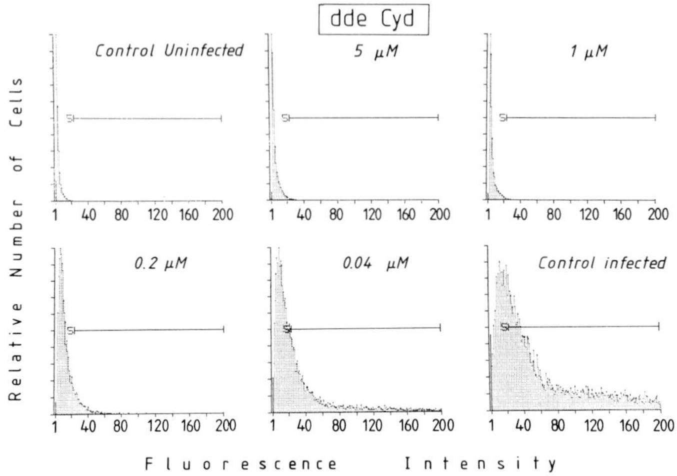

FIGURE 3. Inhibition of viral antigen expression in HIV-infected MT-4 cells by 2',3'-dideoxycytidinene. Antigen-positive cells were measured by indirect immunofluorescence and laser flow cytofluography 4 days after infection, as described in Materials and Methods.

4. DISCUSSION

The potent and selective inhibitory effects of the 2',3'-unsaturated derivative of 2',3'-dideoxycytidine (ddeCyd) on HIV replication, as initially described by Balzarini et al. (9), have been corroborated by various assay methods. Although the most accurate data on the anti-HIV activity of ddeCyd were derived from cytopathogenicity and viral antigen expression, the data of the 'reclustering' assay, being far less time-consuming, proved to correlate very well with the data. This has been confirmed for the more than 100 compounds that have been looked at so far.

The results presented here therefore indicate that microscopic observation of MT-4 cell cluster morphology and 'reclustering' behavior provide valuable indications as to the potential anti-HIV activity of selected antiviral compounds. The assay allows a rapid and sensitive in vitro screening of a large number of compounds.

From the 'MT-4 cluster assay' the necessary information is obtained to optimize the test compound concentrations for potency and selectivity determinations by the more elaborate cytopathogenicity and viral antigen expression assays.

REFERENCES

1. Gallo RC., Sarin PS, Gelmann EP, Robert-Guroff M, Richardson E, Kalyanaraman VS, Mann D, Sidhu GD, Stahl RE, Zoll-Pazner S, Leibowitch J, Popovic M. Science 220:865-867, 1983.
2. Barré-Sinoussi F, Chermann JC, Rey R, Nugeyre MT, Chamaret S, Gruest J, Dauguet C, Axler-Blin C, Vezinet-Brun F, Rouziaux C, Rozenbaum W, Montagnier L. Science 220:868-871, 1983.

3. Levy JA, Hoffman AD, Kramer SM, Landis JA, Shimabukuro JM, Oshiro LS. Science 225:840-842, 1984.
4. De Clercq E. J. Med. Chem. 29:1561-1569, 1986.
5. Horwitz JP, Chua J, Noel M, Donatti JT. J. Org. Chem. 32:817-818, 1967.
6. Popovic M, Sarngadharan MG, Read E, Gallo RC. Science 224:497-500, 1984.
7. Harada S, Koyanagi Y, Yamamoto N. Science 229:563-566, 1985.
8. Pauwels R, De Clercq E, Desmyter J, Balzarini J, Goubau P, Herdewijn P, Vanderhaeghe H, Vandeputte M. Submitted for publication, 1987.
9. Balzarini J, Pauwels R, Herdewijn P, De Clercq E, Cooney DA, Kang G-J, Dalal M, Johns DG, Broder S. Biochem. Biophys. Res. Commun. 140: 735-742, 1986.
10. Baba M, Pauwels R, Herdewijn P, De Clercq E, Desmyter J, Vandeputte M. Biochem. Biophys. Res. Commun. 142: 128-134, 1987.